Eco-Farm

AN ACRES U.S.A. PRIMER

Eco-Farm

AN ACRES U.S.A. PRIMER

Charles Walters

Revised third edition

Acres U.S.A.
Austin, Texas

Eco-Farm

AN ACRES U.S.A. PRIMER

3rd edition, revised
Copyright © 1979, 1996, 2003 Acres U.S.A.

Acres U.S.A.
P.O. Box 91299
Austin, Texas 78709 U.S.A.
(512) 892-4400 • fax (512) 892-4448
info@acresusa.com • www.acresusa.com

Printed in the United States of America

Publisher's Cataloging-in-Publication

Walters, Charles, 1926-
Eco-farm: an Acres U.S.A. primer / Charles Walters. — 3rd ed., rev.
xiv, 462 p., 23 cm.
First ed. published as An Acres U.S.A. primer.
Includes bibliographical references and index.
ISBN 978-0-911311-74-7 (trade)

1. Agricultural ecology. 2. Organic farming.
I. Walters, Charles, 1926-   II. Title.

S605.5.W35 2003   631.5'8

For Anne Walters

Acknowledgments

A great many people have contributed to the evolution of *Eco-Farm —* *An Acres U.S.A. Primer,* including hands that rocked the cradles of a century ago. Some have been cited in the running text. Others, long forgotten, must rest with the assurance that they gave the writer ideas without either giver or receiver being conscious of the spark that then touched tinder. They helped us learn the dimensions of the temple.

This revised edition owes a debt of gratitude to the late C.J. Fenzau, who served as a consultant and an inspiration not only for this work, but for a lifetime that can only be considered an intellectual feast. Other contributors and sources have been acknowledged in the text of this primer. Both Fenzau and the author were students of the late William A. Albrecht, Ph.D., whose wisdom anoints most of these pages.

USDA materials have been amply cited, we hope. Still, we would like to acknowledge the USDA weed drawings by Regina O. Hughes in *Common Weeds of the United States,* which proved invaluable.

There are a number of helpers who deserve to be mentioned: the Acres U.S.A. staffers who typed and retyped the manuscript, and with great tolerance and fortitude presided over the workaday business of bringing the manuscript to print; Paul Cupp who provided most of the art used to illustrate these pages; Anne Walters, who proofed and reproofed the final result; the late John Whittaker, D.V.M., whose constructive criticism resolved many a shortfall; the late Lee Fryer, who gave assist in his specialized areas; Phil Callahan, Ph.D., who read the manuscript critically and provided numerous improving suggestions. Writers and agonomists do not live by words and soil systems alone. They too need compassion and gentleness, both of which were provided by our colleagues above and beyond the call of duty. We thank them all warmly.

CONTENTS

NATURE WILL NOT BE CONQUERED
EXCEPT BY OBEYING

("Natura enim non nisi parendo vincitur")

At first men try with magic charm
To fertilize the earth,
To keep their flocks and herds from harm,
And bring new young to birth.

Then to capricious gods they turn
To save from fire or floods;
Their smoking sacrifices burn
On altars red with blood.

Next bold philosopher and sage
A settled plan decree,
And prove by thought or sacred page
What Nature ought to be.

But Nature smiles—A Sphinx-like smile—
Watching their little day
She waits in patience for a while
Their plans to dissolve away.

Then come those humbler men of heart
With no completed scheme,
Content to play a modest part,
To test, observe and dream.

Till out of chaos come in sight
Clear fragments of a Whole—
Man, learning Nature's ways aright,
Obeying, can control.

The great Design now glows afar;
But yet its changing scenes
Reveal not what the pieces are
Nor what the Puzzle means.

And Nature smiles—still unconfessed
The secret thought she thinks—
Inscrutable she guards unguessed
The riddle of the Sphinx.

FOREWORD

A primer is a first reader.

Eco-Farm — An Acres U.S.A. Primer is a first reader in eco-agriculture.

The objective here is to take it all apart and put it back together again, and to do this so that all can understand.

Accordingly, this book contains nothing that runs beyond the comprehension of a person blessed with average intelligence and a fair education. It assumes no great knowledge of chemistry, botany, entomology or soil physics. It may be that readers will have to look up a few words in the glossary now and then, and handle a few new concepts—transpiration, translocation, cation exchange capacity, biological energy, cellular equilibrium—but these things are also explained adequately in the context of their usage.

Having mastered the several lessons in this primer, the student of eco-agriculture will then be able to read the lines—and between the lines—of articles and books with superb comprehension. Should a passage or chapter prove a bit difficult, the reader is urged to proceed anyway. The overview sought here will come together in any case sooner or later.

This primer takes issue with the National Association of Science Writers, a cautious group that has made the following a part of the organization's code of ethics. "Science editors are incapable of judging the facts of phenomena involved in medical and scientific discovery. Therefore, they only report discoveries *approved* by authorities, or those presented before a body of scientific peers."

Metro farm editors and editors serving industrially owned farm papers obey this injunction and handle no new technology unless it is first blessed by the land grant colleges.

This is scholasticism, the last stage in the decay of the simple and obvious system called the *scientific method*. It seems to say that those who are trained by the universities are not competent to do what they have been trained to do. A measure of self respect requires us to reject this outrageous summary.

In a way, it is fitting that this primer challenge those who would hide behind the complexities of their craft. The economist Clarence Ayres once said it all: "It is easy to maintain sacred fictions in a community to whom every letter is an occult symbol; in a community to whom the printed word has become a common tool, no fiction is shielded from the scrutiny of the people, not even the divinity which hedges kings."

There is nothing occult about science. And we add, there is nothing occult about the rationales constructed by the grant receivers to protect the commerce of their patrons. Indeed, a primer can handle almost everything we need to discuss. And as these lessons unfold, certain principles will emerge, each dazzling in the purity of its challenge. We can state these now and set the stage for proofs in the pages that follow.

1. Simplistic nitrogen, phosphorus and potassium (N, P and K) fertilization means malnutrition for plants, animals and men because either a shortage or marked imbalance of plant nutrients prevent balanced plant health and therefore animal and human health.

2. Plants in touch with exchangeable soil nutrients needed to develop proper fertility loads, structure, and stabilized internal hormone and enzyme potentials, provide their own protection against insect, bacterial and fungal attack.

3. Insects and nature's predators are a disposal crew. They are summoned when they are needed, and they are repelled when they are not needed.

4. Weeds are an index of the character of the soil. It is therefore a mistake to rely on herbicides to eradicate them, since these things deal with effect, not cause.

5. Crop losses in dry weather, or during mild cold snaps, are not so much the result of drought and cold as nutrient deficiency.

6. Toxic rescue chemistry hopes to salvage crop production that is not fit to live so that animals and men might eat it, always with consequences for present and future generations of plants, animals and men.

7. Man made molecules of toxic rescue chemistry do not exist in nature's blueprints for living organisms. Since they have no counterpart in nature, they will not likely break down biologically in a time frame suitable to the head of the biotic pyramid, namely man. Carcinogenic, mutagenic and teratogenic molecules of toxic rescue chemistry have no safe level and no tolerance level.

The summary stacks up like any college syllogism. NPK formulas as legislated (and enforced by state departments of agriculture) mean malnutrition, insect, bacterial and fungal attack, toxic rescue chemistry, weed takeover, crop loss in dry weather, and general loss of mental acuity—plus degenerative metabolic disease—among the population, all when people use thus fertilized and protected food crops. Therefore the answer to pest crop destroyers is sound fertility management in terms of exchange capacity, pH modification, and scientific farming principles that USDA, Extension and Land Grant colleges have refused to teach ever since the great discovery was made that fossil fuel companies have grant money.

Young people today do not understand this profound philosophy. They turn to the farmer for answers, but most farmers no longer understand. They may still remember that nature created life, but they think the test tube and fossil fuel factory have vacated nature's rhythm of life and death.

The authors of this book have worked in agriculture for more than a quarter of a century. And if you ask us, *What does a farmer do?*, we will answer quite differently from most. In agribusiness they say a farmer produces corn, wheat, cattle or swine, or perhaps one of a hundred other crops, and this may be correct as far as it goes. But we and a few farmers see the final product of the farm as human bodies with minds capable of thought and reason.

True, the farming profession requires farmers to bargain with his fellow men for dollars according to some few economic laws. These laws have been covered in *Unforgiven*. Still, there is a more subtle message in *Unforgiven*, and it states that the farmer must also bargain with nature to get human food according to the laws of life and death. This primer tells how.

After reading *Eco-Farm*, you'll be ready for a postgraduate course, one available monthly in the pages of *Acres U.S.A.* and in a few select book titles that can be recommended as being in tune with nature.

PRESCHOOL

Photosynthesis

Land, its management, its ownership, all dominate the lifespan of ecological systems, and therefore economic systems. Thus the primacy of agriculture and the labor of love agriculture requires.

1. margin of life

"Long before man could make a plow or a test tube," writes Gene Poirot in *Our Margin of Life*, "nature was creating life, including man, and providing an environment in which all life could live. She used the resources of air, water, sunshine and soil plant food minerals to make life. If she had created only life, these resources would soon have been tied up in all living things. So she created death. This way resources could be recycled and used again and again. There is a basic law which says, *All life forms must return at death what they took from the resources of the earth during their lifetime.* For example, when any creature dies, the water it took returns to the environment. The air it used in terms of oxygen, nitrogen and carbon dioxide is recycled. And because death occurs either on the soil or in the water, these resources are held within reach of new life forms which also come from the soil surface and from close to water's surface. So to live with the laws of life and death, we must follow a basic rule of nature."

These laws say, return to the environment all that which is taken from it and made in growth, and hold these additions within reach of new life.

There is wonder and awe in Poirot's little story about *Our Margin of Life*. Everything that eats will be eaten. This is easy enough to understand whenever an owl swoops down to snag a field mouse, or a sharpshin hawk takes the life of a rabbit. But is the reverse true, that everything eaten must also eat? Do plants feed? Of course they do. Even the ancient Greeks knew that plants relied on nutrients in the soil. But they did not know how chemicals of the air contributed to plant bulk and food production.

2. jan baptista van helmont

Jan Baptista van Helmont, a 17th century Flemish physician, started getting a handle on exactly what happens when he performed his now famous tree experiment. He simply wanted to know how soil matter was being displaced when plant life grew. No one could measure such a proposition in a field, or in a forest. So van Helmont planted a willow tree in a large earthen tub. The little sprig weighed in at 5 pounds. Soil used in the experiment scaled in at an even 200 pounds. The tub was then covered so that only a small hole for the tree trunk and one for watering remained.

Five years later the tree was not only larger, it now weighed 164 pounds. Obviously, reasoned van Helmont, if the willow tree picked up the difference between 5 pounds and 164 pounds, then the soil remaining in the tub should weigh only 41 pounds, potting material having been reduced to oven dry soil for the postgrowth weighin. The results proved van Helmont hopelessly wrong. After contributing to the tree's growth for five years, the 200 pounds of soil had lost only 2 ounces. Van Helmont pondered the problem in deep consternation. Could it be that all this growth came from the water he had given the tubbed tree all these years? Surely this was the answer.

3. in contact with air

No, said a scientist by the name of Nehemiah Grew. He had used the new tool called the microscope to study the tiny pores of a plant leaf. Here were openings not unlike those observed in the skin of an animal or a man, obviously for the elimination of perspiration. By this time it had become uncomfortable to use the term obviously too much. Grew took a second look and this suggested another possibility. Perhaps the pores were for the admission of air. Van Helmont had completely overlooked air, and the willow tree had been in contact with air as well as water and soil. Even so, he didn't pause to consider that air contained dust particles—debris from volcanoes, pollution from forest fires and industry, soot from town chimneys.

It was one thing to have a vision of how nature works, another to test it out. How, indeed, could one determine whether air figured in the growth of a plant? An English preacher and scientist named Stephen Hales reasoned it out and constructed his famous peppermint plant experiment.

The proposition was hellishly simple. A peppermint plant was potted in a small cistern. Next, a glass was clamped down over the plant to shut off outside air. The water level in the glass that surrounded the plant was regulated simply by siphoning away water in the cistern. Now if the plant made use of the air, why the water in the container would either rise or fall. Hales knew about atmospheric pressure, but his control would handle that variable. The control was a similar setup, except that it contained no peppermint plant. Atmospheric pressure would treat each vessel alike. If a difference developed, this would be sound scientific evidence that the plant had interacted with the air in the vessel.

Hales tabulated his results and watched. One day the plant took on a noticeable wilt. Hales replaced this plant with another, taking care to preserve the old air in the vessel. It took less than five days for the new plant to fade away. Yet a plant introduced into the control vessel lived approximately as long as the first one. Apparently plants were taking something out of the air. Was this something food?

Of course! Once the food in the first vessel had been exhausted, a second peppermint plant couldn't live. Yet a peppermint plant placed in the stale air of the second vessel lived as long as the first plant. Being an astute observer, Hales reasoned that his mint plants were changing the air. How? He died without finding out.

The idea of using inverted glass vessels to follow where logic led didn't die with Hales, however. Joseph Priestley, the British chemist who was later to isolate and describe the properties of elemental oxygen, was fascinated by those trapped bodies of air. At one point in his experiments, he clamped such a glass vessel over a burning candle, and of course the flame went out in a short time. He did the same with a mouse and found it could live only a short time.

A mouse and a candle! What could they possibly have to do with agriculture? Priestley continued his experiments and finally announced a serendipitous discovery. "I flatter myself that I have accidentally hit upon a method of restoring air which has been injured by the burning of candles, and that I have discovered at least one of the restoratives which nature employs for this purpose. It is vegetation."

Indeed, even a candle consumes a gallon of oxygen a minute. People and animals used air with reckless abandon even before the advent of the automobile and the industrial age. Something was renewing the air, and Priestley stumbled on some parts of the answer. The recorded date for this discovery was August 17, 1771. First Priestley consumed the air in a chamber by burning a candle. Then he inserted a peppermint plant. It lived.

Moreover, after only ten days a candle would burn in that same air again. The same principle held for a mouse. Air that had been exhausted until a mouse died could be renewed by the mint plant so that ten days later another mouse could live in it.

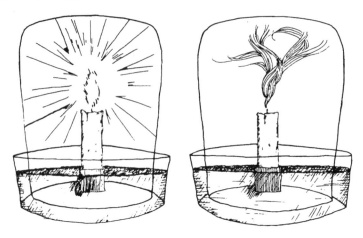

Priestley consumed the air in a chamber by burning a candle. After a candle had exhausted the air, a mouse could not live in it.

After a peppermint plant had renewed the air, a mouse could live in it until the air was again exhausted.

4. this strange miracle

This seems so fundamental now that we wonder aloud how such complete simplicity could stand as a turning point in agricultural science. Certainly Priestley believed plant life was working this strange miracle, but he didn't understand why. At the time he worked his candle/mouse experiments, he hadn't isolated oxygen yet. As a matter of fact, Mendeleyeff's periodic table of the elements wouldn't be published for another hundred years. Because of this, perhaps, a final answer to the riddle being served up by Hales and Priestley had to wait well in excess of another hundred years for solution.

In the meantime, there was Jan Ingenhousz of Holland, a pure scientist and a rare professional for any season in history. Working in a world that saw the ferment and chaos of an American revolutionary war, he first checked Priestley's findings, then added to the existing body of knowledge. Yes, Priestley was right—as far as he went. Plants did purify air, albeit only in light.

5. photosynthesis

Here was a factor the great chemist hadn't even considered. Men like Hales and Priestley didn't work in the dark. They worked in sunlight or with the aid of candlepower, and they didn't give the matter a thought. Yet here was the reality Ingenhousz discovered: plants kept in the dark would not purify the air. As a matter of fact, in the dark all plant parts damaged air much the same as did candles and mice. Moreover, certain parts of plants wouldn't purify the air either. Roots wouldn't. Old stems wouldn't. Only leaves and young stems with green coloring seemed to do this job. Green coloring? The equation seemed to get more complicated as man's puny mind expanded itself. First air, now green coloring seemed to walk hand in hand with that air purification task.

Once this door had been kicked open, it became more difficult to overlook contributing factors—no matter how strange—in the Creator's handiwork. At each step along the way some investigator had a vision of what was happening. If plants didn't get their food from the soil as a finished product, then plants—quite unlike animals—manufactured it. In the scientist's lexicon the term of choice was synthesize. If this manufacturing chore was accomplished only in light, then photo (meaning light) and synthesize described the process.

That is, photosynthesis described what, but not how. There were all those nagging questions. If, in Hale's experiment, the water level only rose, it meant that plants not only gave up something (oxygen), they also consumed something else. After all, those mint plants eventually died. Did they exhaust what they were taking from the air, possibly the same thing plants and animals were giving up?

Mental acuity and sound reasoning suggested a balance in nature. But it took chemistry to answer how this balance was being maintained. Laboratory experiments said that of the various substances released by animals into the environment, carbon dioxide was the one plants needed to grow. Somehow the plants combined water with carbon dioxide to produce food called glucose.

Glucose has 6 atoms of carbon, 12 atoms of hydrogen, and 6 atoms of oxygen—$C_6H_{12}O_6$. Did the oxygen released from plant life come from the water or from the carbon dioxide? Students of agriculture will run across names such as Claude Louis Berthollet, Jean Senebier, Nicholas Theodore de Saussure as they watch the debate rage back and forth. All passed from the scene long before the question was answered, or the answer was explained. A great deal had to be learned about chemistry and the chemistry of life first, about inorganic and organic chemistry.

6. table of known elements

Dmitri I. Mendeleyeff, the Russian chemist, first constructed his table of known elements hardly a hundred years ago. His insight into the Creator's order was so great he supplied spaces where he believed an element should be to comply with the rhyme and reason of the supreme plan. That table still stands. The blank spots have been filled in. And the Periodic Chart of the Elements has provided a simple and beautiful picture of the order in the universe where we live. It has done more. It has opened chemistry and physics as never before, and made it possible for lesser minds to understand the structure of the atom. Moreover, it has provided facts agronomists could not ignore.

Each elemental entry features an inventory of information, starting with hydrogen, the lightest.

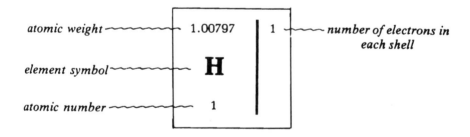

The atomic weight on a chart is almost always expressed as a single figure average when isotopes are involved. Isotopes are atomic brothers, so to speak, atoms of the same element which differ in weight. Hydrogen, as an example, has three isotopes. Protium hydrogen, the lightest form, has one proton and one orbiting electron, thus:

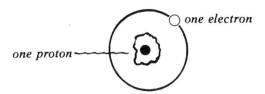

A slightly heavier form is deuterium, sometimes called heavy hydrogen.

The heaviest form of hydrogen is called tritium.

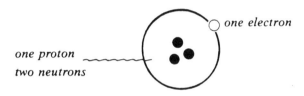

The average of these several weights is 1.00797.

Each of the elements of life and death has an abbreviation symbol. You'll see some on fertilizer bags, in farm literature, and even field them as slang in daily conversations. Here, for easy reference, is what each of the symbols mean.

Element	Symbol	Atomic number	Atomic weight
Actinium	Ac	89	(227)
Aluminum	Al	13	26.9815
Americium	Am	95	(243)
Antimony	Sb	51	121.75
Argon	Ar	18	39.948
Arsenic	As	33	74.9216
Astatine	At	85	(210)
Barium	Ba	56	137.34
Berkelium	Bk	97	(247)
Beryllium	Be	4	9.0122

Element	Symbol	Atomic number	Atomic weight
Bismuth	Bi	83	208.980
Boron	B	5	10.811
Bromine	Br	35	79.909
Cadmium	Cd	48	112.40
Calcium	Ca	20	40.08
Californium	Cf	98	(247)
Carbon	C	6	12.01115
Cerium	Ce	58	140.12
Cesium	Cs	55	132.905
Chlorine	Cl	17	35.453
Chromium	Cr	24	51.996
Cobalt	Co	27	58.9332
Copper	Cu	29	63.54
Curium	Cm	96	(247)
Dysprosium	Dy	66	162.50
Einsteinium	Es	99	(254)
Erbium	Er	68	167.26
Europium	Eu	63	151.96
Fermium	Fm	100	(253)
Fluorine	F	9	18.9984
Francium	Fr	87	(223)
Gadolinium	Gd	64	157.25
Gallium	Ga	31	69.72
Germanium	Ge	32	72.59
Gold	Au	79	196.967
Hafnium	Hf	72	178.49
Helium	He	2	4.0026
Holmium	Ho	67	164.930
Hydrogen	H	1	1.00797
Indium	In	49	114.82
Iodine	I	53	126.9044
Iridium	Ir	77	192.2
Iron	Fe	26	55.847
Krypton	Kr	36	83.80
Lanthanum	La	57	138.91
Lawrencium	Lw	103	(257)
Lead	Pb	82	207.19
Lithium	Li	3	6.939
Lutetium	Lu	71	174.97
Magnesium	Mg	12	24.312
Manganese	Mn	25	54.9380

Element	Symbol	Atomic number	Atomic weight
Mendelevium	Md	101	(256)
Mercury	Hg	80	200.59
Molybdenum	Mo	42	95.94
Neodymium	Nd	60	144.24
Neon	Ne	10	20.183
Neptunium	Np	93	(237)
Nickel	Ni	28	58.71
Niobium	Nb	41	92.906
Nitrogen	N	7	14.0067
Nobelium	No	102	(254)
Osmium	Os	76	190.2
Oxygen	O	8	15.9994
Palladium	Pd	46	106.4
Phosphorus	P	15	30.9738
Platinum	Pt	78	195.09
Plutonium	Pu	94	(242)
Polonium	Po	84	(210)
Potassium	K	19	39.102
Praseodymium	Pr	59	140.907
Promethium	Pm	61	(147)
Protoactinium	Pa	91	(231)
Radium	Ra	88	(226)
Radon	Rn	86	(222)
Rhenium	Re	75	186.2
Rhodium	Rh	45	102.905
Rubidium	Rb	37	85.47
Ruthenium	Ru	44	101.07
Samarium	Sm	62	150.35
Scandium	Sc	21	44.956
Selenium	Se	34	78.96
Silicon	Si	14	28.086
Silver	Ag	47	107.870
Sodium	Na	11	22.9898
Strontium	Sr	38	87.62
Sulfur	S	16	32.064
Tantalum	Ta	73	180.948
Technetium	Tc	43	(97)
Tellurium	Te	52	127.60
Terbium	Tb	65	158.924
Thallium	Tl	81	204.37
Thorium	Th	90	232.038

Element	Symbol	Atomic number	Atomic weight
Thulium	Tm	69	168.934
Tin	Sn	50	118.69
Titanium	Ti	22	47.90
Tungsten	W	74	183.85
Uranium	U	92	238.03
Vanadium	V	23	50.942
Xenon	Xe	54	131.30
Ytterbium	Yb	70	173.04
Yttrium	Y	39	88.905
Zinc	Zn	30	65.37
Zirconium	Zr	40	91.22

It will be noted that a few numbers appear in brackets. This means the number represents an isotope of the element, usually the one with the longest known half-life. These numbers have been struck up by the National Bureau of Standards. The unbracketed numbers are standard according to the International Union of Pure and Applied Chemistry.

All the elements needed for life are listed as the first 53 of 92 natural elements on planet earth. Of these, all except one falls in order among the first 42, and all except two are listed among the first 34. There is also a natural order for abundance of elements, according to atomic weight and number. The heaviest elements are the rarest. Elements with even atomic numbers are more abundant than those with odd numbers in our universe. We don't know why, nor can we even guess.

The table itself is a veritable encyclopedia. There are series with missing electrons. As the eye moves from titanium to zinc, unfilled orbits change, an electron at a time. These transitions take place in natural order, moving across the table. There is also a vertical order to the table weight increasing as each element is listed under the one above. There are groups that figure in biology and signal the entrance and exit of disease. Henry A. Schroeder, M.D., possibly the world's foremost authority on trace elements at the time of his death, wrote in *The Trace Elements and Man*, that a heavier metal can displace a lighter one in the same group in biological tissues and alter the reaction of the lighter one. He went on to say that tissues with an affinity for a certain element have an affinity for all other elements of the same group. Some elements are bone seekers. Some are thyroid seekers. All elements in two groups are liver and kidney seekers.

In terms of plant life, it is too early to say which of the elements are essential, although college texts and agronomy manuals are fond of listing 14 or 16 or 18, sometimes more.

If you look at any Periodic Chart of the Elements diagram, you will note a † symbol, designating a footnote, and the footnote says that atomic weights presented are reliable to ±3 in the last place. Other weights are reliable to ±1 in the last place. These are the isotopes. Isotopes are atoms with the same number of protons, but different number of neutrons. If you run across a notation,

$$^{16}_{8}O, \; ^{17}_{8}O, \; ^{18}_{8}O$$

it simply denotes three stable isotopes of oxygen. The small index 8 means there are 16 - 8 = 8 neutrons in the isotope 16; 17 - 8 = 9 neutrons in the isotope 17; and 18 - 8 = 10 neutrons in the isotope 18.

This means oxygen 18 can be identified, just as in the case of those hydrogen forms illustrated earlier. It has more mass than, say, oxygen 16, and this mass can be identified by the mass spectrometer, a relatively modern instrument now available to scientific investigators. It was the development of this instrument, together with codified knowledge of the periodic table of elements, that finally served up an answer to the question: did the oxygen released from plant life come from the water or from carbon dioxide? Why is it so important to know the answer?

7. shared, rented, occupied

There is a wonderful little book in print nowadays entitled *The Lives of a Cell*. In it Lewis Thomas, M.D., makes quite a point of the fact that the human system is not alone as an operating mechanism. We are shared, rented, occupied, he writes. The very interior of our cells are homesteaded by the mitochondria, small separate creatures that may or may not have entered early precursors of our eukaryotic cells, and stayed on for a few billion years. These little fellows are self sufficient sooners. They have their own DNA and RNA and replicate in their own way. Like the rhizobial bacteria on the roots of beans, they are symbionts. Except for them we couldn't even drum a finger, much less think a thought.

Plant life is not plant life as such either. It too is rented out and occupied. Little one cell individuals are everywhere. They are so small they are invisible to the naked eye. Except for powerful microscopes, we would never lay an eye on one, and yet plant and human life depend on them, suffer disease because of them, live and die according to how life and death are balanced among them.

8. the "?" at the end

Certain bacteria handle the business called photosynthesis. In plant cells, chloroplasts work with the sweep of the sun to manufacture the oxygen we breathe. There is also a purple bacteria that carries out photosynthesis much like chloroplasts in plants, using carbon dioxide as a raw material. Unlike the chloroplasts, purple bacteria use no water. They use hydrogen sulfide, H_2S, a gas compound that smells like rotten eggs. (Parenthetically, it might be noted that recent research has also uncovered the presence of purple protein inside the placenta which supplies the beginning colostrum nutritional needs of developing embryos. Umbilical systems for nutrition are not the only conduit for embryo requirements.)

It remained for an American biologist, C. B. Van Niel, to fit into place a last stone of the mosaic started when van Helmont planted his willow tree. Van Niel suspected that purple bacteria were doing exactly the same thing chloroplasts were doing in green leaves, using carbon dioxide plus hydrogen sulfide instead of carbon dioxide and water. The equation for green plants read $CO_2 + H_2O \rightarrow$ glucose $+ O_2$.. The equation for purple bacteria read $CO_2 + H_2S \rightarrow$ foodstuff $+$?. That purple bacteria made another foodstuff, not glucose, didn't matter. What mattered was whether the "?" at the end of the purple bacteria equation read oxygen or not. A positive answer meant that oxygen released during photosynthesis came from carbon dioxide. If the "?" meant sulfur, then, indeed, the oxygen byproduct of photosynthesis came from water.

Van Niel knew the answer in terms of purple bacteria: it was sulfur. This was confirmed by isotope studies hardly more than three decades ago. By using water with oxygen-18, and carbon dioxide with oxygen-16, and exposing the one cell Chlorella plant to the water, it was possible to measure oxygen released. Mass spectrometer reading revealed that the released oxygen indeed was oxygen-18, the isotope found in the experimental water.

Picture the irony. It had taken 2,000 years to scratch the surface on how plants fed, and another 200 years to solve a relatively simple question— where the oxygen byproduct of photosynthesis came from. In the process, the world learned something about the absurdity of single factor analysis, about the tragedy of legislating what is true or false in science, about the curse that descends on science when practitioners are not well grounded in logic.

9. limiting factor

There is a limiting factor in almost everything that concerns farming, and a limiting factor often has absolute power. Light is a limiting factor in photosynthesis. Temperature, water, certain earth minerals, all serve up limitations. The green pigment called chlorophyll gives off a reddish light when leaves are placed in acetone, and the chlorophyll in chloroplasts goes into solution. Why is this? No reddish glow can possibly be seen when chlorophyll remains in the chloroplasts. The answer is absurdly simple— and complicated!

There are energy exchanges involved here. Molecules are formed, as is the case with glucose—$C_6H_{12}O_6$. Orbits of electrons are different when chlorophyll goes into an acetone solution from when it remains in the chloroplasts. There is an energy differential when electrons are excited. As electrons are raised to higher orbitals and fall back to lower ones, green plants manage to trap the energy into energy-rich chemical compounds. Light figures in creating this excitement that splits water molecules and presides over energy-rich compound production. These many energy-rich compounds in effect power food production once light no longer has a role to play.

It is the function of plants to produce more than they themselves require. That is why the farmer can harvest directly, or harvest to feed animals for later human consumption.

People who make computations on such things say that only about one part in 2,000 of sunlight reaching the earth is ever absorbed by plants. This seems a very low efficiency ratio. This equation becomes even more awesome when we consider that grass, trees and farm crops account for only 10% of the photosynthesis going on. At least 90% of this activity is carried on by algae, which are small single cell plants living in the lakes, streams and oceans of spaceship earth. However, once chlorophyll molecules in a green plant get a hammerlock on light energy, fantastic efficiency is invoked. Over half the absorbed energy is locked up into the energy rich molecules we've been discussing.

As Gene Poirot so aptly put it, nature creates life using air, water, sunshine and earth minerals. What is even more astounding is the fact that a typical crop of corn—one of the simpler monocot grasses—is 95% air, water and sunshine, and only 5% earth minerals. Nature furnishes air, water, sunshine with little or no variance, apparently leaving the farmer to concentrate most of his management capabilities on 5% of the nutrient traffic. Yet it stands to reason that we need to deal with function and nutrition in the 95% range, where results can be proportionately better than in the 5% range.

Thomas A. Edison was once asked what he thought was the greatest invention. The wise old inventor pondered the question, then answered, "A blade of grass." And then he added, "Do you really understand what makes it grow?"

FIRST SEMESTER

The Forgiveness of Nature

Grass is the forgiveness of nature, her constant benediction, wrote old time Kansas Senator John J. Ingalls in his famous paean to blue grass. "Sown by the wind, by the wandering birds, propagated by the subtle horticultural touch of the elements, which are its ministers and its servants, it softens the rude outline of the world. Its tenacious fibers hold the earth in its place and prevent its soluble components from washing into the sea. It invades the solitude of the desert, climbs the inaccessible slopes and forbidden pinnacles of mountains, modifies climates, and determines the history, character and destiny of nations. Unobtrusive and patient, it has immortal vigor and aggressiveness."

Ingalls was more poet than scientist, yet he correctly saw grass as the most important plant to man. All our breadstuffs—corn, wheat, oats, rye, barley, plus rice and sugar cane—are grasses. So are bamboo shoots on that plate of delicacies in a Polynesian restaurant.

1. grasses

All grasses have stems with solid joints plus two ranked leaves, one at each joint. Leaves have two parts—a sheath that fits around the stem like a tube that has been split, and a blade. Even seed heads have a character all their own. Flowers exist on tiny branchlets, sharing a crowded residence, always paired like the leaves.

Most grasses flower each year. There are exceptions. Some perennials are spread with the aid of underground specialized stems, rhizomes and rootstocks, and fail to flower regularly. Withal, grasses are specialists at simplification.

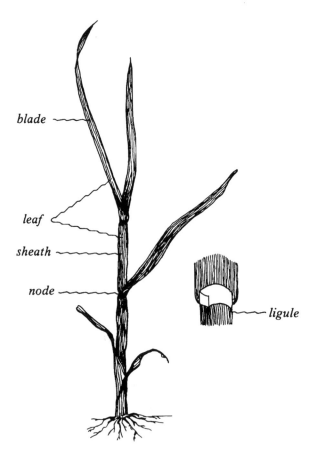

2. between the scales of a seed cone

Obeying the biblical injunction to increase and multiply is the name of the mandate in nature. Hidden deep in the ovary of the mother flower or between the scales of a seed cone is the ovule. This contains an embryo sac and a tiny egg. The egg must be fertilized by a sperm cell from a pollen tube before it can start to develop into an embryo, and thus perpetuate the parent's life.

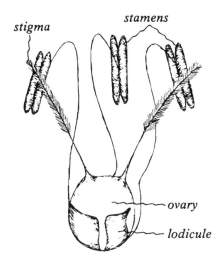

stigma   stamens

ovary

lodicule

Much like any infant, the embryo needs a special store of food, a formula on which to live after it has become separated from the mother plant. That's why every seed has its ration of carbohydrates, proteins, fats and minerals. Just what the package contains is programmed by the computers of nature according to the kind of seed. Corn, as might be expected, is heavy in starch. Flax and sunflowers specialize in oils and fats. Peas and beans like protein. Seeds like to hide out their food stores in a diversity of places. Some stash the reserves inside seed leaves. Some place the goodies away in tissues developed from the embryo sac, in the endosperm, or in tissues developed from cells of the ovule that surrounded the embryo sac.

Nature has even programmed a distribution system for seeds. Some travel on the wings of the wind, fitted out with a feathery pappus that serves as both a sail and a parachute. Sticktights attach themselves to animals and hitch a free ride. There are the tumblers, and the passengers in alimentary tracts which are impervious to digestive fluids and gizzard grinding. All seeds need proper temperature vacillation and environment or they won't grow.

3. a built-in computer

Each seed a farmer deals with comes with a built-in computer. Nature's programming tells the plant when to sleep, when to wake up, how to translocate nutrients, and how far to go in the food production business. As long as enzyme systems work properly, the seed does what it is supposed to do. It takes trace element keys to activate these enzymes. If the farmer does anything to upset this fine-tuned computer system, strange abnormalities start appearing in plant, animal and man.

Even so, seed producing plants—or spermatophytes—are merely an end product in a long chain of development. Affected by the process has been not only the reproductive, but also the vegetative structure of plant life.

4. botany divides

Botany divides the plant kingdom into four divisions, the pecking order being from the simplest to the most complex. The simplest are the algae and fungi in the soil, microscopic plant life that can deliver great benefits and hand out great damage to farm crops. Next are the liverworts and mosses, parasites all, living as symbionts with higher forms of plant life. Above the mosses and liverworts are ferns and fern allies. Highest of all in the evolutionary scale are the seed plants. It is well to have a least a primer knowledge of the plant kingdom before we move on into soil system management, seedbed preparation, tillage, and the grand diversity known as eco-agriculture.

Algae and fungi have sexual reproduction, but the sex organs and spore producing structures are one-celled and very primitive. Almost all algae live in water—in rivers and ponds and in upper layers of soil—and depend on water for function and distribution.

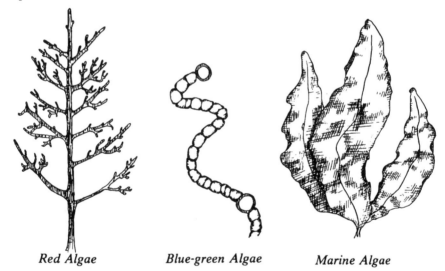

Red Algae   *Blue-green Algae*   *Marine Algae*

Fungi are classified into too many orders and families for instant comprehension. Yet everyone is familiar with molds common to old bread or rotting oranges, or the growth inside damp logs. Mushrooms are a fungi of a different stripe. Many can be propagated under controlled conditions, and made a profitable delicacy for gourmet tables. Others, kinsmen of toadstools, are quite poisonous. Friendly fungi in the soil are as much a work force for the farmer as are plant nutrients. Unfriendly fungi—*Aspergillus flavus*, for instance—are sometimes cancer causers and production robbers. Rhizoctonia is a fungus that causes damping off of-seedlings and some older plants. Antracnose is not only a serious fungus problem for sycamore trees, it is also destructive to beans and commercial plants. Alternaria, a serious blight of tomato plants, is another example of fungus problem. To illustrate the point, here are several fungi.

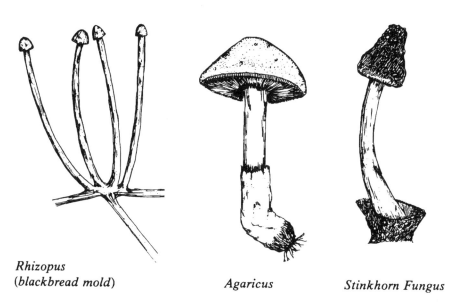

Rhizopus
(*blackbread mold*)   *Agaricus*   *Stinkhorn Fungus*

Not all plants grow in the soil. Some grow on trees, even on telephone wires, or on the scaly surfaces of rocks. Take lichens. Sometimes pale green, sometimes pavement gray, they resemble the worn out hair of an old man's beard. Sometimes lichens put on their Sunday best and add an orange colored skin to tree branches and rocks. A rootless creation, Spanish moss obtains no nutrients from the surface on which it grows. Several decades ago it was discovered that Spanish moss took its nourishment via leaf and stem, knowledge that was a harbinger of leaf feeding technology now commonly used in eco-farming. Incidentally, it is a mistake to clear Spanish moss off live oak trees as advised by most tree experts. Live oak and Spanish moss go together like ham and eggs. The moss doesn't harm the tree. It serves as a sponge to soak up rain and store it for later drip irrigation on the roots of the tree, thus keeping it watered over a long period of time.

Liverworts and mosses are land plants. Here the zygote, or fertilized egg, is retained in the female sex organ. There it divides to form a mass of cells. Mosses have sporophytes—that is to say, an asexual or vegetative part of a plant as opposed to the gametophyte, or sexual portion. Scientists define this to mean diploid generation, wherein cells have twice the gametic number of chromosomes. Don't worry about all this heady stuff now. We'll touch on just enough of it to make nature's pattern come clear, then back away and hand off the ball to you for any amount of further study you care to pursue. For now, here are some of the mosses you ought to know about.

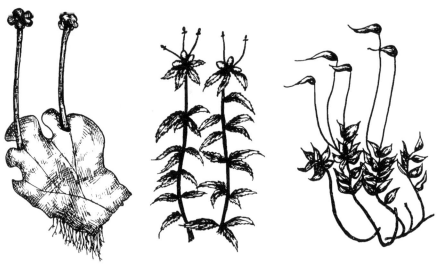

Marchantin Moss *Sphagnum Moss* *Common Moss*

Ferns and fern allies represent a big leap forward in the evolutionary process. They have a vessel system. They have plumbing which permits water and nutrients to move rapidly through roots, stems and leaves. Still, they depend on asexual reproduction. When conditions are favorable, small gametophytes—sex cells—are simply disseminated as part of the Creator's propagation plan.

All the rest are seed plants, either parasitic plants such as dodder and mistletoe, or cycads (symospermous plants that reproduce by means of spermatozoids) and conifers, firs and cedars. The flowering plants are either monocotyledons, monocots for short, or dicotyledons, dicots for short.

5. monocots and dicots

As the name implies, monocots have seeds which contain only one seed leaf. Dicots have seeds containing two seed leaves. All are flowering plants, and all rely on sexual reproduction, probably nature's greatest invention, rivaling only death as an evolutionary mandate.

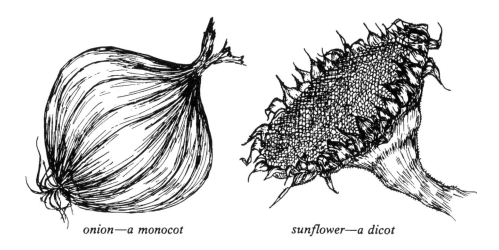

onion—a monocot   *sunflower—a dicot*

Among the important monocots and dicots in world agriculture are the following:

Monocots	Dicots
Bananas	Beans
Corn	Rubber
Canes	Cocoa
Grasses	Coffee
Lilies	Oranges
Orchids	Breadfruit
Onions	Yams
Pandanus	Croton
Palms	Crotolaria
Sedges	Most garden flowers and vegetables
Taros	
Vanilla	

6. the seedbed becomes a rootbed

Seedbed is one of those words we wish had never been invented. At first glance, it seems appropriate enough—a prepared bed for placement of seeds for optimum growth and production. But nothing is static in any form of agriculture. The seed doesn't stay a seed. Very soon the seedbed becomes a rootbed, and what may be suitable for a seed eating up its sustenance simply won't do when tap roots go down and lateral roots venture out hard on the hunt for nutrients.

When a seed germinates, a tiny root peeps out. It is called a radicle. As it ventures deeper and deeper, it often becomes a taproot, thus changing its role from being an opener to being an anchor. At the opening to the seed made by the radicle, two more roots venture forth. Styled *lateral roots*, these adventitious roots have roles that differ according to whether a plant is a dicot or a monocot. The taproot is front burner stuff when the plant is a dicot. Not so with the monocot, which relies more on the adventitious roots for stability against stress.

Much like an army patrol, the root tip pushes ahead, down into the soil, as if paving the way for a whole host of fine white hair troops looking for a meal. Fine root hairs not only seek and find a whole cafeteria of food, they also carry home the water needed for plant life.

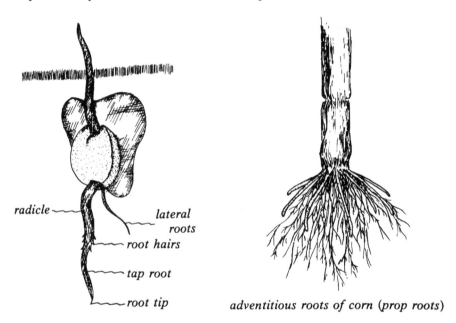

radicle

lateral roots

root hairs

tap root

root tip

adventitious roots of corn (prop roots)

Watching a corn plant grow takes on the color of a miracle in slow motion. Little swellings appear along the sides of larger roots, and these form root eyes. Here lateral roots will form. If the stress of life breaks away a root, a new root will appear as if by magic out of one of the root eyes. After the soil really wakes up in spring, one can sometimes hear corn growing, so rapid and fantastic is the scenario of a seed sprung to life.

An old paradox applies to almost any discussion of plant life. It says, all generalizations are false, including this one. One must always keep this point in mind when summarizing. Roots nevertheless have several things in common. They anchor the plant in the ground. They take up plant nutrients and participate in water transpiration. They store and hold plant food. And—here goes the generalization out the window—they propagate some species of plants, the sweet potato, for instance.

The sweet potato is really a swollen root plant. The familiar Irish potato is a tuber that got mixed up in the evolutionary process and now grows its fruit on an underground stem.

There are plants, of course, that have no root hairs, and there are plants that can be propagated with cuttings, such as grapes, in which case nutrients enter through the thin root epidermis.

7. a typical plant

From the ground up, here is a typical plant—corn.

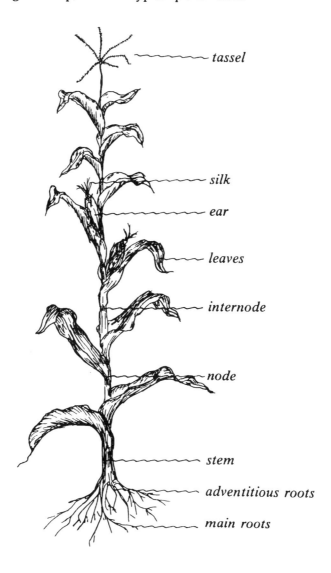

tassel

silk

ear

leaves

internode

node

stem

adventitious roots

main roots

The soybean plant has a different nomenclature.

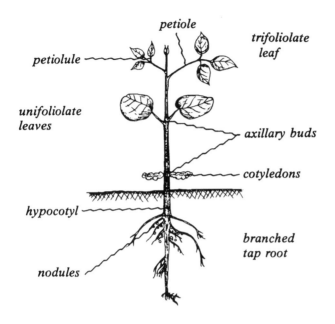

Here are wheat, left, and rice, right.

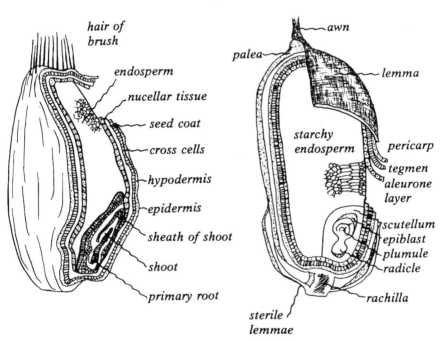

8. seeds and stems and leaves

Seeds and stems and leaves have their several functions, stems carrying water and nutrients, holding up the structure of the plant and the fruit it produces. Seeds warehouse life and preside over the genetic makeup, the stamina, the health and the production capacity of the commercial plant.

It is the leaf, however, that now attracts more attention than it has heretofore. Not that the nature of the plant has changed. Those foliage leaves at the tips of stems and branches still handle the chlorophyll payload that permits plants—during daylight hours—to make carbohydrates from starch and sugar from oxygen, hydrogen and carbon. A leaf has its own nomenclature. The skin is called a cuticle. It has tiny holes called a stomata. The stomata allows a leaf to breathe.

The leaf blade is held out, stretched and shaped by a midrib, a sturdy fiber that takes on the function of a barn's roof timber. Leaf veins run across the leaf like secondary roads at the end of each section in the countryside. Sometimes they make a network, like trails and traces lacing country roads together in Vermont, and sometimes they keep their straight lines, like an Iowa section line. Dicot leaves form up at the stem via a narrow leaf stalk—the petiole. Monocot leaves attach themselves to the stem with a sheath or leaf base.

Here is a dicot leaf, in this case the willow.

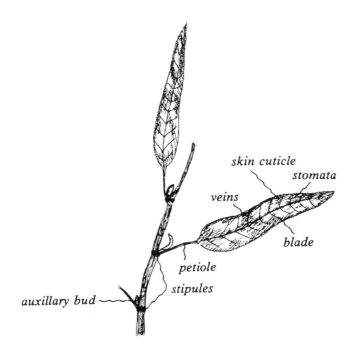

skin cuticle

stomata

veins

blade

petiole

stipules

auxillary bud

The arrowhead leaf is typical of the monocot plant.

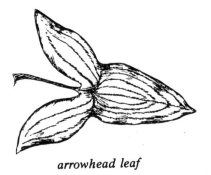

arrowhead leaf

It is probably an oversimplification, but leaves come in two shapes—simple and compound. A leaf system is compound if it is composed of two or more small leaves.

Simple

white oak

sugar maple

Compound

buckeye

black locust

9. plants capture solar energy

All energy comes from the sun. Tiny spines called trichomes probably focus the sun's energy onto the surface of the leaf. The curved trichomes on small veins of wild grape leaves, for instance, are hollow. They are also coated with wax exactly as are the sensilla spines on insect antennae—a happenstance we will discuss in the lesson on insects. Likely, the leaf trichomes should be credited with the efficiency mentioned earlier, that of plants needing only one part in 2,000 of the sun's energy for efficient growth.

In any case, plants capture solar energy in their leaves, as we have seen in lesson 1, and use this energy to make sugar. They then go out to construct cells of all kinds, using sugar, air, water and earth minerals as building blocks. The first part of the equation is the photosynthesis story everyone from van Helmont to Van Niel has spent so much time unraveling—the business of plants using sunlight to synthesize a new substance, namely sugar. After that, sugar and oxygen serve up carbon dioxide, water and energy. From a chair of a theory of energy, chlorophyll (the green coloring matter of vegetation that is built around a single atom of magnesium) is the original capitalist. Energy capital is there for the taking. Plants always have been and always will be the chief key to natural energy because the chlorophyll is the principle transformer of solar energy into the kind of power plants, animals and human beings can use.

Plant leaves thus manufacture plant food. They store food, run a refrigeration device for the plant, and do the plant's breathing. Knowing this much, scientists kicked open the door to even more knowledge during the 1950s. Sylvan H. Wittwer, Ph.D., of Michigan State University found that the efficiency of foliar fertilizers was 100 to 800% greater than fertilization with dry materials applied to the soil. The results of this pioneer research were made a matter of record in an audio film styled *The Non-Root Feeding of Plants.*

Stems, too, have a character and nomenclature all their own.

Rootbed Cutting

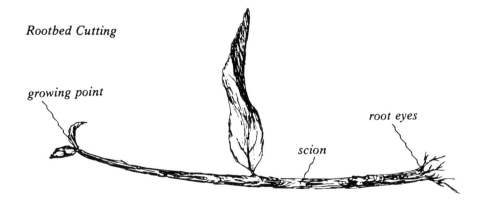

growing point

root eyes

scion

10. at the terminal bud

Growth takes place at the terminal bud. If the terminal is pruned away, the next bud in the pecking order takes its place and becomes the leader. Many dicot plants—but not all—have lateral branches or suckers. Some have a bud that can give shots that ultimately carry flowers and fruit on the side of the stem.

As noted earlier, root eyes appear on many plants. It is from these eyes that adventitious roots grow to supply prop roots, such as one sees in a corn field.

Another distinguishing feature that separates dicots from monocots should be noted.

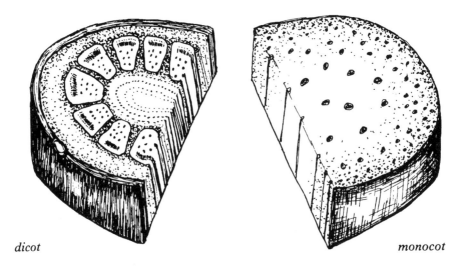

dicot   *monocot*

Taking a cross section of each stem, it will be seen that dicots have clear rings, a cambium layer, bark and a distinguishing core. The monocots seem to have vascular bundles scattered more or less at random. Dicots such as the sweet potato, soybeans, and rhubarb are usually woody. Monocots such as corn and asparagus are soft and fibrous.

The cambium layer is simply growing tissue. It grows wood and veins inside and pushes the bark out. This process continues year after year, hence the growth ring. Cut the cambium layer, and it will heal, leaving a scar. This is the reason naval stores—made into rosin, pitch and turpentine—are tapped without going through the cambium layer, otherwise the tree trunk would become a mass of scars.

The monocot stem is quite different. There are no cambium growth rings, just the fibrous materials running vertically through the flesh of internodes. There are the vascular bundles mentioned earlier. Here is the plumbing system that carries plant food and water from soil to leaves, and sometimes water from leaves to the soil.

The dicots have veins in the growth ring. The cambium is creating new tissue constantly. In monocots, veins are run like network through the entire stem. The stem grows bigger as old tissues swell and become vascular bundles and also expand.

11. rhizomes

Stems can be the source of new plants as cuttings are taken, or runners reach out for a new anchor. Some plants—Bermuda, for instance—have underground stems called rhizomes. These send out roots from some of their nodes. The sweet potato does the same, as does the watermelon plant. A grape plant can be trained to re-root in exactly the same way.

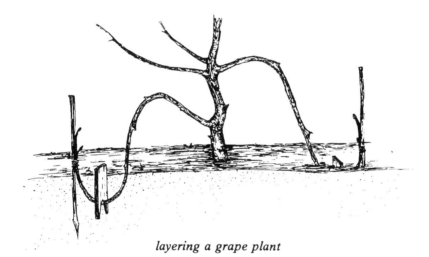

layering a grape plant

12. non-root feeding

As suggested by *Non-Root Feeding of Plants* experiments, some plants absorb dust particles from the air to obtain minerals. Some plants—like pepper plants—send out roots that attach themselves to soil and other plants, but they obtain no food or moisture this way. They simply catch and hold rain and dew with their foliage. Such plants are styled *epiphytes*.

Flowering plants are of main concern to the farmer since most commercial production—leather ferns and mushrooms excepted—have to do with sexual reproduction, pollination, and the man-directed experiments called hybridization.

13. sex in plants

The Arabs were the first to recognize male and female sex in plants. They realized that boy trees had to be planted with girl trees or there would be no date crops. Still it remained for an Austrian monk to put plant sex studies on a scientific basis.

Gregor Johann Mendel entered the order of Augustinians at Brunn in 1843, age 21. In the monastery garden he grew peas—tall with dwarf, yellow seed with green seed, whatever—making all the combinations ingenuity and logic could account for. Some 22 years later Mendel read his famous paper on genetics before the Natural History Society at Brunn. Needless to say, those who heard Father Mendel's epoch making work on how peas transmit their traits failed to comprehend Mendel's Law, or the new world of knowledge it had opened to them.

Mendel directed attention to the plant flower.

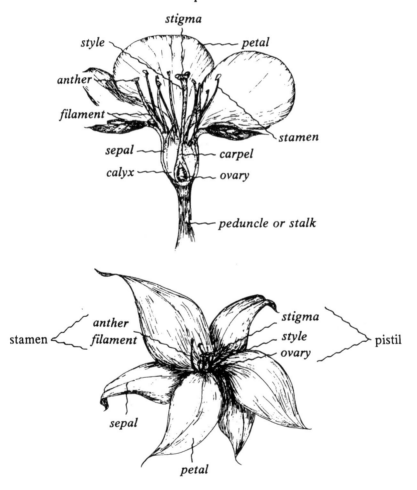

As noted above, the cut of floral leaves is called the calyx. Each of the individual leaves is called a sepal. These are sometimes joined together. Sometimes they are not. In the middle of any flower is the style, and further down, the ovary. Ovary, style and stigma are called pistil. Note that the pollen sacs attach at the ovary and shoot out and up, reaching almost to the stigma. The pistil is the female part of the flower. The pollen sacs are the male part.

A sweet nectar characterizes many flowers, as does a pleasant perfume. These pheromones are simple molecules featuring eight or ten carbon atoms in a chain, but they telegraph messages that put ITT to shame. They program the bees and insects, telling them to come or stay away, hand out instructions on how to behave toward the opposite sex, and how to help the flower in its birth-oriented program. Bees move about, gathering and coming, testing and going, all to feed their own young and obey the injunction to increase and multiply. As they move and fly, they distribute the pollen to the stigma—pollination!

14. pollination

Everyone knows the bee can't fly according to the laws of aerodynamics. But the bee does not know this, so it does fly, delivering 225 wingbeats per second. These wingbeats help the wind stir pollen and make their contribution to the sex act in plants.

When pollen reaches a stigma, it causes a tiny, thin root to grow—the pollen tube. This tube plunges down through the stigma, down the style, into the ovary. There it joins a tiny organ called ovule, and from all this will grow a seed if fertilization is right with God and the world.

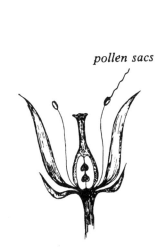

pollen sacs

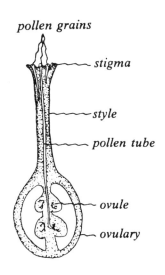

pollen grains

stigma

style

pollen tube

ovule

ovulary

Bees are not alone in their work. Insects also pollinate, as do the winds.

All insects are not equally effective in pollination work because of wingbeat and size and distance capability. Here, nevertheless, is a table giving wingbeats per second and how far their efforts take them per second.

	Wingbeats Per Second	Meters Per Second
Medium butterflies	8-12	2-4
Damselflies	16	1-2
Scorpion flies	30	.5
Large dragonflies	25-40	7-15
Noctuid moths	30-60	4
Hawk moths	50-90	5
Hoverflies	120	3.4
Bumblebees	130	3
Houseflies	200	3
Honeybees	225	2-3
Mosquitoes	150-500	-.5
Midges	1,000	-.5

15. the grand diversity

Any plant that can achieve fertilization with its own pollen is said to be self-compatible. Many plants do not accept their own pollen—the so-called self-incompatible plants.

The grand diversity in nature again reveals itself. Some plants produce fruit, albeit no seeds, notably the banana. Some plants are sterile—that is, they produce no fruit in one case, or seed in another. These plants can be reproduced only from cuttings or other parts of the mother plant.

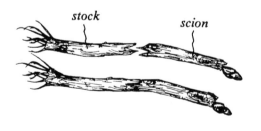

grafting

cutting tubers into pieces

A few plants are hermaphrodites. That is to say, they are both male and female complete in the same flower. In botany they say the flower is perfect. Plants also commonly have male and female parts on different flowers. Cucumbers and squash provide good examples.

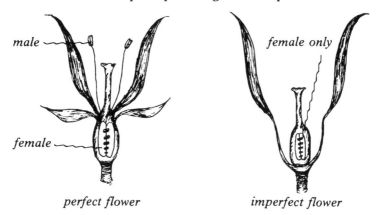

perfect flower　　　　　*imperfect flower*

When male and female are in fact on the same plant—pumpkins, corn, squash—the plant is called monoecious, from the Greek for "one house." When male and female are in fact on different plants in the same area, the term is dioecious. It is a dioecious plant. This is what the Arabs discovered when they found they had to plant male and female date trees in the same grove.

With a few more terms, the lexicon is complete. A plant that can pollinate itself is self-pollinated. A plant that is fertilized with pollen from another plant is cross-pollinated. When a flower from one variety is fertilized with pollen from another variety, it has been hybridized. As new seeds are produced from this mating, they are called hybrids. It isn't exactly the same as with the mule, which has no pride of ancestry or hope of progeny. Hybrids can be planted back, albeit without good results. Hybridization makes it possible for seed companies to keep farmers coming back for machine counted, high priced seed instead of growing their own.

16. hybridization

Control of seeds has long been an economic objective. Thomas Jefferson once risked the death penalty to smuggle an upland rice seed out of northern Italy. Then as now someone wanted to protect a monopoly. History tells us that the trigger mechanism for Mexico's revolt from Spain was sprung when a provincial padre was told he couldn't plant grapes for altar wine, this also to protect the monopoly in Spain.

Of all the techniques ever used to control seed production, none have rivaled hybridization.

OPEN POLLINATION

Some of the successes that have attended corn production may have been the result of distorted accounting procedures. In terms of nutrients, open pollinated still has an enviable record. Adolph Steinbronn of Fairbanks, Iowa put the matter in perspective by having two samples of corn tested for ingredients usually added to commercial feeds. One was a sample of hybrids he had grown. The other was OP corn. The OP corn contained 19% more crude protein, 35% more digestible protein, 60% more copper, 27% more iron and 25% more manganese.

Compared to some 4,000 samples of corn tested in ten midwest states in a single year, Steinbronn's OP corn contained 75% more crude protein, 875% more copper, 345% more iron, and 205% more manganese. The same trend has also been seen in the content of calcium, sodium, magnesium and zinc. It can therefore be said that OP corn could contain an average of over 400% more of these nutrients.

Ernest M. Halbleib of McNabb, Illinois confirmed the failure of hybrid corn to uptake certain mineral nutrients. In comparing Krug OP corn and a hybrid in the laboratory of Armour's Institute of Research, Chicago. spectrographic testing revealed the hybrid short of nine minerals. The hybrid failed to pick up cobalt and any other trace minerals. Both varieties had the same chance to pick up a balanced ration.

The reason I mentioned cobalt," wrote Halbleib, "is that we found (on the 16 farms in test) that no hybrid picked up cobalt, and in all the tests the hybrid was short seven to nine minerals, always exhibiting a failure to pick up cobalt."

The core of vitamin B-12 is cobalt. Ira Allison, M.D., and others have found that a lack of cobalt is implicated as a cause of brucellosis and undulant fever, and cobalt is part of the cure.

In the opinion of many eco-farmers, hybrid corn merely masks poor farming by producing bins and bushels without the nutrient goodies that are really corn's reason for being.

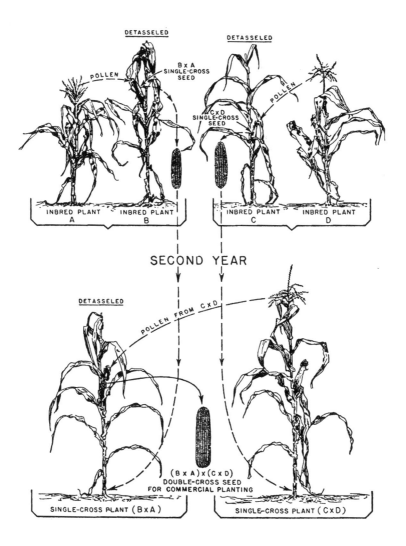

FIRST YEAR

SECOND YEAR

This diagram reveals the method of producing double cross hybrid seed together with representative ears of the crop. B x A or C x D would be a single cross; (B x A) x C or (C x D) x B, a three way cross; and (B x A) x (C x D), a double cross. The fourth is a top cross—a cross between an inbred line and an open pollinated variety.

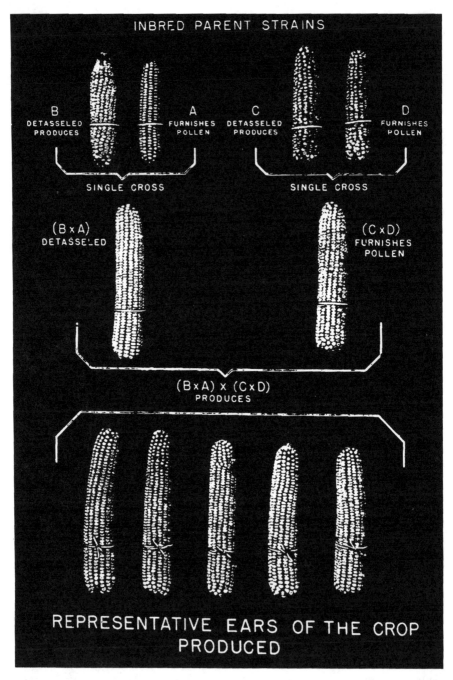

Method of producing double-cross hybrid seed corn and representative ears of the crop produced from hybrid seed.

Hybridization has to do with combination of desirable characteristics found in two plants. The hybrid can be a cross between varieties, and sometimes between species. The details need not detain us. Suffice it to say that to produce purebred strains of, say, corn, by inbreeding, the breeder selects plants with known characteristics. These are heavy yielding plants with strong roots, stiff stalks, good ears, wide leaves and other desired qualities. These plants are inbred—in other words, forced to self fertilize.

The silk of an ear cannot be fertilized by pollen until it grows out through the husk. To inbreed, ears are covered by a paper bag. This prevents wind pollination, sometimes called open pollination, nature's process whereby pollen from many corn plants is carried by air to settle down on receptive silks. There it germinates, sending down a pollen tube through which the sperm reaches the egg—or fertilization! Each kernel of corn has its own thread of silk and will not develop unless this thread has been pollinated.

In the inbreeding process pollen from the tassel of the same plant which bears the ear to be pollinated is shaken into another paper bag. The bag covering the ear is then quickly replaced with the bag containing the pollen. The corn is thus self-pollinated.

Hybrid seed corn is a first generation cross or hybrid between two or more inbred strains of corn.

The USDA diagram presented here illustrates the method used in crossing four separate inbred plants to obtain two single crosses. Note how the single crosses are crossed to produce a double cross hybrid seed. Once a desirable hybrid combination has been determined, it can be reproduced repeatedly, simply by maintaining and recrossing the same inbred lines which have already been inbred to the point where each reproduction will give the same results.

Of all the commercial farm crops, corn is the least difficult to hybridize. Usually six rows of the seed-bearing parent, a ratio of one pollen parent to three seed bearing plants is the general rule. High school and college kids are usually employed to remove tassels of the six rows. As a consequence, the seed set will be hybrid.

Today corn, celery, carrots and onions are grown with hybrid seeds, corn and onions easily leading the way. There is a reason for this. Many hybrids are produced by hand crossing. This means emasculation, or removal of the anthers of flowers before pollen has spread, and at the same time protecting these flowers from insect and wind-spread pollen. Not many plants can stand the ordeal. As a consequence, a lot of emasculation and pollination is carried on in a greenhouse. Sometimes flowerbuds—cabbage, for instance—are opened artificially and pollen applied to the stigma. Bud pollination is used on self-compatible radishes.

There is always a question about how pure a seed crop can be kept. Commercial plants that are easily cross-pollinated by insects and wind are heterozygous for some characteristics. They can't be put into sealed cages, at least when grown on a large scale.

The world has listened long and hard to science telling about hybrid vigors, about dwarf wheats to feed the world, about the much desired male sterility in plants—it prevents unwanted pollination—about those extra chromosomes in potatoes—tetraploid, meaning four, diploid, meaning two—about control of disease via genetics, about green beans with less pod fiber. There is even research afloat to produce a gasless bean. (There is such a bean, of course, one evolved by nature, but it yields poorly, and so far consumers have elected to endure pain in the intestines more than pain in the pocketbook.) In general, however, hybrids simply have less of a burden to carry—that is, reproduction of their species in the seed. This has accounted for more bins and bushels, often with greater protein and trace mineral deficiencies.

Nevertheless, the hybrid concept has swept most of the agricultural scene. When D. F. Jones of Connecticut Agricultural Experiment Station developed the double cross corn concept, the door flew open. From that moment on seed breeders worked long and hard with high-yielding, single cross parents. In 1933 only .3% of the American corn acreage was planted in hybrid seeds. By 1960 this figure had jumped to 95%. It is probably in excess of 98% at this writing.

Cytoplasmic male sterility hasn't been limited to corn alone. It has reached out and touched pearl millet. Hybrid sorghum has largely replaced open pollinated varieties. Soybeans have largely replaced cowpeas for hay and green manure purposes, and hybrids became the name of the game during the 1930s—hybridization and backcrossing (followed by selection).

Even the buffalo grass has not escaped genetic engineering. The plant has staminate and pistillate on separate plants. This specialty has been used to produce the hybrid Mesa. In other words, seed fields are interplanted with selected clones of two dioecious parents.

All this is to simply illustrate that crop potential is determined largely before a farmer puts the first tool to the soil, even before he starts his seedbed and rootbed preparations. Many farmers hone close to low-priced seeds, seeds without the genetic potential needed to deliver a healthy, abundant crop. This is being penny wise and pound foolish, to use that folklore saying.

Seed selection has to be made while the production year is still on the drawing board. Climate figures, so does land. Germination time must be understood. Temperatures that can be endured, and water requirements—all must be made part of the decision process.

The decision is further complicated by the fact that most seeds can handle and germinate in a wide range of temperatures—usually slowly at the lower range, more rapidly as the soil comes alive at higher temperatures. Temperatures that run too high, on the other hand, may impede good germination.

IMMUNE PLANTS

Much reliance is put on the belief that by selecting and propagating certain plants of a crop we eventually find those which will tolerate diseases like smut, rust, root rot and others. Much is said about freeing resistant crops, or those which will tolerate such troubles. We fail to see the germ diseases as attacks by these invading foreign proteins in their struggle to get their necessary protein while parasitically taking over organically elaborated materials of that kind as starter compounds from which to synthesize their own. We fail to see that immune plants are those getting enough soil fertility support for creating their own protective proteins or antibiotics in the same way as fungi make theirs to protect themselves from each other and to protect us similarly when we take their antibiotics into our blood stream.—William A. Albrecht, Ph.D.

WEATHER MODIFICATION

Without expressing it this way, farmers know instinctively that there is a highly elastic envelope around spaceship earth. Scientists have used this knowledge of this energy continuum to engineer rain and drought, wind and hail, lightning and snow-pack. Weather modification in fact emerged from WWII as a byproduct of smoke screen research. It marched off into the New Mexico desert, where Irving Langmuir, Ph.D., released nucleating materials to bring down rain—both locally and in the eastern states—at regular intervals. Cloud seeding materials cause release of latent heat, each gram of silver iodide capable of releasing a trillion calories of latent heat, thereby generating high winds.

Unless seedings are engineered with precise mathematics, the likely result of cloud seeding is removal of snow and rain by forcing moisture aloft as ice crystals, and sending them downwind. Langmuir's 88 weeks of research revealed that seedings in New Mexico could produce rains and floods in the Ohio River Valley five days after release. Indeed, Stokes Law decrees that tiny ice crystals can take a week to fall, this after being swept along by high velocity winds.

A single gram of silver iodide can bring down 500,000 acre feet of rain. Unfortunately this rain can't be pinpointed.

Private, uncoordinated cloud seeding and inept public programs are riding a bonanza. With airplanes and ground generators, chemicals and promises, they seed everything in sight. The results have been the worst droughts since the dust bowl, and the worst floods, high winds and tornadoes in history. No mystical Act of God theory can assign blame for these uncomprehending acts of men. Farmers may not be able to do much about the weather—nature's weather, or man's weather—but they owe it to themselves to understand cause and effect.

LESSON 3

The Light Factor

Plants ought to enjoy perfect health. They keep regular hours. They don't smoke or drink hard liquor. They're programmed by their genetic codes to grow and produce on a certain timetable. Yet somehow nature's clock often goes wrong, the fruit fails to arrive, the plants may even lodge or die, with the farmer's efforts going down the tube.

1. guerrilla war

Why is one farmer successful in producing crops, another a failure, even when they have the same resources and the same soil? What is the best way to farm? Does the fertility management really underwrite the health of a crop, and therefore the health of animal and man? These are questions farmers ask each day, and the answers are usually wanting. Nor do the great myths of agriculture come to the rescue. The Food and Drug Administration started carrying on a guerilla war with the organic gardening people many years ago as a part of a refurbish-the-findings-of-medicine drive. In the process the FDA developed some fairly naive ideas about farming, soil systems, and why the plant ought to grow properly (even though it doesn't). Some of these ideas were even published in the *Federal*

Register as findings of facts to be accepted under the pain of law, if not mortal sin.

Item. "There is no relation between the vitamin content of foods and the chemical composition of the soil in which they are grown."

Item. "Scientifically, it is inaccurate to state that the quality of soil in the U.S. causes abnormally low concentrations of vitamins or minerals in the food supply produced in this country."

It was probably not enough for the organic people to answer the above by saying, perhaps correctly, that FDA people had been on white bread a little too long. The bureau people in fact leaned on the 1959 *Yearbook of Agriculture,* in which USDA blandly proclaimed that "Lack of fertilizer may reduce the yield of a crop, but not the amount of nutrients in the food produced." It seems that up in Ithaca, New York, some scientists took a worn out piece of ground and ran a test. They fertilized one plot with nitrogen, phosphorus, and potassium. Otherwise the two plots were handled alike. After testing the results they found both crops more or less alike in nutrient content, except that the fertilized plot produced more. Milk from animals fed production from the two plots revealed no significant difference. In due time the experiment became an entry in the *Journal of the American Medical Association*—thus scientific evidence, if you will, that poor soil will produce quality food.

Not many people can accept this breath-taking absurdity as science. Even workaday farmers know that not all soils have an ample supply of calcium, nitrogen, phosphorus and a long list of elements, including the trace elements. Earth minerals are required to synthesize amino acids, proteins, vitamins, enzymes, lipids, octacosanols, phosphatides and the rest of the building blocks used to construct plant life. Plants and microbes— even those in the cow's gut—synthesize the amino acids that make up proteins. "Both plants and animals assemble their proteins to provide their reproductive functions," noted Professor William A. Albrecht, "since these are the only compounds through which the stream of life can flow."

The stream of life in plants has to do with more than providing bins and bushels. Plants may not smoke or drink hard liquor, but they suffer from lack of moisture and endure nutrient shortages or imbalance because of the physical, biological and chemical climate in which they live. When plants suffer, it shows up in the vitamin content and mineral content of food.

2. health and light

Many plants grow poorly as long as the soil fails to warm up. During this period in early spring, very little nitrogen is served up by proteins in the soil's organic matter. One can see pasture grasses literally starved for want of nitrogen. Then one day the soil wakes up, and the landscape is painted green overnight.

SINGLE FACTOR ANALYSIS

A caution is in order as you continue to read and study this primer. To be sure, light affects plant growth, synthesis of food materials, and seed production according to the type of plant involved. And it may be—as John Ott says—that "possibly the basic principles of photochemistry in connection with the process of photosynthesis do carry over from plant life into animal life, but in a greatly improved way. If the basic chemistry of the human body responds to glandular actions controlled by the pituitary gland responding to light energy, then—as with plants—the characteristics of the light energy would be a very important factor."

Obviously, these observations call for answers in an arena where the questions haven't even been framed yet. And, indeed, science cannot get the answers until it knows the questions .

In any case, light is only one factor—one of the several environmental factors that preside over crop response in the field. Any plant in a natural environment has to live with natural temperature, high or low soil moisture, during bright, dark or cloudy weather, under polluted skies, whatever. The plant does not respond to any of the environmental factors separately. It responds to all factors as they are blended into the grand mosaic of the composite environment.

Ambient air temperatures also figure in plant performance. Photosynthesis comes to a halt at night because the chloroplasts settle down for a sleep of sorts. There is a word for this—*photoperiodism*—and scientists who use that word speak of daily rhythms, biological clocks, and the like. USDA scientists H. A. Bortwick and S. B. Hendricks found, as early as 1948, that red and far-red light—that is, visible light almost in the infrared part of the spectrum—regulates plant growth. (See page 267 for an illustration of the full spectrum.) They found a protein which runs the light switch in plants, so to speak. They proved that the molecule styled phytochrome is triggered by light. It presides over the plant life process—its germination, flowering, growth. Even in sleep, respiration and burning of sugar continues. Good Iowa farmers will tell you that they can hear the corn grow at night, and they can. Entomologist and philosopher Phil Callahan has watched bamboo growing. "After a good electrical storm you can sit down level with the fresh little bamboo shoots and actually see them get longer," Callahan said. But when air temperatures soar beyond endurance, respiration and sugar burning are affected simply because too much sugar is lost.

"Light exerts a profound effect on plants and on all animal life," John Ott once told *Acres U.S.A.* readers via the medium of a taped interview. His two books, *My Ivory Cellar* and *Health and Light* furbish and refurbish this thesis.

"Sunlight is a broad, continuous spectrum peaking a little in the blue-green. It then cuts off abruptly in the ultraviolet at about 2.900 angstroms because of the filtering effect of the earth's atmosphere," Ott wrote in *Health and Light*. An angstrom is a unit of length so small it demanded a name of its own. Technically speaking it is one ten-billionth of a meter, and is used in optics and to measure light, something a farmer hardly concerns himself with. Yet a farmer has to be concerned with light.

Using time-lapse photography, Ott has been able to show the streaming of protoplasm with cells of a living plant leaf. This has to do with the photosynthesis we discussed in the first lesson of this book. Air, water, sunshine and a few earth minerals make it possible for plants to create food energy in the presence of an appropriate temperature.

But when the sun sets, photosynthesis stops. Long periods of cloudy weather and faltering sunlight intensity not only affect plant life, they help nature decide which plants can survive in climes with long and short days. Modern technology, particularly weather modification and industrial development, has a profound effect on sunlight availability, ergo crop production. The consequences of inserting carbon black and other nucleating agents into the atmosphere by climate and weather modifiers escapes instant comprehension. But the result of heavy and concentrated industrialization has been a matter of record. The Smithsonian Institution in Washington, D.C. has reported a 14% loss of overall light intensity over the past 60 years. Mount Wilson observatory in California has published figures to the effect that all farm acres have lost 10% of average sunlight intensity during the last 50 years, and 26% reduction in the ultraviolet part of the spectrum. Some virus problems and aphids brought on in part by inadequate sunlight have in fact been controlled by placing light reflective aluminum foil on the ground beneath the plants.

Poultry growers have long realized that the light in a chicken's eye stimulates the pituitary and thereby increases egg production. Ott says the pituitary is the balancewheel of the entire glandular system, not only in chickens, but in men as well.

Indeed, Ott's experiments tell us something we ought to know, even if we can do little about it. In one experiment he photographed chloroplasts within Elodea grass as they responded to different wavelengths of light energy. When the Elodea grass was exposed "to the full spectrum of all the wavelengths of natural sunlight, all the chloroplasts would stream in an orderly fashion around and around from one end of the cell to the other. However, if the sunlight was filtered through ordinary window glass that blocked most of the ultraviolet, or if an ordinary incandescent microscope light which is lacking in the ultraviolet part of the spectrum was used, some of the chloroplasts would drop out of the streaming pattern and remain immobile near the center or off in one corner of the cell of the leaf." Under red light, chloroplasts would drop out of the streaming pattern or take a shortcut, not touching all the bases. But when the color filters were removed, the chloroplasts would go back to their normal streaming procedure.

3. light sensitive

Most plants are extremely light sensitive the moment they break out of their shell, some even before. Bermuda grass, bluegrass and lettuce are light sensitive even before they break the seed coat, assuming suitable conditions of moisture and temperature. The absence of light determines the rate at which a stem elongates during those early break-out moments, and this makes it important to sink a seed at the optimum depth when planting.

The scenario of a seed come to life contains real life-death drama. Germination takes place in the dark, sending a stem to the surface. At this point elongation is checked by sunlight. At night, in the dark, the rate of elongation increases. This becomes the rhythm. In the presence of high light intensity, stems are regulated to be short and sturdy. If a cloudy pall overhangs the scene for too long, stems stretch out to become long and spindly.

Light continues to figure even after a young plant extends itself above the level of the soil. Obviously a seed can contain only enough food to keep the plant, say, a week or two. After that survival based on seed food storage is difficult because of inadequate food supply. Leaves that are grown in the dark do not expand fully. But if seedlings get adequate light before the food supply in the seed is exhausted, they rush to the task of manufacturing their own sustenance. The process can be dramatic in the extreme, taking only a matter of hours. Once carbohydrate synthesis starts, the plant no longer has to rely on that faltering supply of food in the seed.

Nature and nature's evolution has served up plants for every level of light intensity and duration. There are short-day plants and long-day plants.

4. short-day plants

Just to illustrate the point, here are a few plants that can be characterized as short-day plants.

Maryland Mammoth tobacco (*Nicotiana*).

SOIL FERTILITY AND ANIMAL HEALTH

The provision of proteins is our major food problem. Carbohydrates are easily grown. Any growing plant is synthesizing carbohydrates from mainly the elements of the weather by sunshine energy. For the output of these energy foods very little soil fertility is required in terms of either the number of chemical elements or the amounts of each. But, quite different, for the plant to convert its carbohydrates into proteins by its life processes and not by the sunshine power, calcium, nitrogen, phosphorus, and a long list, including the trace elements are required. Plants and microbes—even those in the cow's gut—synthesize the amino acids. They only collect them from the plants and assemble them into their proteins of milk, meat, eggs and other body-building foods. Both plants and animals assemble their proteins to provide their reproductive functions since these are the only compounds through which the stream of life can flow.

It is in the protein synthesis and in the reproduction of life, then, where the control by the soil of the nutritive quality of food is pronounced. Our ignorance of this control is suggested when we classify as proteins anything that gives off nitrogen on its burning in sulfuric acid. By this we include nitrogenous compounds that are not proteins. Yet we recognize about two dozen different amino acids as components of the proteins. We know that life is impossible without providing the complete collection of at least eight of them. When even the trace elements, manganese and boron, applied to the soil at rates of but a few pounds per acre for alfalfa increase the concentration of these essential amino acids in this crop—especially those amino acids deficient in corn—there is evidence that the nutritive quality of this forage is connected with the fertility of the soil.

The assessment of the contributions by the soil through only the ash analyses of the crops, has left us ignorant of the numerous roles played in the plant's synthetic processes by the elements of soil origin. In believing that we need "minerals" according to such analyses of our bodies and our foods for their inorganic contents, we consider the soil as the supply of these and the plants as haulers of them from there to us. We conclude therefrom that limestone fed to the cow in the mineral box is the equivalent in nutritional service for her as that from lime as soil treatment coming through the plant with all the effects by calcium en route in combined action with all other fertility.

Like for the protein analyses, so have we been content to accept and useaverage figures for ash analyses. In the same year and in the same state, for example, the protein of wheat has varied from a low of 10 to a high of 18% of the grain. Ash chemicals may double or treble their concentration in the crop on one soil over that on another. Such variations go unappreciated for their provocation of variations in crop services in nutrition if we are content to believe that "plants are good feed and good food if they make abig crop."

Crops that are doing little more than to pile up carbohydrates, as was demonstrated with soybeans, make big yields of bulk. But when fertilized to produce proteins, the hay yields are smaller. To be content with the above simple faith is to be as agronomically gullible as the youngster content with the knowledge of reproduction that credits this process to the delivery services by the stork.

Our reluctance to credit the soil with some relation to the nutritive quality of our feeds and foods is well illustrated by the belief persistent during the last quarter of a century, namely, that the acidity of the soil is injurious and that the benefit from liming lies in its help in fighting this acidity when, in truth, it lies in its nourishment of the plants with calcium and its activities in their synthesis of proteins and other food essentials. To say that we don't believe there is a relation between the nutritive values of feeds or foods and the fertility of the soil is a confession of ignorance of all that is to be known of this fact and is not a negation of it.

We may well ask whether the soil in its fertility pattern is of no import relative to nutritive quality of what it produces when we grow cattle and make beef protein more effectively today in the former bison area; when that area is now growing the high protein wheat; when we fatten cattle farther east on the more weathered soils and combine this speculative venture with pork production that puts emphasis on fat output by carbohydrates and the lessened hazard by marketing these smaller animals nearer their birthday; when soil fertility exhaustion has pushed soft wheat westward; when the protein in corn has dropped, because of soil exploitation, from an average figure of 9.5 to

8.5%; and when the pattern of the caries of the teeth of the Navy inductees in 1942 reflects the climatic pattern of soil fertility. Such items related to the national pattern of soil fertility suggest that many of our agricultural successes (or escapes from disaster) have been good fortunes through chance location with respect to the fertility of the soil when we have too readily, perhaps, credited them to our embryo agricultural science.

When a crop begins to fail we search far and accept others if they make bulk where the predecessor didn't. We credit the newcomer with being "a hay crop but not a seed crop." If it cannot guarantee its own reproduction via seed, we call it feed for the cow. With the cow's failure to reproduce under such poor nutritional support we, apparently, economize on the bull's energy by resorting to repeated artificial inseminations. The grazing animals have been selecting areas according to better soils. They have been going through fences to the virgin right-of-way. They have been grazing the very edges of the highway shoulders next to the concrete to their own destruction on the Coastal Plains soils. All these are animal demonstrations that the nutritive quality of feeds is related to the soil that grows it. But to date, the animals rather than their masters, have appreciated this fact most.

Shall we keep our eyes closed to the soil's creative power via proteins, organo-inorganic compounds, and all the complexes of constructive and catalytic services in nutrition? When the health and functions of our plants, our animals and ourselves indicate them all crying to be better fed, isn't it a call for agricultural research to gear production into delivery of nutritional values related to the fertility of the soil rather than only those premised on bulk and ability to fill? By directing attention to the soil for its help in making better food, we may possibly realize the wisdom in the adage of long standing that tells us that "to be well fed is to be healthy," and that good nutrition must be built from the ground up.—*William A. Albrecht, Ph.D.*

Biloxi soybean (*Soja Max*).
Ragweed (*Artemisia*).
Aster (*Aster linarifolius*).
Cosmos (*Cosmos sulphureus*).
Chrysanthemum (*Chrysanthemum morifolium*).

5. long-day plants

Good examples of long-day plants are hibiscus, goldenrod, timothy hay, radish, ryegrass, chrysanthemum (*Chrysanthemum maximum*), and Cineraria.

Typical plants not sensitive to the length of the day are asparagus, cyclamen, tomato, narcissus, primrose and foxglove.

6. at the mercy of the gods

As can be seen, the several varieties of the same plant can respond differently to light intensity. Note that one type of chrysanthemum flowers only in a relatively short light period of late summer. Much the same is true of soybeans and radishes. This is why a soybean might grow well in one area (south) and not so well in the north.

Plant breeders have used this intelligence to bring plants into flower at the right time, to make crosses and follow genetic engineer patterns not normally available under natural conditions.

Withal, light is not well understood. Intensity is not always the key, nor is light the only key. Indeed, plants do not respond to light and heat and water and other environmental factors separately, but to all the factors—many of them still unknown—as they are blended together. Admittedly the farmer can do almost nothing to govern crop light conditions in most cases. If the crop has a sufficient value, he might stretch cheese cloth so as to furnish partial shade, as is common in Connecticut tobacco fields, or among herb producers who grow exotic plants such as ginseng and goldenseal, which require shade. Greenhouse operators could do well to study John Ott's works so as to know what glass to use. But the row crop farmer is at the mercy of the gods and the men who play god as they pollute nature's clear sky.

MEMO ON LIGHT

Light affects hormones. That is how light regulates all life processes and why it is a dominant factor in development of yields. The ability of a plant to utilize light available to it is a prime limitation. After that, the question is—how much light is available, and how warm is the temperature? Light and temperature go together when they figure as limitations. All of the cell activity within leaves is either fast or slow, depending on how warm or cold it is. The colder it is, the slower it happens and the warmer it is, the faster it happens. In July and August, the cornbelt gets its highest temperatures, and cell activity in living plants is greatest. For cell activity to be producing something every day, it has to have raw materials flowing in. A plugged up plant in July and August functions like a reduced plumbing system. It does not allow the nutrients and the raw material from the root system to come up into the leaf area where it can match the sunlight.

LESSON 4

Temperature Connection

We often speak of biotic geography, but we must never forget that a bit of fine-tuning called bioclimatics is in order. Values for latitude, longitude and elevation often override geographical considerations in determining the performance of a plant at flowering and growing time.

Andrew Delmar Hopkins both coined the term and wrote down the principles of bioclimatics under USDA auspices over three decades ago.

1. For each degree of latitude north or south of the Equator, flowering is retarded four calendar days.

2. For each 5 degrees of longitude, from east to west on land areas, flowering is advanced four calendar days.

3. For each 400 foot increase in altitude, flowering is retarded four calendar days.

1. light and temperature

As suggested in the first lesson of this primer, light and temperature affect the water requirement, and temperature itself affects every single chemical and physical process involved with plants—solubility of minerals, absorption of water, gases, and mineral nutrients, diffusion, syn-

thesis—as well as growth and reproduction.

Temperature defines crop areas. Those hallmarks of biotic geography—the cotton belt, the cornbelt, the winter and spring wheat areas, the great fruit belt of Michigan, all are defined by temperature. The limiting parameters are rolled into place by nature because mean temperature decides for the farmer whether a season is too short for crop maturity or whether it is unfavorably high or low for economic and ecological crop development. Temperature presides over death and injury, either because it is too hot or too cold. Cold can injure and kill dormant plants, or pave the way for insect pests.

Average dates of the last killing spring frosts in the United States. Locate the line nearest to the locality in which you live, note the date on that line; first figure indicates the month, the second the day.

killing frost liable annually

killing frost liable in half the years

no record of killing frost

The Creator has taken all this into consideration. There are very few places on spaceship earth where it is either too hot or too cold for some form of plant life. There are blue-green algae that live in hot springs where the water is constantly near the boiling point. There are arctic plants that survive -90 Fahrenheit temperatures. As winter approaches nature has such plants brace themselves by going into a rest stage. There are trees that can withstand -65 F. in winter and yet perish when temperatures dip a few degrees below freezing during the growing season.

In China farmers often hand-till the soil, turning it up to the winter on the theory that hard freezes will kill spores and pathogens. The senior author once mentioned to a Nanking experiment station worker that spore stages of some fungi could survive temperatures of liquid air and others could survive heat at 266 F., and that all the hard work was probably an exercise in futility. Of course the Chinese farmer had his orders from Peking (now Bejing) and obeyed official folklore much as American farmers obey official folklore out of USDA.

Suffice it to note that each species and variety has a temperature below which growth is not possible. This is called the minimum growth temperature. There is also a maximum temperature for plants. Above this temperature, growth stops. The ideal, of course, is the *optimum growth temperature*, the precise point at which growth is best of all.

From the minimum to the optimum, the growth curve follows van't Hoff's Law. This means for every 18 F. the rate of growth approximately doubles. Above the optimum, it falls off rapidly. At the maximum toleration of temperature, growth stops. Optimum and minimum are further apart than optimum and maximum.

2. cool season crops

Cool season crops such as oats, rye, wheat and barley have typically low cardinal growth temperatures. The minimum is 32 to 41 F., optimum 77 to 87.8 F., maximum 87.8 to 98.6 F.

3. hot season crops

As the name implies, hot season crops—melons, sorghums, cotton, sugar cane—have much higher cardinal temperatures. Minimum is 59 to 64.4 F., optimum 87.8 to 98.6 F., maximum 111.2 to 122 F.

4. the entire range

There are other crops that function within the entire range of growth temperatures—hemp, for instance. It grows within the minimum of cool season crops and grows within the maximum of hot season crops.

There are other complications. Cardinal temperatures for growth vary according to the stage of plant development. Germination, seedling and maturity each may take a varied temperature range. Nor can it be implied that the optimum growth temperature is always best for the disease-free crop production. If growth is too rapid because growth temperature is always optimum, fruiting may be delayed or prevented. Plants may be produced structurally weak, and prime targets for insect and fungal invasion, or incapable of standing the stress of high wind, hail or heavy rain.

AVERAGE DATE CORN PLANTED

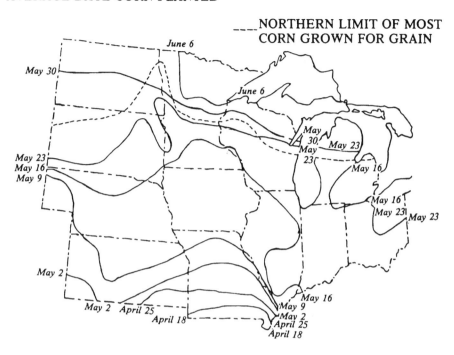

----- NORTHERN LIMIT OF MOST CORN GROWN FOR GRAIN

AVERAGE DATE 50% OF CORN SAFE FROM FROST

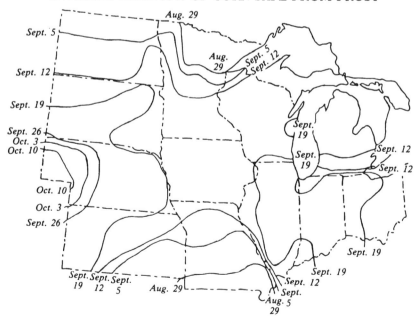

At the same time, dramatic departure from optimum temperature levels will so reduce growth as to cancel out crop profitability.

Agronomists in the land grant colleges have put all this on computers to determine "degree hours," degree days and the like, a degree hour being a 1 degree temperature above the base for an hour. *Growing Degree Units for Corn in the North Central Region* is a publication that summarizes the general growing degree unit concept (*North Central Regional Research Publication No. 229*). They have isobars and hand off things like average date of corn planting, when corn is safe from frost with several margins for safety, the objective being to single out the prime planting and growing times according to temperature.

Unfortunately nature can deliver her own whammies, and all the charts in the world won't make up for the fact that an average is merely an average, and not a spot situation for this year, this month, this week. Remember 1966, when corn started without a single stress limitation due to weather and moisture, and it disqualified all statistics. Corn tasseled in late June—earlier than ever before—received ideal moisture during August, and corn belt farmers produced the greatest yield ever. Heat degree days needed by plants to flower was actually greatly more efficient because soil and seedling nutrition had not been stressed by nature. The first 30 days of nutrient availability is more important to plant potential than any charted fertilizer prescriptions.

5. low temperatures.

Survival of the fittest could easily become an entry in each of the lessons of this primer. As part of the great design, nature has taught plants to survive heat and cold, imbalanced nutrients and lack thereof, even starvation to a degree. But nowhere is the battle plan of life as intricate as in coping with low temperatures. Tender annuals escape the freezing point by completing their life cycle—seed to seed—between frost dates. Herbaceous perennials die back but keep the life stream functioning in roots, bulbs, tubers, rhizomes. Once temperatures become more favorable, new tops emerge. In the meantime these organs are protected in their underground bunker. To this end snow cover, leaves, mulch, all protect the perennials during a harsh winter.

Many cold-hardy plants genetically develop resistance to low temperatures within their tissues. Many plants, such as cabbage, can withstand ice on the leaves for a time. As might be suspected, trees, shrubs, vines, all endure cold best, all the while keeping the secret of their physiological adaptation. Plant breeders constantly try to run a tap into this knowledge in order to genetically engineer cold resistant plants.

Farmers who are hard on the hunt for greater production efficiencies often push crops into geography that is outside their temperature range.

Cold injury is a common result since fringe days are forever barking at the heels of such a maneuver. This is not to say that cold temperature injury is not a problem for the entire nation. There is no area that does not have at least occasional frosts.

Hardest hit are the tender crops—beans, melons, tomatoes. Cold snaps move in and kill or ruin fruit on trees in the citrus areas, sometimes taking trees with them. Deciduous fruits and nuts take a whipping when frost hits buds, flowers, young fruit. The flowers, shoots and leaves of ornamental trees, shrubs and perennials are often annihilated by cold weather at the wrong time.

There are many dimensions to be considered when cold damage is inventoried. For example, not all plant parts are killed or injured when there is frost damage. Given an assist with some of the technology to be discussed in a later lesson, recovery can be dramatic—and complete. At the same time injury is not always at once apparent. Like an old battle wound, cold damage can surface after buds are open and fruit is set, or in late summer.

In case of herbaceous plants such as winter wheat, grasses, alfalfa, clover, strawberries, winter cold over a long period of time can outright kill them. Milder cold damage can kill buds or damage crowns or roots.

The effect of cold damage is a study all in itself, and could easily fill a book the size of this one. Indeed, an inventory of damage to woody plants could take exhaustive treatment. Terminals may be killed back by cold. Internal injury—"black heart" being an extreme example—can be detected by discoloration. This stands to reason. The cambium—that region between the wood and bark where new wood and bark cells are formed—is the toughest and most resistant part of a woody plant. This is where the new rings of sound wood are laid down. It is nature's way to cover discolored layers of wood with new layers.

Some parts of the country have sunscald. This happens when the sun shines brightly, warming one side of a tree well above the temperature of the shady side. When evening falls, these warmed tissues freeze rapidly—hence sunscald. Winterburn is a name sometimes given to evergreens that bow to the rigors of extreme cold in the same way.

This is not to suggest that all low temperatures are harmful. A lot of plants—the deciduous fruit trees included—have a rest period. No growth takes place during the rest, even if external conditions are favorable. Cold helps preserve the rest period.

In the case of winter wheat, a low temperature during germination and early seedling is a natural necessity, or there will be no normal development. Farmers who sow winter wheat in spring find that it will not head. It is possible, on the other hand, to slightly germinate the seed, then hold it for a couple of months at near freezing temperature before spring sowing. This way spring sown winter wheat will produce a crop. Agronomists have a name for this—*vernalization*. When farmers can't get into the fields in fall, they can still plant many cool climate crops in spring with the principle of vernalization in tow.

6. heat

It is difficult to separate light intensity, high temperatures and water requirements because the optimum growth temperature is inescapably linked to transpiration and the sweep of the sun. Above the optimum growth temperature, plants become dwarfed.

When plants are pushed out of their natural temperature zones, production results usually falter. A Temperate Zone plant will produce vegetative bulk in the tropics, but very little fruit. As a matter of fact one can almost generalize that a crop out of place will produce poor fruit, poor flavor, poor color and poor keeping quality. High temperatures can affect crops grown in the right place just the same. The lexicon has many names for this kind of damage: tipburn (lettuce and potatoes), scale fruit (strawberries and gooseberries), heat cankers (flax and fruit trees). The specifics vary but the general effects are defoliation, premature fruit drop, and death of the plant.

Water Requirement

Water, or lack thereof, is often named by angry farmers when crops fail to deliver those bins and bushels. Abundant literature and folklore findings have convinced them that the plant's life processes, centered in the proteins, are most disturbed by great heat and lack of water. According to van't Hoff's Law, such processes are doubled in their rate of activity with every 18 degrees Fahrenheit increase in temperature. Under practical farming conditions, however, many life processes are interrupted long before the protein is coagulated or changes meet the naked eye.

Let's consider the chicken's egg in an incubator at 100 F. At that temperature a chick will hatch. A few degrees higher even for a short period of time, physiological processes are so disturbed that a normal hatch becomes impossible. Yet the protein of the egg does not need to coagulate or even coddle to upset the hatching process. Life processes in the plant are not all that different. They deal with proteins within the plant cell. They are mightily concerned with enzymes that encourage the processes of life. And enzymes depend on trace mineral keys.

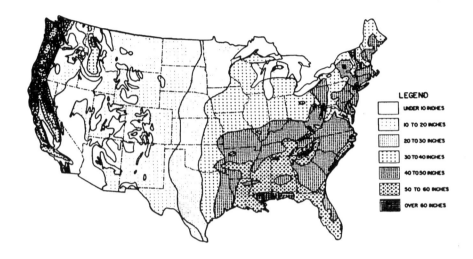

LEGEND

UNDER 10 INCHES

10 TO 20 INCHES

20 TO 30 INCHES

30 TO 40 INCHES

40 TO 50 INCHES

50 TO 60 INCHES

OVER 60 INCHES

Distribution of mean annual precipitation in the United States.

1. transpiration

Water is often no more than an alibi when plants fail to grow and produce. In folklore and conjecture, water is perceived to be the transportation system that takes nutrients from the soil to the several parts of the plant, and for this reason fertilizers—in those legislated formulas—are made soluble, "a damn fool idea," according to Professor William A. Albrecht.

As a matter of fact, the transpiration stream can be empty, as when seeds are planted in moist sand. Nutrients can go from plant back into the soil while a transpiration stream of water is moving the other way. In desert situations, moisture condensed on plants at night moistens the soil around plant roots by reversing the stream of transpiration, and all the while fertility is moving the other way. The tamarugo plant grows in a part of Chile where there isn't rain in as often as once a decade. The water table—if there is one—would be 300 feet deep. Yet the plant grows 15 feet in ten years, produces pods high in protein, and literally saturates the matted root area with water, all of it taken from the air. Crop production and relative humidity have a relationship that will emerge more clearly in the remaining pages of this primer.

Last, there can be the situation where the transpiration stream is not flowing either way. Yet with enough carbon dioxide and sunlight, plant growth and nutrient movement from the soil continues. Indeed nutrient movement is quite independent of the water supply and has little to do

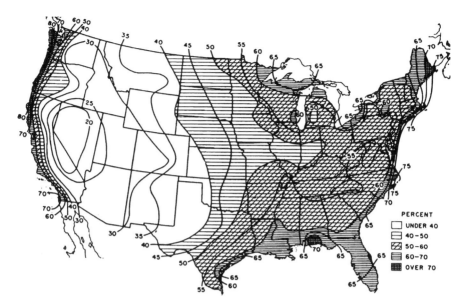

Average relative humidity in the United States at noon during July, as measured through a cool-dry and hot-dry cycle.

with water solubility for plant soil utilization.

2. the bank account

It is the soil system that has the bank account. Roots enter for a withdrawal, not to go joy-riding. Deep rooting crops will deep root if there is something to deep root for. Fine roots act as a scout, growing through infertile soil to find a payload. How do the roots know where the nutrients are? The answer can come quickly—"In search of water"—but this is wrong. In many experiments and farm situations water was available for 30 feet or more, yet a root chose to grow toward a clump of manure or a sewer tile break—remaining a thin thread all the while—then becoming thick once contact was made with the fertility load. Roots will tell you a lot about the fertility of the soil. Thin roots mean fertility shortage, roots searching and finding very little. Big, massive root systems loaded with miles of root hairs mean the plants are finding plenty to eat and have the capacity to supply the factory up in the leaves.

3. water requirements

Water requirements can't be considered independent of fertility, and yet we are obliged to consider moisture availability the greater determiner of bins and bushels.

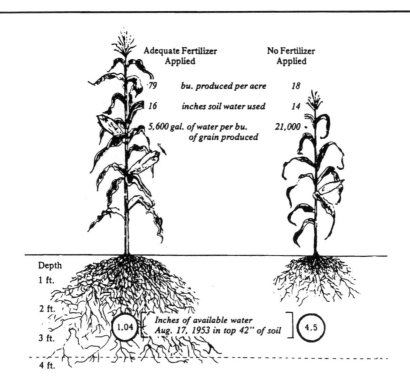

Adequate Fertilizer Applied		No Fertilizer Applied
79	bu. produced per acre	18
16	inches soil water used	14
5,600 gal. of water per bu. of grain produced		21,000

Depth
1 ft.
2 ft.
3 ft.
4 ft.

1.04 — Inches of available water Aug. 17, 1953 in top 42" of soil — 4.5

Corn on soil with inadequate nutrients cannot develop a root system capable of utilizing subsoil moisture. This fact makes fertility, not water, the key to successful production, and poor fertility an open sesame to a burned out crop in dry weather. The general principles of this concept were worked out at McCredie in central Missouri during a hot-dry phase in the most recent 100 year cycle.

Dwight D. Smith, a research associate working with William A. Albrecht at the University of Missouri, found that corn on soil with proper nutrition—grown in a four year rotation of corn, wheat and two years of meadow, grain and hay crops removed—produced 79 bushels an acre with 16 inches of soil moisture. Some 5,600 gallons of water were required to produce a bushel of corn during a drought year even with ample available nutrients.

Corn grown without soil treatment—in a corn, oats rotation with only grain removed—produced only 18 bushels an acre, and required 14 inches of soil water. On a per bushel basis, 20,000 gallons of water were required to produce a single bushel of corn. The point in all this surfaced long before harvest. Corn grown on soil with ample fertility developed a better root system and penetrated deeper on the hunt for water. Root development proved sufficient to take in water at a greater rate. And since roots penetrated to greater depths, the plants had a greater supply available from which to draw water than did the shallow root-ed corn grown on unfertilized soil. By early August in a very dry-cycle year, plant roots in the fertility area had penetrated below a 3.5 foot depth.

Water removal from the unfertilized land was about 1/8 inch daily throughout the season. This was also the approximate water removal rate from fully fertilized soil for the first one-third of the season. After that the removal was more than 1/4 inch daily. Evaporation of moisture from the soil was an important factor during the early growth stages. As the corn grew the crop required more water, but shading of the soil cut evaporation losses. Corn grown on soil with full treatment shelled 79% compared to 70% for the corn on untreated soil. Fertility treatment in this college experiment included lime, rock phosphate, muriate of potash, and starter, the usual fertilizer fare at experiment stations. Obviously, if there is deficient nutrition in the few top inches of soil, then a plant can't carry any of its nutrients deeper as those same roots hunt for more water. The most important lesson that emerged was that soil without nutrients does not provide plants with the lease on life a dry season requires. Plant roots can sometimes carry their nutritional requirements with them as they tunnel deeper, but starved plants run out of food when forced to hunt water early in the growing season. As Albrecht pointed out elsewhere in his work, "Drought may be merely the soil situation in which we have no soil fertility deep enough to feed the plants when they are compelled to have their roots go deeper to get stored water."

On the basis of moisture alone, plants can be divided into three general groups. Hydrophytes are either water plants or water loving plants. They are large celled with thin walls and thin epidermal cover. As might be suspected, the root systems are often poorly developed, serving merely to anchor the plants while water does much to hold them in position. Rice is the prime hydrophyte in American agriculture. Certain weeds and herbs also qualify.

Mesophytes are plants best served by medium moisture conditions. Most farm crops are of this type. The mesophyte plant needs water, but it also needs air around the root system. Here cells are medium sized, with epidermal covering well developed and thick so as to prevent moisture loss. Pores in the leaves regulate transpiration. When water intake is cut off or restricted, these stomata close shop and go into a water rationing routine to prevent excessive leaf wilting. This means mesophytes can stand great swings in moisture availability, the range of these swings apparently governed by nutrient availability.

Xerophytes are common to the desert. They can handle drought and come up smiling. To enable them to do this, nature has built in certain water conservation mechanisms. These plants have very small leaves. The epidermal coverings are thick and often waxy. Stomata are almost always extremely small and sometimes hidden away in pits rather than on the surface of the leaf. Cells are small and thick-walled. Root systems are fantastically complex and large. When moisture is plentiful, xerophytes grow slowly. When drought hits, they go into a holding action pending arrival of more moist days. With rare exceptions, xerophytes are of little importance in agriculture.

Mesophytes frequently take on the characteristics of xerophytes when drought conditions threaten, and they respond somewhat like xerophytes when rains continue beyond what is normal, adjusting leaf size and cell structure accordingly.

In either drought or rainy season, agricultural plants are largely water. On top of that, fantastic amounts of water pass through the plant on the way back to the atmosphere. Pores that admit carbon dioxide gas also preside over the departure of water vapor.

4. root systems

Water is channeled into the plant primarily via the root system, and we are now coming to appreciate the real potential of a good root system. At Taichi, China they grow corn with roots as deep as the corn is tall. Loess soil has been filled into gullies and ravines, and inoculated with human excreta, providing a loose and open soil. It is a matter of record that roots of a single lima bean or cabbage plant can ram their way through 200 cubic feet of soil. Many grasses and legumes have root systems 16 feet deep.

Wheat in the loess soil of the Texas panhandle has been dug out with roots even longer.

Soils that hamper root movement do more than shut off water. They also effectively slam the door to the nutrient vault. This may be because of a lack of tilth, because a plowpan barrier has developed, or the soil system has a marked imbalance of nutrients that complex each other, tighten the soil and cause starvation in the midst of plenty.

It has been estimated that the average mineral soil contains between 20,000 and 40,000 pounds of potassium per acre. Of this amount, hardly 100 pounds is available for plant use in a given year. The same situation has been observed for other plant food requirements—an abundance securely locked up, and very little available for plant use by way of iron, calcium, boron, sulfur.

Scientists aren't even agreed on what it takes to feed and grow a plant. Some 16 elements are commonly listed essential for plant growth, yet over 56 different elements have been detected in plant life. Some stimulate plant growth, and yet they are not being characterized as essential. In other instances, nature has set up a substitute system. When potassium has been benched as missing or unavailable, sodium moves in to substitute in sugar beet production. Cobalt, magnesium, silicon, sodium—items often not mentioned—all have this role at times.

Nutrients required in relatively large amounts are called macro, meaning 1 to 100 pounds per acre. Macronutrients are carbon, oxygen, nitrogen, phosphorous, potassium, calcium and sulfur. Also macro is hydrogen—not a nutrient—but essential as a hyperactive ion used by the plant to trade off for certain nutrients. All this will be explained later. Micronutrients include magnesium, copper, zinc, molybdenum, boron, chlorine, iron—and many others in trace form.

Roots of plants are surrounded by soil particles, soil solution, and biotic life. Think of what we're talking about in terms of the following focal points.

Water is the common denomination for each dimension. It presides over life, chemistry and physics.

GROUNDWATER IN THE U.S.

Patterns show areas underlaid by aquifers generally capable of yielding individual wells 50 gpm or more of water containing not more than 2,000 ppm of dissolved solids (includes some areas where more highly mineralized water is actually used.)

✕ *Watercourses in which groundwater can be replenished by perennial streams*

Buried valleys not now occupied by perennial streams

Unconsolidated and semiconsolidated aquifers

Consolidated-rock aquifers

Both unconsolidated and consolidated-rock aquifers

☐ *Not known to be underlaid by aquifers that will generally yield as much as 50 gpm to wells*

The Cell

The disturbing thing about modern agriculture is that it has not kept up with its science. The old N, P and K fertilization ideas seemed superb when they were first discovered, but now we know that they worked only by accident when they worked. Powerful electron microscopes have unfolded for all to see that marvel of microminiaturization known as the cell.

Let's review a few of the items we've covered before we move on into the business of metabolic activity, or the chemistry of growth. The seed, it will be recalled, "comes alive" by imbibing water. At this time the desiccated colloidal content becomes soaked. The seed swells. Fantastic forces are put into motion. Perhaps 2,000 miniature atmospheres actually function. Sailors know something about this, especially if they've ever shipped on a leaky tub carrying grain in its hull. The housewife knows it if she's ever made the mistake of soaking seeds in a closed container.

1. metabolic activity

It also takes air to trigger metabolic activity. Without oxygen, growth stops. This is induced dormancy. It doesn't mean the seed can't grow. It simply means nothing will happen until there is a reintroduction of oxygen. If oxygen is cut off for too long, on the other hand, the seed will die. Anaerobic metabolism will prove lethal.

The point in all this is that the chemistry of growth involves the cell, first, last and always. Likely, the farmer will never see a cell because the average farm operation has no microscope. Yet what happens at the cellular level will determine the bottom line of the income statement at the end of the year.

2. the shapes of cells

This reality is further complicated by the fact that biologists can't generalize about the shapes of cells. Cells, indeed, have an endless variety of shapes, each governed by the special functions the Creator has assigned to them. They can be platelike, elongated, concave and disclike. They can be spherical. And they can't hold their shape because of close packing. In biology texts—and for the purpose of this primer—it is a common practice to portray a generalized cell. The diagram of a generalized cell is both a fiction and the truth. Think of it as an abstraction, and think of it as a bird's-eye view of Creation.

3. the catalytic nature of enzymes

Before the discovery of the electron microscope, it was perhaps satisfactory to consider as definitive the idea that plant activity could be controlled by random saturation of cellular substrate nitrogen pools with N, P and K. This was understandable, considering the state of the arts and the level of knowledge. Now, however, it is known that "the precise regulation of substrate or metabolite pools resided with enzyme structure and activity due to the catalytic nature of enzymes." The quoted words have been taken from a paper by Mark E. Jones, one of the new breed of microbiologists commercially engaged in environmentally sound agriculture.

4. the center of life

A close look at the center of life is certainly in order before we go on our way. This look reveals that "cells carry out thousands of chemical reactions per second, and that the substances involved in these rapid reactions move by diffusion," to use the summary of Andrew Zaderej, an Ukrainian refugee scientist who possibly has done more to define soil science in terms of the cell than any other individual.

"In terms of the biochemical tasks that it must perform," wrote Zaderej in a privately circulated paper, "the cell approaches the theoretical limit of miniaturization. A simple cell requires at least a few hundred enzymes for functional purposes, many of them being molecules. Other large (DNA)

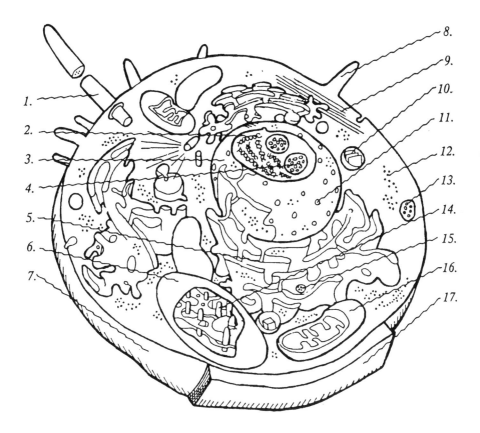

1. Cilia (locomotion); 2. Chromosome (contains DNA); 3. Nucleus (heredity bearing core); 4. Nucleolus (center of synthetic activity and storage); 5. Endoplasmic reticulum (pathway to nucleus, bears ribosomes); 6. Chloroplast (photosynthesis center of cell); 7. Cell membrane; 8. Microvilli (promote absorption); 9. Microfilaments (maintain cellshape); 10. Lysosome (scavenger organelle); 11. Pore (allows diffusion to and from nucleus); 12. Ribosomes (contain RNA); 13. Food storage gland; 14. Chlorophyll; 15. Secretion gland; 16. Mitochondria (regulates cellular activity); 17. Plasma membrane.

molecules store genetic information. The remaining biochemical machinery occupies a certain volume, as do stored food and water. The DNA molecule is built something like a spiral staircase. The phosphates and sugars form the twisted frame of the stair, and the bases form the steps. Each step is made up of two bases joined in the middle, but it is not random pairing. The sequence of nucleoticles in the DNA molecule is a kind of molecular alphabet that spells out the formula for manufacturing a specific protein."

The farmer deals with this cell, its components, its functions, and his failure to work with nature's own computer finally translates itself into something all can understand when the crop goes to market. When a farmer decides to use synthetic nitrogen instead of protein nitrogen, his decision runs headlong into the cell. When soil or leaf nutrients are either withheld or complexed, or supplied in the wrong form, the cell is the first to field the shock. When toxic genetic chemicals are sprayed over plant life, it is the cell that winces with the violence.

In the cell, plant, animal or human, there are chromosomes which carry almost all of the information needed to direct that cell's growth, division and production of chemicals such as proteins. These chromosomes are composed of information-bearing genes. Radiomimetic chemicals (chemicals that ape the character of radiation), radiation itself, and many of the chemicals used in agriculture can injure the chromosomes either by altering the chemistry of a single gene so that the gene conveys improper information (called *point mutation*) or by actually breaking the chromosome (called *deletion*). The cell may be killed, or it may continue to live, sometimes reproducing the induced error. If it continues to live, its self-regulatory powers are often affected. Some types of cell damage may cause genetic misinformation that leads to uncontrolled cellular growth. The phenoxy herbicides all operate on this principle. In human beings the same general phenomenon is called cancer.

Cellular damage because of malnutrition or the invasion of toxicity can cost a farmer either part or all of his crop. The same damage in human beings can cost a nation its heritage and its future. Damage to the sperm or ova in a human being can cause malformation or mental retardation in future generations. It can also contribute to degenerative metabolic disease. Indeed, the United Nations Scientific Committee on the Effects of Atomic Radiation has noted: "It is generally accepted that there is a genetic component in much, if not all, illness."

5. let andrew zaderej explain

Again, let Andrew Zaderej explain. "Although proteins are synthesized from about 20 different kinds of nucleotides: adenine is joined with thymine to form a stem and quinine is joined with cytosine to form another." The mathematics of it all is that simple, and that complicated.

"At first it might seem impossible that a four letter (A, T, C and G) alphabet could govern the protein code that uses 20 different letters. But the Morse Code uses only two symbols, a dot and a dash, to represent our 26 letter alphabet, and a binary computer uses only a zero and a one to perform its enormous repertoire of calculations. Each word of the DNA code is a sequence of three nucleoticles. Each code may follow each other endlessly in any order: AT, GC, AT, CG, GC, TA, etc. A single gene might be a part

COMPLEMENTARITY IN BIOLOGY

That biology is "mechanistically indeterminate" has long been a seat-of-the-pants judgment of the eco-farmer. The same general concept is now emerging—albeit reluctantly—in academic circles. The finest statement on this subject to date has been published by Johns Hopkins Press, Baltimore, under the title, *Complementarity in Biology*. This book by James P. Isaacs and John C. Lamb is not primer stuff. In fact, it is light years ahead of the science being taught at the university level, an exception being the thinking of people like Phil Callahan (see index). In *Complementarity in Biology*, the writers set down a theoretical foundation for new laws of biology, many of which are were adopted by consultant C. J. Fenzau.

There is a predication in *Complementarity* that a type of radiation will be found to be emitted from living cells. Isaacs and Lamb call this "bionic radiation." The implications outdistance the imagination. Such radiation could underwrite a new spectroscopy, and this could be used to identify tissues, bacteria in infections, processes in aging, malforming changes in embryogenesis, on and on.

In the words of Isaacs and Lamb, "Bionic radiation may have an electromagnetic nature, but not necessarily so. If it is electromagnetic, it is of anomalous character. It is probably coherent in nature with peculiar distance-dependence. The wavelengths are probably mainly in the infrared. Some type of directed molecular motion is implied."

Post graduate students will want to leap forward into such intellectual fare. There is a yawning gulf between what the abstract entity called science knows, and what individual scientists manage to learn. Often, today, practical farmers outdistance institutionalized scientists in that "know" department.

of these steps, the arrangement of TA's, AT's, GC's and CG's giving each gene its special character. The amount of DNA in a living organism and the complexity of the organism seem to be somewhat correlated. The coiled DNA stairway in the chromosomes of a simple cell (virus), for example, might be 1/2,000th of an inch long and contain 170,000 steps in the virus cell."

The figures involved in nature's tiny computer system outdistance the imagination. A DNA might be no more than 1/300th of an inch long and hold several million steps. Zaderej figures a single human cell contains as much information as several sets of encyclopedias. Hereditary material in a cell must be able to reproduce itself. What, then, accounts for genetic forgetfulness, mishaps, missing links in this memory bank? A number of influences figure in the equation, including the earth, air, fire (light) and water of the ancients. A change in environment might favor a certain set of genes. This is how nature favors the survival traits most adaptive to the environment, and it is the prime reason why the gardener's penchant for pushing plants out of their natural environment is almost always unsuccessful.

The gateway to the cell is a thin skin about 100,000th part of a millimeter thick. Through this membrane must arrive the nutrients that say grow or die, increase or fail to multiply. Particles of nutrients have to cross over on their own steam, so to speak. Obviously, ionized chemicals called fungicides and insecticides hold their own key to cellular entrance.

6. trans-membrane trip

All this new found knowledge is being used to better understand nutrient transport into cells, and what energy has to do with this trans-membrane trip. Add a word to your vocabulary here: electrogenic. It defines the pump that pushes home the goodies that cells—in plants as well as animals and human beings—need to grow. Just as orbiting electrons constitute the smallest definable part of the atom, electrogenic energy rates attention as the handle on the nutrient pump. That's what microbiologist Mark E. Jones meant when he said, "precise regulation of substrate or metabolite pools resided with enzyme structure and activity due to the catalytic nature of enzymes." Enzymes are delicate life-like substances found in all living cells. Some experts consider them to be protein. They are in fact small on-scene engineers in the cell building business, commanding an atom of boron here, zinc there, directing phosphorus to the fruit or the bone, and so on.

This primer will talk a lot about enzyme systems in the pages to follow. After all, the enzyme systems are in command, and they didn't go to college or hold office on a fertilizer board. For our part, we choose to trust them, just as we choose to trust the lodgers in our own cells, the mitochondria, and those other little animals that are holding us upright, sorting things out for us, bolting us together—all while homesteading in our cells. Odd that plants are in the same general fix. As we have noted in lesson 1, they couldn't be plants without chloroplasts, those little fellows who run the photosynthetic factory and manufacture oxygen for the rest of us.

7. until science understands

The marginal people who generally end up running state affairs do not think like environmental farmers. They have a kindergarten understanding of survival of the fittest, and this means might must best might. They compute things like megadeaths, but they overlook the element of heat should so and so many atomic bombs go off. Heat affects weather and tides. It has been computed that a tide one kilometer high would circle the globe should those bombs explode at even "acceptable" megadeath levels. The marginal people simply can't think in terms of enough variables.

All this has prompted Lewis Thomas to make a modest proposal in *The Lives of a Cell*. He suggested no further action until science understands fully at least one living thing. He suggested study of the protozoam *Myxotricha paradoxa*, a small critter that inhabits the inner reaches of the digestive tract of Australian termites. This goal—like reaching the moon—could be reached within a decade. "When this is done, and the information programmed into all our computers, I for one would be willing to take my chances," Lewis wrote.

Much the same is true of absolute pronouncements from universities, fertilizer boards, or legislatures imposing their wisdom on biology. As soon as any of the above fully understand a single cell, then the computerization of the answers might permit absolute lawmaking on the matter of how plants eat.

HEALTH AND THE SOIL

Our health depends on the soil. Bodies are really built from the ground up. All of us must build our lives and bodies from the ground up in more ways than one.

"In terms of the moon and the stars," inquired the Psalmist of old, "what is man that thou art mindful of him?" Man, the Bible tells us, if a mere handful of clay into which the Creator has blown his warm breath. Even in such early thinking some consideration was given to the possibility that the soil has something to do with the construction of the human body.

Chemical Analysis of the Human Body in Comparison
with That of Plants and of Soil

	Human Body	Vegetation	Soil
	per cent	*per cent dry matter*	*per cent dry matter*
Combustible			
Oxygen	66.0	42.9	47.3
Carbon	17.5	44.3	.19
Hydrogen	10.2	6.1	.22
Nitrogen	2.4	1.62	
Ash			
Calcium	1.6	.62	3.47
Phosphorus	.9	.56	.12
Potassium	.4	1.68	2.46
Sodium	.3	.43	
Chlorine	.3	.22	.06
Sulfur	.2	.37	.12
Magnesium	.05	.38	2.24
Iron	.004	.04	4.50
Iodine		Trace	
Fluorine	Trace	Trace	.10
Silicon	Trace	0-3.00	27.74
Manganese	Trace	Trace	.08
Water	65.0		

	Human Body	Vegetation	Soil
	per cent	per cent dry matter	per cent dry matter
Protein	15.0	10.0	
Carbohydrates		82.0	
Fat	14.0	3.0	
Salts	5.0	5.0	
Other	1.0		

To the soil chemist, that which may be interpreted as merely allegory in scriptural language is in reality a great truth. Viewed in chemical terms, the adult human body of 150 pounds contains but about 5.5 pounds of ash, or the noncombustibles that came from the soil. This is the handful of clay into which all the processes of creation serve to blow the warm breath of sunshine, of water, of air and all else from above the soil, to build our bodies by way of the foods we eat.

Many are the failing bodies that reveal the weaknesses in their structures and in their functions. Medical science is moving from cure to prevention. Many leading physicians point to poor bone structure commonly called rickets, and to disturbed body functions under the more understandable term of malnutrition, traceable back to the soil. Conservationists are joining with them and going even a bit farther back to fundamentals in thinking and searching for the causes of poor body structure in deficiencies in the very handful of clay from which we are made.

Man's body composition demands much calcium and phosphorus. Note in the accompanying table of chemical contents of the body, of the vegetation, and of the soil.

We dare not, however, focus attention wholly on but these two chemical elements, calcium and phosphorus, when 14 of them are demanded for plant construction, and when 16, or two additional ones are required for animal body building. Four of these—carbon, oxygen, hydrogen and nitrogen—which constitute the bulk of all plant and animal forms, are not of soil origin but are supplied by air and water.

When the dwindling fertility makes protein-producing, mineral-providing crops "hard to grow," we fail to undergird them with soil treatment for their higher nutritional values in growing young animals. The soil fertility as help towards more protein within the body, as protection against microbial and other invasions, has not impressed itself. Instead we have taken to the therapeutic services of protective products generated by animals, and even microbes, in our bloodstream as disease fighters. The life of the soil is not attractive. The death of it has no recognized disaster. Hence, it may seem far-fetched to any one but a student of both the soil and nutrition to relate the nutritive quality of feeds and foods to the soil.—*William A. Albrecht, Ph.D.*

LESSON 7

The Lives of Plants

For a moment, let us consider the biological division of the soil system.

A good soil is alive with microbial life. There are the protozoa, which are the smallest forms of animal life. And there are the nematodes, little eel-like animals that are more commonly thought of as parasites. The top few inches of a soil system also hold bacteria, yeasts, fungi, actinomycetes and algae. Researchers have calculated that a doublehandful of biologically active soil will contain more units of life than there are people on the face of the earth. Aerobic life must have its air. Anaerobic life—below the top few inches—has to have an air-free environment. That is one reason eco-farmers do not like the moldboard plow. That instrument is unkind to the farmer's unpaid labor force. It puts aerobic life in airless chambers, and moves anaerobic life up to where there is air.

1. the top two to five inches of soil

The top five inches of almost any soil system contain 95% of the soil's aerobic life. These bacteria, together with crop residue and organic matter, need oxygen. Humus, as the end-line result of decayed organic matter, is concentrated in the top two to five inches of soil. And humus is the life of

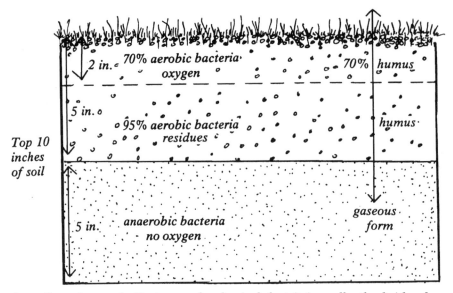

the soil, so to speak. Its tag reads physical, but it is really the bridge between the biological and the chemical. Since most nutrient systems are prolific in the top 5 to 7 inches of surface soil, and quite the opposite in the subsoil, field samples of soil for laboratory analysis must be taken at a uniform core depth. A separate subsoil sample directly below the surface sample will reveal numerous differences.

The fantastic scope of microbe activity is neither appreciated nor understood fully. One laboratory cited in Brian J. Ford's *Microbe Power* has collected 13,000 different organisms that have the ability to break down the cellulose molecule, releasing energy. And that laboratory may have only scratched the surface.

2. the help of microbes

We have seen how plants, via photosynthesis, structure elaborate carbohydrate molecules using sunshine, air and water. We have made the point that this is really 95% of the crop. Unfortunately, animal and man can't thrive on the 95% alone, the so-called "grow" foods. Also required are the "go" foods, the 5%. And it is in dealing with the 5% that microbes come into their own. Fully 80% of the atmosphere is nitrogen, all of it unusable by green plants. This raw nitrogen has to be combined with other elements to make ammonium and nitrate compounds before plants can incorporate it into proteins. The pithy put-down that says the plant doesn't know the difference between nitrogen forms rates attention only as a curious form of kindergarten nonsense. We will discuss nitrogen at length in the special lesson devoted to that subject. For now, it is enough to understand that many species of microbes make the natural nitrogen cycle work, writing checks on the supply in the atmosphere. The star of this team is Rhizobium. Pull up a clover plant and shake away the soil ever so gently. You'll see wart-like nodules. These contain colonies of bacteria that can take nitrogen from the air and effect the chemical combination required for plant use. Rhizobium can in fact exist without that well known symbiotic partnership with roots. Yet it is the special favor that it confers on legumes that makes it so valuable.

Brian J. Ford tells about the shipworm, a creature that lives in an environment that has too little nitrogen. Yet in the intestine of the marine worm is a bacterium which can fix nitrogen. This is interesting. Even more interesting is the fact that this bacterium feeds on cellulose and half liquified cellulose in culture.

There are many species of blue-green algae, a very small form of microbe life. It lives on coral atolls and reefs, and has recently found application in agriculture as a nitrogen fixer. Certain species of Hormothamnion and Calothrix can indeed fix nitrogen on par with controlled agricultural production using symbiotic plants.

Nitrogen isn't alone in requiring the help of microbes before it can be made fit for plant use. The anion plant nutrients—phosphorus, sulfur, chlorine, boron, molybdenum—all have to have microbe help before they attain the refinement needed for plant consumption. Just as nitrogen must first be converted to nitrate, phosphorus must be converted to phosphate, sulfur to sulphate, chlorine to chloride, boron to borate and molybdenum to molybdate, and so on.

THE NATURE AND PROPERTIES OF SOILS

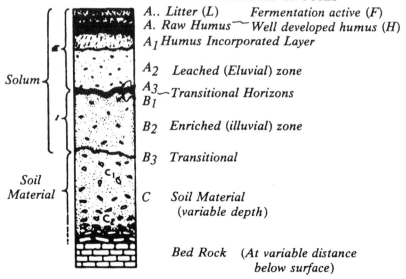

A.. Litter (L)   Fermentation active (F)
A. Raw Humus — Well developed humus (H)
A₁ Humus Incorporated Layer

A₂ Leached (Eluvial) zone

A₃ — Transitional Horizons
B₁

B₂ Enriched (illuvial) zone

B₃ Transitional

Solum

C Soil Material
   (variable depth)

Soil Material

Bed Rock (At variable distance
               below surface)

An ideal soil profile showing all of the horizons usually distinguished.

3. nutrients found to be essential

The general requirements of the soil microorganisms are the same as those of plants: energy, nutrients, water, suitable temperatures and an absence of harmful conditions. Specifically, those nutrients found to be essential in microbial metabolism are as follows:

Essential for all microbes	Essential for some microbes
Carbon	Manganese
Hydrogen	Zinc
Oxygen	Copper
Nitrogen	Boron
Phosphorus	Molybdenum
Sulfur	Iodine
Iron	Silicon
Calcium	Inositol
Magnesium	Hemin
Potassium	Choline
Sodium	Pimelic Acid
Biotin	Glutamine
Nicotinic acid	Vitamin K
Pyridoxine	
P-aminobenzoic acid	
Riboflavin	
Thiamine	

Carbon and hydrogen are available in the form of organic compounds such as carbohydrates, proteins, etc.

The great difference between plants and microorganisms is the source of energy suitable for each. Green plants derive their energy directly from sunlight, and the majority of soil microorganisms obtain their energy either directly or indirectly from the decomposing products of plants. As a consequence, a soil's microorganism population is controlled by the rate at which energy containing materials synthesized in plants are added to the soil.

This energy supply, as well as the necessary nutrients and most of the minerals, becomes available through the decomposition and fermentation of plant materials.

Energy captured in organic matter differs fundamentally from the nutrients in it. Nutrients can be microbes. Energy cannot be used this way. It is indestructible as matter, but when transformed to heat it cannot be used by microorganisms or any living thing. Thus the supply of nutrients and energy become the chief factors in determining a soil population. It is therefore essential that all products added to the soil have these nutrients and energy capabilities.

According to the microbiologists, soil microorganisms can be classified according to their nutritional requirements. The vast majority depend primarily upon organic products for their metabolic needs. Soil microbes can be classified into three simple groups on the basis of their use of nitrogen: those capable of using ammonia nitrogen, those requiring amino acids, and those requiring more complex nitrogenous (organic) compounds.

The organic nitrogenous products so essential to soil microorganisms are produced by the process of fermentation. Most of the known sugars, proteins, amino acids, minerals and growth factors necessary for microbe metabolism are produced and made available as end-products by the decomposition of organic compounds. It follows therefore that the addition of a compound that contains these essential products, when made available to soil microbes, will improve the environment, increase their activity and enhance the population. The effects of this stimulation will benefit the soil and substantially improve the growth of the higher plants. To quote William A. Albrecht, "The plant always eats at the second sitting, the plant only gets what the microbes give it."

"The productivity of a soil is in direct proportion to the number, activity and balance of the soil microorganisms," wrote Stanley Wedberg of the University of Connecticut.

4. microbes at work

Examples of microbes at work could fill hundreds of pages: the miracle of bread and cheese and beer; the leather top for a pair of shoes; even the white cliffs of Dover. Without the microbes, starch rich grains would remain unappetizing and hard to digest. We can understand how yeast is a requirement for bread and beer, but we seem to forget that food crop harvests wouldn't arrive in the first place without the intervention of microbes—microbes in the soil, microbes called chloroplasts in leaf cells, all harnessed to the work of trapping solar energy.

Oh yes, those white cliffs of Dover! These are the skeletal remains of microbes that flourished some 100 million years ago. The seas were just right for great proliferation of life, one million tons of these microbes dividing to become 16 million tons in just 240 minutes. As a matter of fact, most sedimentary rocks are what's left of microbes.

When we think of microbes, we think of disease, and yet most microbes are beneficial in nature. Most of the fungi help, not hinder, the farmer. A good example—more typical than we pause to admit at times—is the predacious fungus.

This is an artist's conception of the predacious fungus in action. In obeying the laws of life and death, it seeks to live and multiply. As a rotifer or eel-worm goes about its handy work in the soil, it might encounter the lasso of the predacious fungi. Like a microscopic snake, this fungus holds its death grip, in time absorbing the body contents of its victim.

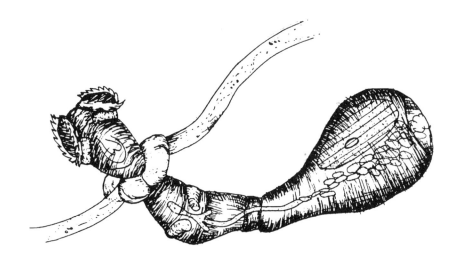

The term *mycorrhiza* was first used about 90 years ago. It refers to the many fungi that are found in close contact with—and entering into—the plant roots growing in virgin soil, or in soil with plenty of organic matter. Organic gardeners have long considered the mycorrhiza a friend. Not a few scientists interested in farm technology have considered them a foe.

Just as there are more kinds of plants that grow underground than there are on the soil's surface, so too are there more kinds and numbers of livestock hidden away in the shallows and depths of a soil system than ever walk the surface of the earth. These tiny underground plants and the little critters that live on them make possible the growth of higher plant life. This underground living complex decomposes dead organic materials, making soils fertile so that higher plants can grow. Reserve mineral elements are made available by life in the soil. Most important, these life systems enter into symbiotic relationships with roots of higher plants and supply them with critically needed compounds.

S. C. Hood of Hood Laboratory, Tampa, Florida once caused these lines to be printed in a company brochure, and so far we have found little in the scientific literature to equal them. "It is probable," wrote Hood, "that this symbiotic relation began when the first primitive plant forms left the primordial sea and took to the land. There were primitive forms of fungi and algae, both of which had developed in water. When cast on dry land, as separates, both were helpless. The fungi could not make carbohydrates. The algae could not secure mineral nutrients from the rocks. But united in a partnership, both could survive. The algae made carbohydrates for both, and the fungi extracted from the rocks the mineral elements needed by both of them."

There is nothing to suggest that this relationship does not persist to the present, especially in the lichens, the first builders of soil. In their development of complicated structures, higher plants kept a part of this early relationship. "They are still dependent on their associated fungi for development, especially chemically," is the way Hood put it.

5. underground plants

These filamentous, underground plants form a cobweb-like growth throughout the soil and over roots. They are so slender that should we twist together 500 of the larger ones, we would have a rope no longer than a human hair. There are some fungus hyphae so small that it would require 3,000 strands to make a fine hair-sized rope. This is the study of them which has been neglected and why their importance has only been recently recognized.

Further, once these are recognized, the fantastic quantities of mycelial fiber and surface area of the fungus in a limited amount of soil around even one plant, the importance of mycorrhiza in symbiosis with higher plants comes clear.

Some hint at this complexity can be found in the scientific literature. We have in mind one entry that has been largely forgotten, if it ever rated attention in the first place. Writing in *A Quantitative Study of the Roots and Root Hairs of Winter Rye Plants* (*American Journal of Botany* in 1937 and 1938), H. Dittmer reported on a single rye plant. He found a root length of 377 miles. Fully 80% or 275 miles of these roots were feeders. The root hairs on that single plant numbered 14.5 billion, having a fibrous length of 6,214 miles. The surface area alone was calculated at more than a tenth of an acre. Combined, the roots and root hairs had a length of 6,990 miles with a combined surface area of 63,784 square feet—close to 1.5 acres. And this was just one plant.

It is true, winter rye has a massive root system and very fine root hairs. And it may be that Dittmer had a very robust plant on his hands. But the point is that all plants have fantastic figures involved when these measurements are taken. In a single acre of winter rye or meadow grass, the area of roots and root hairs may exceed 30,000 acres. At least one-third of this is covered by a net of fungus mycelium, and this provides additional area for soil contact.

The friendly fungus must be kept in mind when we move on in our primer to discussion of chemical requirements, salt fertilizers, toxic rescue chemistry, and other facets extension of these topics would duly account for.

Mycorrhiza in association with root systems isn't a one-way street. Let us refer to a scientific paper translated in 1961 from the Russian under the auspices of Israel. In *Soil Microorganisms and Higher Plants*, N. A. Krasilnikov put together the findings of some 20 investigators and served up some breath-taking data on exudates from plant roots. The Russians found growing roots to exude inorganic elements, sugars, many amino acids, a host of organic ones, vitamins, biotics, antibiotics and a number of organic compounds. A man named Denidenko was cited as having found a single corn plant which—during the vegetative period—exuded 436 milligrams of organic substances when the nutrient solution remained unchanged. When it was changed seven times during the growth period, 2.3 times more—or 1,136 milligrams—of organic substances were exuded. Fantastic. Certainly. But this has been known and ignored for a long time.

What does this mean? Apparently the root surfaces of higher plants are used by fungi as feeding ground. Are these fungi friend or foe?

6. the important fungi

Apparently Fusarium, Trichoderma, Gliocladium and Basidiomycetes are the important fungi in this fungus-plant symbiosis, the mycorrhiza complex. Moving from richer virgin soil, where fungus is ever-present in both species and number, to soils with less organic matter, fungus growth is greatly reduced in both amount and kind. The Basidiomycetes are the first to disappear. As conditions worsen, one group after another vanishes. Finally, when the corpse of a soil is all that is left, only an occasional Fusarium remains in evidence.

When soil has been reduced to a barren waste, plant species of a weed nature take over. They can grow well without a fungus association. Indeed, many cultivated plants can do this also, suggesting that fungi are useful but not indispensable. Yet logic requires us to look a bit further.

Without a full complement of mycorrhiza, lowered quality and yield result. Lowered quality is the chief reason salt fertilizers are not entirely satisfactory. Still, inferior quality—lowered protein, less vitamins, poor mineral content—finds acceptance in the market simply because the naked eye can't see the difference as long as bins and bushels remain. It is only when yields falter that the farmer recognizes the problem.

And yet the problem can be stated briefly and succinctly. Perhaps the Fusarium genus can provide us with the key for much needed understanding. *Fusarium oxysparum*, for instance, is very versatile. Whenever investigators look for fungi, they invariably find *Fusarium oxysparum* or other groups of that genus—*F. salani*, *F. rodeum*, and so on. Generally this genus is a peaceful homesteader in the soil and a beneficial symbiont on plant roots. *Yet when this fungus finds a root that is poorly nourished, a plant with low resistance, it quickly becomes pathogenic.* If the farmer permits plant malnutrition to continue, pathogenic potential really comes into its own, and the fungus rates attention as an active parasite.

This is why the biochemistry of immunity is seated in fertility management, and not in having Dow Chemical, Monsanto and other oil company satellites create more lethal molecules of poisons to combat fungus attack. This is what William A. Albrecht was talking about when he charged that "We are exhausting the quality of our soils. As we do so the quality of our plants goes down."

7. first place at the table

At one time *Acres U.S.A.* ran an article called *The Two Fires of Agriculture.* Here was a picture of a Texas farmer burning his wheat stubble because he didn't want the decay process stealing nitrogen when a second planting was being contemplated. Next this same farmer took another slice out of his humus supply by knifing in anhydrous ammonia. Some farmers in California burn their rice straw because (they say) it is not biodegradable. Nothing is biodegradable if soil microorganisms have been bombed out entirely with harsh chemicals, intolerable forms of nitrogen, and certain types of salt fertilizers used under the wrong conditions.

It is true, microbes eat more simply than all other life forms. Like a banker on the hunt for higher interest rates, they have a one track mind. Also like a banker, they have the first place at the table and take the first helping.

Microbes have to live in a symbiotic relationship with plants because plants build carbohydrates, using the power of photosynthesis, of course. Plants live in the sun and air. Microbes live, eat and die in the dark. Since they cannot manufacture the chemical compounds they need through photosynthesis, they have to wait for the death of plants to have the needed compounds passed on to them.

It is then that microbes burn or oxidize these organic compounds found in dead plants and animals. This is the supreme function of microbes, to act as decomposers, as a wrecking crew in the business of tearing apart more complex molecules and returning them to elemental parts from whence they came—air, water, soil, or other points of origin.

This is nature's way. Energy from the sun is released for microbial life service via the process of digestion and metabolism of organic matter by those same microbes. It really isn't all that different from what human beings do. We eat, digest and metabolize, excrete, all as part of the struggle for calories and nutrition. The microbes do their thing, and we call it rotting.

Crop trash turned under by a tillage instrument delivers organic matter to microorganisms for food. Microbes indeed burn or oxidize this organic trash, and this can cause problems when organic matter is turned under only shortly before a new crop is planted. The farmer who burned his wheat stubble knew this and chose to waste his organic matter rather than suffer poor crop performance. He did this perhaps without knowing that poor crop performance would be his legacy in any case—in the fullness of time.

8. "go" foods and "grow" foods

Microbes are not as finicky as human beings. They do not require their nitrogen in complete proteins as do humans. Just the same, they are exacting in their nitrogen requirements. As Albrecht explained it half a hundred times in *Acres U.S.A.*, thousands of times in his classrooms, they need both "go" foods and "grow" foods. They require a balance between nitrogen and carbohydrates.

A lot of plants give up their protein contents in the process of seed making, and this makes such plant residues unbalanced food for microbes. Livestock below the soil's surface can't grow well on this trash, which is mostly carbohydrate "go" food, and largely deficient in "grow" food.

Thus when a farmer turns in a lot of sawdust, straw and the like, the microbe is given a meal high in carbon or energy and low in body building foods. Remember, the microbes are there first. They have ready access to clay catchpens where other nutrients are warehoused for plant use. They take all the ammonia nitrogen, all the potassium, all the phosphorus, calcium, magnesium, molybdenum, whatever, that they need for growth and to balance off that sawdust and/or straw or stalk diet. That is what we mean when we say the microbes eat first. When the soil is too low in fertility to serve both microbes and plants at the same time, it will be the plant that suffers first.

9. microbial fires

Fortunately for the farmer, this natural game plan is short lived. While the microbial fires are at work, excessive carbon moves out into the atmosphere as carbon dioxide. Nitrogen and inorganic nutrient elements remain in the soil. As carbon rolls off, the carbon to nitrogen to inorganic elements balance narrows. The ratios in fact even out with those in microbial body composition, a level close to protein. At the start, straw residue might read carbon-nitrogen at a ratio of 80 to 1. As the microbes live and die, this edges down to what we call humus, a carbon-nitrogen ratio more nearly 12 to 1. Since new trash isn't turned in every day of the week, microbes soon continue to live by "consuming their predecessors or the humus residues of their creation," to use Albrecht's words for the process.

Humus is low in energy values, high in body-building values. It is unbalanced, to be sure, but not critically unbalanced as are straight energy foods, carbohydrates. After all, "grow" foods, proteins, can also be burned for energy. That is why proteinaceous organic matter—legumes, for instance, with a high content of nitrogen, calcium, phosphorus, magnesium, potassium, and a wide range of inorganic elements—puts microbes on a narrow carbon-nitrogen ratio. Such a diet still has microbes taking the first helping, but it is a helping that has a surplus of nitrogen and minerals. The surplus is not built into microbial bodies. It is in fact liberated in simpler forms and left as fertilizer for crop production.

Obviously, what a farmer turns under determines what will be left over for crop use. Poor soil will produce poor trash, and poor trash will do little by way of lifting soil fertility out of the doldrums. The mandate for manure, especially composted manure, ought to come clear at this point. Only a sound cycle of returning quality organic matter to the soil will support continuing agriculture. If soils are mined of their nutrients and pushed to low levels of fertility, the microorganisms that once worked for the farmer become a strike force working against him.

10. this uneasy war

Fertility management must be pursued with this uneasy war forever first in mind. Salt fertilizer nutrients inserted into the soil can purchase crops, but only at the cost of considerable shock to the fungal and microbial complex in the soil. Such shock continued over a period of time takes its toll. And still it is possible to put inorganic fertilizers and minerals into the soil below the zones of microbial activity. Roots, powered by the sun, can find these nutrients in an arena where the plants rather than the microbes eat first.

The farmer who hopes to farm ecologically has to become concerned about his organic matter. He has to learn to incorporate crop residue into the top four to six inches of soil, into the area where microbes operate, certainly not deep. Organic matter should be recycled rapidly because any delay in recycling nutrients contained in residues means leaching losses.

11. management of organic matter

Thus the first order of business for an eco-farmer is to concern himself with the management of organic matter. Even public policy ought to be concerned with this aspect of agriculture. If a nation is to look at farming as a system good for thousands of years, instead of a mining operation good for 50 or 90 or 200 years, it has to deal with organic matter and its maintenance. The National Resources Defense Council is currently using a figure of 3 billion tons of soil per year eroding by water from American croplands. That's just equivalent to under .25 inches per year off 100 million acres, or 8 inches from those same acres in 32 years. This means subsoil is being called topsoil, and it won't stay put because microorganism growth has been stopped because of a lack of organic matter.

12. the nitrogen cycle

You hear the term organic farming a lot nowadays, and it means different things to different people. To us it means the primacy of organic matter. Organic matter can make the natural nitrogen cycle work. If organic matter falls below a level required to free enough nitrogen from microorganisms, it cannot be made to work for the next crop year. Yet the country is full of people with .5% organic matter who are having a try at organic farming. They might just as well walk straight into the bankruptcy court without the bother of going broke.

Success has to do with making the natural cycle work.

Making the nitrogen cycle work depends on the right level of organic matter. If this can be managed between 4 and 6%, the soil system will have a tap into 4,000 to 6,000 pounds of nitrogen.

Organic matter shouldn't be too low, but it shouldn't be too high either. At a 2.5% organic matter level, leaching cannot be controlled. Below a 2% level, there are not enough existing metallic trace elements to form good enzyme systems in a plant, and above 6% there is too much competition

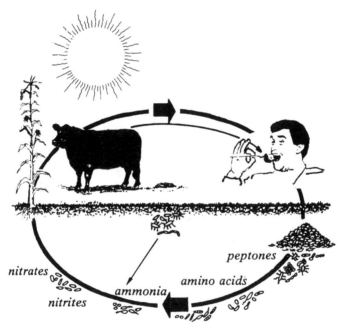

The nitrogen rotation in nature.

between plant and soil organisms, and this might cause deficits for the enzyme systems of the plant.

13. the carbon cycle

Making the carbon cycle work is also dependent on a well managed decay system. Introduction of nitrogen into a soil system when there is a

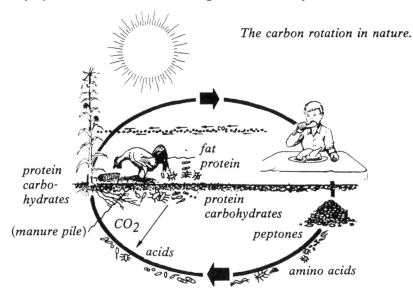

The carbon rotation in nature.

fat
protein

*protein
carbo-
hydrates*

protein
carbohydrates

(manure pile) CO_2

peptones

acids

amino acids

low amount of carbon forces robbing of the humus colloids, and this is also why production is achieved at the cost of considerable chemical shock to the soil system.

Organic matter becomes humus in the fullness of time if the farmer regulates conditions properly. Humus and organic matter plus life mixed with clay account for porosity and tilth—but not quite!

14. protein agriculture, synthetic agriculture

A sandy soil is one made up of more than 50% sand. The same general ratio applies when it is a silty soil. Soil, however, takes on clay properties when 20%, no more, is made up of a fine clay separate. In terms of organic matter, and nutritional service to microbes and plants, the situation is decidedly improved when organic matter stays in the 5% range for clay and sand, with clay and sand helping define tilth. The nation would be thrice blessed—as would microbes, plants, animals and men—were all farm acres to be maintained at 5% organic matter.

This, of course, is clearly impossible in many geographical areas. And it is this reality that precisely defines whether farming can be pursued entirely in terms of a protein agriculture, or whether it must be in part at least a synthetic agriculture.

Any productive soil must have its minerals. These must be broken free from rocks via a process of erosion and disintegration. Mineral elements go into action by going into solution or becoming absorbed to the clay and humus components of the soil system. No soil can be productive by being little more than a pile of pulverized rock with water added. There has to be organic matter to support the microbes. In short, it has to be a living soil.

In the everything-relates-to-everything-else equation called eco-agriculture, tilth relates to nutrients and nutrients in the final analysis take the sting out of drought, cold snaps, and production woes in any difficult year.

As we continue our first reader lessons, consider the soil as a stomach into which you put nutrients and water to make plants grow. The soil is a living system, a biological environment. It can only be maintained through the proper receipt of organic material, whether it be humus, manure, or crop trash mixed with air and water. It is the farmer's job to set in motion the environment in the soil that enables him to manage and retain his organic matter.

The value of a living soil can be illustrated with this diagram.

100 POUNDS OF DRY SOIL

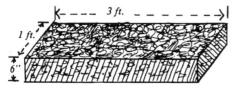

4% to 5% organic matter can hold
165 - 195 pounds of water
equal to a 4" - 6" rain.

1½% to 2% organic matter can hold
only 35 - 45 pounds of water
equal to a ½" - 1½" rain.

...below 2½% organic matter, elements leach out.

A block of 4 to 5% organic matter soil weighing 100 pounds, taking a space of 3 feet by 1 foot by 6 inches deep, can hold 165 to 195 pounds of water, according to USDA. This means a field with such soils could absorb a 4 to 6 inch rain in an hour. Unfortunately the cornbelt of the United States has organic matter ranging from a mere .5% to perhaps 2.5%, and 100 pounds of such soil can absorb only about a half inch of rain.

Organic matter is the reservoir for nitrogen, phosphorus, sulfur, boron, zinc—in short, it is a general catchpen for nutrients. Ehrenfreid Pfeiffer, who refined and practiced Rudolf Steiner's biodynamic principles, once estimated that about 5% of the total organic matter in a soil system is present as nitrogen in various compounds. In terms of 2% organic matter, this would mean about 40,000 pounds of organic matter to the acre. Approximately 5% nitrogen in this 2% organic matter would mean the acre had a 2,000 pound nitrogen reserve. Nitrifying or ammonifying microorganisms—the Azotobacters, Beijerinckia aerobes, or Clostridium anaerobes—transform a small fraction of this stored material to nitrate or ammonia—enough in a living soil to sustain healthy plant growth. If we feed these organisms and otherwise care for a proper balance, the biological process in soil will make these microorganisms release nitrogen to the tune of 80 to 120 pounds per acre per year, according to Pfeiffer. Clemson University workers have constructed a similar equation, citing about 85 pounds per acre per year.

Crops require no more than nature can and does provide in a well-managed eco-system. Even the advocates of anhydrous and salt nitrogen are admitting this. The great overloads of 200 and 250 pounds of nitrogen are being viewed in many circles as distorted technology. A mere 40 pounds seems to be optimum, according to Minnesota researchers.

The reason plants fail to enjoy perfect health and fall victim to insect and fungal attacks is simply that a lot of soil systems cannot supply necessary plant nutrients in sufficient quantities for high yield crops because of physical, chemical and biological complications. These complications account for malnutrition, imbalanced hormone and enzyme systems within plants, fungal and insect attacks, dry weather losses, undue damage due to cold snaps, and general loss in vitamin and mineral content in thus produced food crops.

15. ancient and venerable errors

There are many ancient and venerable errors afloat in agriculture today, most of them accepted as facts for many years. Many still serve as premises for pseudo-investigations. Some of them are so securely implanted into scientific thought that even when better science reveals them to be in error, they go on being propounded as though nothing better had been learned. Indeed, errors refuted more than three and four decades ago still appear in classroom texts as though they were gospel. Many have to do with the biological division of the soil system. Even more have to do with the physical and chemical divisions.

The linchpin of these ancient and venerable errors has been called the von Liebig Complex. In 1846 Justus von Liebig wrote *Die Chemie in An Wendung aus Agriculture and Physiologic.* In this landmark text von Liebig blandly announced that plants needed no more than proper levels of nitrogen, phosphorus and potassium in a water soluble form for optimum growth. Liebig understood almost nothing about microorganism—in fact, he leveled ridicule at Pasteur when that great scientist named pathogens as agents in the cause of disease, and friendly bacteria as unpaid workers in wine fermentation. He probably never heard of John Tyndall,[*] the Irish scientist who helped Pasteur and wrote a classic treatise on fermentation.

There is still another error associated with the von Liebig Complex. It is the claim that nitrogen bearing materials in a soil system must be changed to nitrate or ammonium salts before their nitrogen can be taken up by plants. As a codicil to all this there is the claim that only small and simple molecules can pass through cell walls of roots.

Here for instance is the *Reader's Digest*-type fare put out by a chemist-agronomist at an experiment station: "Before the plant foods contained in compost, manure and other organic matter can be used by plants, they must be broken down by bacteria into simple mineral compounds which the plants assimilate, and there is no difference between these minerals and those processed in fertilizer factories." That this conventional wisdom is naive is at once apparent, even to those not trained in chemistry, biology and plant botany. Such breath-taking ignorance is aided and abetted by state fertilizer codes, which make water solubility a criteria for fertilizer units, all based on the folklore belief that water solubility is an index to availability, or absorption, by plant roots.

What about these myths?

William A. Albrecht has answered each of them, one at a time. Writing in *Organic Matter for Plant Nutrition (The Albrecht Papers),* Albrecht noted that "Textbooks of botany and bulletins of some years ago reported the uptake of organic compounds for their nutritional service. Among those reported as improvers of plant growth were such complex ones as coumarin, vanillin, pyridine, quinoline, asparigin, nucleic acids and, in fact, some in each group of carbohydrates, organic acids and nitrogenous compounds—aimed to be systemic insect poisons within the plants—were taken into the plant roots from the soil. Apparently such reports have not registered as cases of

---

[*] Tyndall also proved infrared radiation by inventing the infrared spectrophotometer, and observed penicillin fully 60 years before Sir Alexander Fleming.

soil organic matter serving as large organic molecules moving into the plants from the soil."

It remained for P. C. de Kock of Scotland to give eco-farmers a real handle on this super and continuing fiction about inorganic salts and small molecules. In experiments with chlorotic or iron deficient plants, de Kock offered those plants the chelating agent ethylene diamine tetraacetic acid, commonly known as EDTA. The same material is used nowadays to leach lead and excess calcium out of human beings. EDTA was taken up by the plant, and it mobilized iron from the soil. Next de Kock substituted organic for EDTA. Organic matter gave the same service—that is, moving essential nutrients by chelation into plants, just as had been demonstrated for EDTA. This seems to be the role of humic acids contained in humates, now one of the fertility management aids offered by commerce to the eco-farmer.

How long does this chelating agent serve? In *Iron Chelates in Soybean Exudates*, researchers L. O. Tiffing and J. C. Brown [*Science Magazine*] concluded that the inorganic element fed in union with EDTA was no longer under control of the chelating agent by the time it was moving up through the stem of the soybean plant.

Unfortunately classes of instruction have separated inorganic chemistry from organic chemistry. This has seemed to make more important the status of inorganic plant nutrients. Yet nature does not make this distinction. Nobel Prize winner Melvin Calvin proved as much. In *The Path of Carbon in Photosynthesis* [*Science Magazine*] he illustrated the results of certain research. Somehow phosphorus turned up in one of the sugars produced by photosynthesis. In other words, the so-called anion phosphorus from the soil was chelated and combined into an improved molecular biochemistry of growth which later split back into six-carbon sugar heretofore believed to be the first stage of photosynthesis. Phosphorus was released to perform the same work all over again.

We have the same general thing with cationic elements: magnesium in chlorophyll, iron in hemoglobin of animal blood; copper in the life fluid of crustaceans such as the lobster.

The water solubility bugaboo has now become encased in law like a pillar in concrete. The myth has hardened there and defies mature science to unseat it. Yet water solubility is no index of availability to plant life. It is not the prime task of the water transpiration stream to move nutrients. "Calcium, magnesium, nitrogen, phosphorus, potassium, and all the other essentials are not swept into the plant because they are applied to the soil in water soluble forms of fertilizers and flooded in, as it were," concluded Albrecht in the paper *Insoluble Yet Available* (*The Albrecht Papers*). Yet reliance on fertilizers that deliver chemical shock to the fantastic life complex in the soil is considered the modus operandi in American agriculture, and this procedure is based on blind faith in what the grant receivers at the institutional level say.

16. *other voices in other rooms*

It is now a matter of record that the land grant colleges get grant money from the fertilizer industry based on the amount of agriculture in each of the states. It is perhaps because of this reality that we must look to other voices in other rooms to answer the questions and explain the answers.

Over a half century ago, A. W. Blair of New Jersey noted, "It is well known . . . that by judicious use of lime and vegetable matter on the soil, reserves of locked-up mineral plant food may be made available." At that time others spoke of maintaining soil fertility and a permanent agriculture by cycling organic matter together with rock minerals to the soil. These pioneers did not visualize a permanent agriculture based on soluble salts. They had no petro-chemical industry to color their vision.

Unfortunately, man's puny mind can't comprehend the many ramifications involved here without fencing off each topic and handling it as if in isolation. We will do this, of course, but always at the cost of some misunderstanding. At the end of these compartmentalized lessons, it will take a bit of memory to bring it all together again. We won't have time for much single factor analysis after that. It is, in any case, the curse of the amateur. We realize that the single variable has been and remains the lodestone of much college research, but it has all too little to do with production on a living, vibrating, dynamic farm. It ruins true science by making specialists out of young minds, killing off their imaginations.

MIDTERM BREAK

LESSON 8

The Soil-Life Connection

To enlarge upon the outline suggested by the previous chapter might require a shelf swayback with texts. Nevertheless, the cause and effect of any farm model can best be understood in terms of the livestock in the soil.

Few people use the fingerprint left in the soil to identify the drum rolls of history, and yet the connection is so obvious that only the simple-minded, the boasting dishonest or the rank opportunist can manage to ignore it. Goethe touched on the genius of Joseph, who saved Egypt from starvation by foresight and wisdom, at the same time putting the Pharoh into possession of the land by unprecedented speculation. Goethe had his Mephistopheles do the opposite, creating inflation rule, as have government economists in centuries 20 and 21.

We are told that the Xhosa and Zulu of the African continent once enjoyed the lush savannahs of the area now known as that carbon-less Sahara. The bottom line that adjusts history is always the food supply and man's witless destruction thereof.

Even without recent studies, USDA cannot pretend ignorance of this fact. In 1938-1939, Walter Lowdermilk, formerly assistant chief of the Soil

CONQUEST OF THE LAND

Agriculture had its beginning about 7,000 years ago and developed in two great centers — the fertile alluvial plains of Mesopotamia and the Valley of the Nile. We shall leave the interesting question of the precise area in which agriculture originated to the archaeologists. It is enough for us to know that it was in these alluvial plains in an arid climate that tillers of the soil began to grow food crops by irrigation in quantities greater than their own needs. This released their fellows for a division of labor that gave rise to what we call civilization. We shall follow the vicissitudes of peoples recorded on the land, as nations rose and fell in these fateful lands. . . .

In the Zagros Mountains that separate Persia from Mesopotamia, shepherds with their flocks have lived from time immemorial, when "the memory of man runneth not to the contrary." From time to time they have swept down into the plains to bring devastation and destruction upon farming and city peoples of the plains. Such was the beginning of the Cain and Abel struggle between the shepherd and the farmer, of which we shall have more to say.

At Kish, we looked upon the first capital after the Great Flood that swept over Mesopotamia in prehistoric times and left its record in a thick deposit of brown alluvium. The layer of alluvium marked a break in the sequence of a former and a succeeding culture, as recorded in artifacts. Above the alluvium deposits is the site of Kish.

— *W.C. Lowdermilk,*
in Conquest of the Land Through 7,000 years

Conservation Service, toured the Middle East, North Africa, Cyprus and parts of Europe to study food production and discern what separated desert from fertile soil. The lands he inspected had been cultivated hundreds and thousands of years. He wrote that "in the last reckoning all things are purchased with food." He went on to propose that food buys the division of labor that begets civilization. He discerned land and farmer and soil life as the work foundation of our complex social structure.

The farmers of 7,000 years ago could not have known what we know now. But they must have had some appreciation of fertility. Ancient artifacts reveal slaves wringing the sweat from their garments for a soil amendment. In Egypt as well as in Mespotamia, Telus learned how to grow wheat and barley, giving rise to a renewable civilization. Flood irrigation and silt from the Nile charged and recharged the soil, giving a fix of nutrients for prolific soil life, year after year. It was perhaps in the Valley of the Nile that a genius of a farmer learned how to disturb the soil with a yoke of oxen and a plow, unwittingly re-establishing nature as a mandated balance between bacteria and fungi.

Bible students will recall that King Solomon nearly 3,000 years ago made an agreement with Hiram of Tyre to furnish cypress and cedars for the construction of Tyre's temple. We are told that Solomon supplied 80,000 lumberjacks to work in the forests and to skid the logs to the sea. Only about 40 acres remain of a forest that was once 2,000 miles square. Obviously, clear-cutting annihilated the microbial population, especially the mycorrhiza. Apologists for man's debauchery cite climate change, intervention of the gods, the cycles of life and death, whatever.

Lowdermilk's message was clear. Man's intervention prevailed. In Babylon he pondered the ruins of Nebuchadnezzar's canals. At the ruins of Jerash, one of the ten cities of Decopolis — once populated by 250,000 people, now 3,000 — he wondered aloud about cities under erosion and silt. He was told that the French archeologist Father Mattern counted at least 100 dead cites in Syria alone.

The Sahara is expanding in excess of 30 to 40 miles a year. The Aswan Dam, a mechanical marvel and an ecological disaster, will silt over in 500 years. The common denominator everywhere is the death of life in the soil. Man proposes, but God disposes.

Often, analysts became lost in their own metaphors. The Seattle Indian with that same name may have been the first to make clear that what we do to the earth, we do to ourselves. In fact there is no food chain; rather there is a food web, a mesh of life in the soil, this according to Elaine Ingham, Ph.D. of Soil Food Web, Inc., formerly with Oregon State University, Corvallis. Ingham wrote a sizeable chunk of *Soil Biology Primer*, the most useful booklet published by USDA since that agency gave its *impri-*

matur to Walter Lowdermilk's *Conquest of the Land Through 7,000 Years* well over a half century ago.

1. a connection

When life in the soil becomes a consideration, it is no longer time to indulge in single-factor thinking. The irrigation pump may deliver fluid, but the impact on root organisms could be devastating. Microorganisms that live rent free in nature's settings often die or leave the scene not only when the weather changes, but also when salt fertilizers or rescue chemistry pelt the land. Only recently has university science assembled the data base and the insight necessary to identify Ingham's food web. Hints for the direction trail back to the beginning of the last century — as illustrated in the previous chapter — but definitive answers are as new as the present edition of *Acres U.S.A. Primer.*

What then are the right food webs needed to support wholesome field-ripened crops without reliance on inorganic fertilizers and/or toxic rescue chemistry? How can the grower identify the organisms that power crop production?

Poverty acres support weeds, as Albrecht pointed out, because bacteria dominate, the way mycorrhizae dominate woodlands. Grass systems seem to have two times more bacteria than forage. Row crops, in turn, require an eight to one ratio, forage to bacteria. The Wisconsin ginseng grower who expects open prairie under wooden slats to approximate the environment of shaded woods is either ignoring Ingham's food web or is still ignorant of the concept.

Perennial crops, vines, blueberries, blackberries, strawberries — all require more fungi than bacteria. The ratios vary. Indeed, the grand mosaic of nature's whole is an exponential infinity of variations. Deciduous trees demand at least ten times more fungi than bacteria. Without the ratio, growers are forever spraying and waxing fruit to preserve a cosmetic look. Conifers simply won't survive without 1,000 more fungal life forms than bacteria, all according to Elaine Ingham's research.

Investigators have categorized the twenty or so microorganisms we refer to as soil life. Their names — genus and species — are of interest in the same way postage stamps are of interest to collectors. The names create arrays under heads such as algae, fungi, protozoa, nematodes, micro arthropods, earthworms, vertabraes and, not least, plant roots. All of the above eat. All move through the soil. They filter water, decompose organic matter, sequester nitrogen, fix nitrogen, preside over aggregation and porosity. They prepare nutrients for assimilation, they battle crop pests, and, with biblical dedication, present themselves as food for above-ground animals.

A FEW IMPORTANT BACTERIA

Nitrogen fixing bacteria form symbiotic associations with the roots of legumes like clover and lupine, and trees such as alder and locust. Visible nodules are created where bacteria infect a growing root hair. The plant supplies simple carbon compounds to the bacteria, and the bacteria convert nitrogen (N_2) from air into a form the plant host can use. When leaves or roots from the host plant decompose, soil nitrogen increases in the surrounding area.

Nitrifying bacteria change ammonium (NH_4^+) to nitrite (NO_2^-) then to nitrate NO_3^-) — a preferred form of nitrogen for grasses and most row crops. Nitrate is leached more easily from the soil, so some farmers use nitrification inhibitors to reduce the activity of one type of nitrifying bacteria. Nitrifying bacteria are suppressed in forest soils, so that most of the nitrogen remains as ammonium.

Denitrifying bacteria convert nitrate to nitrogen (N_2) or nitrous oxide (N_2O) gas. Denitrifiers are anaerobic, meaning they are active where oxygen is absent, such as in saturated soils or inside soil aggregates.

Actinomycetes are a large group of bacteria that grow as hyphae like fungi. They are responsible for the characteristically "earthy" smell of freshly turned, healthy soil. Actinomycetes decompose a wide array of substrates, but are especially important in degrading recalcitrant (hard-to-decompose) compounds such as chitin and cellulose, and are active at high pH levels. Fungi are more important in degrading these compounds at low pH. A number of antibiotics are produced by actinomycetes such as *Streptomyces*.

Source: Soil Biology Primer, *United States Department of Agriculture*

"All food webs are fueled by the primary producer, the plants," — may be one of the more profound statements to have emerged from the last half century. Lichen, moss, bacteria, algae — all are actors in a drama that fixes carbon dioxide from the atmosphere. In the main, energy and carbon are supplied by compounds found in plants, waste byproducts, etc. A few organisms have very special needs. These are the chemoautotrophs, and they take their energy from nitrogen, sulfur and iron compounds, not from carbon compounds or a solar source.

Nature has decreed that all organisms attempt to increase and multiply. They liberally feed each other and in turn feed the plants, for which reason the objective of fertility management should be "feed the soil" so that in turn it can feed the plants. Plants, of course, provide the larder for soil organisms.

The old cowboy song had it, "You've got to know when to hold 'em, and know when to fold 'em," and that becomes the eco-farming imperative, as a fuller knowledge of soil life requirements reveals itself. A few years ago Edward Faulkner wrote a book, *Plowman's Folly*. Albrecht answered with an article, *Plowman's Wisdom*. There are times when the disturbance called plowing is indicated. There are also times when moldboarding and discing are dead wrong. In the tellers of time, databases will propose and ask the farmer to dispose.

For now it is enough to call attention to the bacterial-fungal requirement, and to invite readers to examine the chapter on mycorrhiza in *The Albrecht Papers, Hidden Lessons in Unopened Books*. The basic answer, of course, is compost, which is discussed in Lesson 19 of this primer.

Farmers tend to think of fungi in the same way most people think of a virus. Cell biologists tell us that once a viral agent enters a warm-blooded body, it stays on for the life of that body. The key is the immune system. A strong immune system simply overpowers a virus. Much the same is true of a soil system. Fungi will always be present in the soil, evil fungi included. A healthy soil system requires the presence of organisms that inhibit, compete with, and consume disease-carrying organisms. Sadly, decisions made by the farmer over the kitchen table either inhibit or invite the pathogen.

The lessons that fill out the rest of this book deal chiefly with the correctness of those kitchen-based decisions. If an action is decided that results in compaction, *send not to know for whom the bell tolls*. It tolls for organisms, usually for the right ones. The soil has to be kept aerobic. It has to permit intake of water and capillary return.

Now that scientific proof is in place proving that so-called conventional agriculture merely mines the soil by annihilating its biotic life, the choice seems clear. So is the necessity for use of a tool introduced to agriculture in the mid-1980s.

2. refractometer

The refractometer had been on line in the grape industry for decades, but its use by farmers apparently was limited to the clients of Carey A. Reams. His chart of values appears as a boxed entry in this chapter.

The food web in a desert soil has the earmarks of poverty. One million sounds like an awesome figure, yet a good corn eco-system has 500 million bacteria per gram. That's what it takes to grow grass, wheat, barley or corn.

In an old-growth forest, the number annihilates the imagination — 1,000 million units of life per gram of soil. This count has been ratified by scientists who are well into the business of changing agriculture's paradigm.

Still, the deserts Lowdermilk wrote about continue to expand. In the first six chapters of *Soil Biology Primer*, USDA No. PA-1637, Elaine Ingham unravels the mysteries behind those refractometer numbers Dan Skow talks about, albeit only if one ties the two concepts together.

Those soils being turned into desert in California's Imperial Valley have only 5,000 or 6,000 species, this compared to 15,000 species per gram in a productive soil capable of producing wheat, barley, hay or corn. Most of these species are as nameless as Nameless Cave on Colorado's west slope. This namelessness becomes the name and the role played out in the theater of farming.

The point having been made, we can abandon this line of inquiry for now. Still, life in the soil will continue to surface as we expand the inquiry and milk the soil for eco-agriculture.

3. the brix index

Bacteria, fungi, arthropods, protozoa (grazers), organisms that live in symbiotic relationship with mycorrhizae, nematodes, earthworms, and photosynthesizers, all populate the environment of plant life, soil and its plants being a residence. There are hundreds of crops and equally hundreds of food webs that anoint optimum conditions. The republics of learning have developed databases for a few definitive answers, all of them costly.

Fortunately, an on-the-spot answer is almost always available, one that signals what can be done "in the growth process."

As noted earlier, photosynthesis is the term assigned to the business of expanding rather complex biochemical reactions covered by the general formula:

$$CO_2 + H_2O \ \frac{\text{sunlight}}{\text{green plants}} \ \longrightarrow \ \underset{\text{carbohydrates}}{(CH_2O) + O_2}$$

COMPARISON CHART FOR BRIX READINGS

PLANT	POOR	AVERAGE	GOOD	EXCELLENT
Alfalfa	4	8	16	22
Apples	6	10	14	18
Asparagus	2	4	6	8
Avocados	4	6	8	10
Bananas	8	10	12	14
Beets	6	8	10	12
Bell Peppers	4	6	8	12
Blueberries	6	8	12	14
Broccoli	6	8	10	12
Cabbage	6	8	10	12
Carrots	4	6	12	18
Canteloupe	8	12	14	16
Casaba	8	10	12	14
Cauliflower	4	6	8	10
Celery	4	6	10	12
Cherries	6	8	14	16
Coconut	8	10	12	14
Corn Stalks	4	8	14	20
Corn, young	6	10	18	24
Cow Peas	4	6	10	12
Endive	4	6	8	10
English Peas	8	10	12	14
Escarole	4	6	8	10
Field Peas	8	10	12	14
Grains	6	10	14	18
Grapes	8	12	16	20
Grapefruit	6	10	14	18
Green Beans	4	6	8	10

COMPARISON CHART FOR BRIX READINGS (*continued*)

PLANT	POOR	AVERAGE	GOOD	EXCELLENT
Honeydew	8	10	12	14
Hot Peppers	4	6	8	10
Kohlrabi	6	8	10	12
Kumquat	4	6	8	10
Lemons	4	6	8	12
Lettuce	4	6	8	10
Limes	4	6	8	12
Mangos	4	6	10	14
Onions	4	6	8	10
Oranges	6	10	16	20
Papayas	6	10	18	22
Parsley	4	6	8	10
Peaches	6	10	14	18
Peanuts	4	6	8	10
Pears	6	10	12	14
Pineapple	12	14	20	22
Raisins	60	70	75	80
Raspberries	6	8	12	14
Rutabagas	4	6	10	12
Sorghum	6	10	22	30
Squash	6	8	12	14
Strawberries	6	10	14	16
Sweet Corn	6	10	18	24
Sweet Potato	6	8	10	14
Tomatoes	4	6	8	12
Turnips	4	6	8	10
Watermelon	8	12	14	16

Neither a formula nor a glossary of life-supporting plants tells an entirely meaningful story; nor do terms like *anabolism* or *catabolism*, both terms for metabolism.

Yet there is one term that codifies the anatomy of plant life. That word is *sugar(s)*. In the process of photosynthesis, sucrose takes its place in the leaves. Osmosis becomes operative, and cells with a high sugar content achieve turgor.

This is the phenomenon Carey Reams harnessed when he constructed the sugar index presented here (*see page 108*). Only a few plants have been included. All plants could fill out this index if full extension of this idea were to be accomplished.

Once this table became perfected, Reams could announce that he had discovered why the anatomy of insect control was seated in fertility management and not in fabricated poisons. The answer was securely locked in each brix reading.

Reams discovered that a high brix reading on a refractometer meant a lower freezing point for fruit. Reams noted that soft rock phosphate best affected brix readings favorably, but the entire spectrum of anion-cation compiled fertilization served equally as well. The reason was simple. Sugar also meant minerals. The refractive reading being a good common denominator, a low brix reading simply means a likely scenario of bacterial, fungal and insect attack. An average or good reading still indicates stress. However, once an excellent brix rating has been achieved, it means soil life is in place, proper fertilization has been achieved, and general immunity to bacterial, fungal and insect attack have been achieved, and weed proliferation — especially poverty weeds — has been avoided.

Timely use of a well-calibrated refractometer can govern progress during the growing season, usually with the use of subtle foliar help. At harvest, a high brix fruit will simply dehydrate rather than rot. An alfalfa that achieves a brix reading of 50 will confer stamina on a race horse and probably determine the winner of the race.

There is a saying that all generalizations are false, including this one. The tomato will rot even with a great brix number, but its shelf life will have been extended greatly.

Dry grains and produce can be measured by reducing moisture content to zero, then hammering the part of the crop that is sold into a fine talcum powder. Reconstituted with distilled water, the solution will provide a brix reading and a window into the soil below.

The brix reading is simply the concentration of carbohydrates in 100 pounds of juice, stated as a percentage. The nomenclature *sugar index* is a misnomer. In fact, the higher the carbohydrate mineral content, the higher the oil content and the protein quality, none of which can be achieved without the food web in place and strong. Finally, the brix index provides a whole biography for the plant, if we have the wit to read it. Always, falling brix levels indicate low phosphate, cation imbalance and general unfamiliarity with the other lessons in the primer.

One parting shot remains before we move on. When a refractometer is properly focused, the instrument has the capacity for bringing insight. View a sharp demarcation; it telegraphs the fact that the crop is low in calcium. When the line becomes diffused, the acid is low, calcium high. High plant calcium with sugar will deliver a sweet taste. This explains why a low brix, high calcium plant tastes sweeter than a low calcium plant, even though both have the same brix reading. It takes art and science to mine the motherlode contained in the brix reading.

Biotic Geography

No two farms are alike. The soil systems, the cropping systems, the people on the farms, their machinery programs, their capital resources, all vary from farm to farm. Take the soils. At one time, in the 1938 *Yearbook of Agriculture*, entitled *Soils and Men*, USDA mapped out some 60 basic soils in the United States, Puerto Rico, Hawaii, the Canal Zone, and the Virgin Islands. These range from Berkshire-Worthington to lava beds, soils that sustained or failed to sustain all types of agriculture. Despite the fact that local conditions were submerged into this common denominator treatment, the map looked something like a county map of the entire country.

1. a fair index

Before vegetable agriculture went to monoculturing and open-air hydroponics—growing most of the vegetables for the nation in no more than four states—soil profiles were a fair index to human health expectations. At the end of the WWII era, we all heard of the town without a toothache, Hereford, Texas. The Texas panhandle has southern dark brown soils, much of it wind deposited loess. This soil is blessed with a good amount of apatite, which chemically is phosphorus, calcium and fluorine.

When apatite decomposes by weathering within the soil, fluorine is the most soluble. It has but a single valence and therefore moves off into the ground water. Calcium has a double valence, or higher combining power, and is therefore less soluble and less active. Phosphorus has a triple valence, and is the least soluble. As fluorine goes out, calcium and phosphorus combine to become available to crop plants. It is because of this excellent crop support that the panhandle produced such sound teeth and healthy skeletons. Needless to say, industries with great overloads of sodium fluoride to get rid of had their scientists conclude that it was fluorine in the water that made Hereford, Texas a town without a toothache—and thus was born the drive to fluoridate the nation's public water supply.

Yet the bison never drank well water. This beast drank surface water and for years provided the fertilizer industry with bones for phosphorus even after it had vanished from the scene. In one study, William A. Albrecht found that dental cavities increased and decreased according to soil types across the nation. Poor Ozark region soils produced more draft rejectees per 1,000 registrants than any other soil in the state of Missouri, according to this map in Albrecht's files.

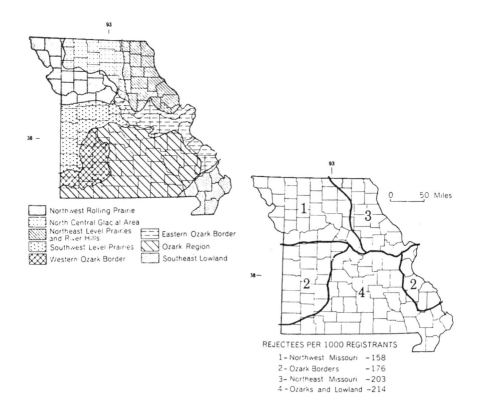

Northwest Rolling Prairie
North Central Glacial Area
Northeast Level Prairies and River Hills
Southwest Level Prairies
Western Ozark Border
Eastern Ozark Border
Ozark Region
Southeast Lowland

REJECTEES PER 1000 REGISTRANTS
1 - Northwest Missouri   -158
2 - Ozark Borders   -176
3 - Northeast Missouri   -203
4 - Ozarks and Lowland   -214

THE WILDCAT OF HALOGENS

Fluoride is an uniquely potent enzyme poisoner, in fact the most powerful one of all elements. There are several reasons for this. In the table of elements, fluorine belongs to the halogen family, sharing chemical properties with its close relatives, chlorine, bromine and iodine. As ions, reacting with other particles, they all carry one negative charge. As the halogen having the smallest atomic weight, fluoride is naturally the most active. It is extremely active in combining with any element or molecule having a positive valence, such as the mineral ions (enzyme co-factors). It decomposes water to form hydrogen fluoride which readily attacks glass. It actively replaces its sibling halogen, chlorine, in any solution, including the hydrochloric acid within our stomachs, or any chlorine-containing molecule within our blood or our intracellular fluid. Fluoride's negative charge and atomic weight of 19 is almost identical to the negative charge and weight of the hydroxyl group (OH), 17.008, which is vitally important to the chemical composition of innumerable substances throughout the human organism. It is, in fact, such interchangeability with the hydroxyl group that is cited as the reason for increased hardness of the apatite crystal of tooth enamel when fluoride is involved. Unfortunately, and all too obviously, this structural change is not confined to teeth, but occurs elsewhere in the body as well. Fluoride poisons enzymes. The book, *Fluorides*, published by the National Academy of Sciences, 1971, lists nine enzymes involved in the breakdown of sugar (glycolysis process) that are fluoride sensitive. The halogen inhibits many enzymes by tenaciously binding with the metal ions they require in order to function. It inhibits others by a direct poisoning action of their protein content. But the ultimate shocker is the toxic effect fluoride has on genes. Painstaking research at the International Institute for the Study of Human Reproduction, Columbia University College of Physicians and Surgeons, as well as at the University of Missouri, have proved beyond doubt that fluoride is mutagenic, i. e., it damages genes in mammals at doses approximating those we humans receive from artificial fluoridation exposure.—*John R. Lee, M.D.*

Here is a map of 48 states based on Navy inductees in WWII. It illustrates how dental caries varied according to climatic development of soil fertility.

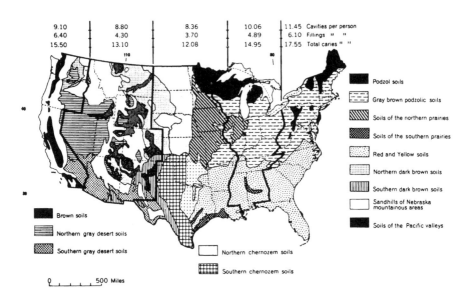

2. *rainfall*

The reasons for this should now be at once apparent to readers of this primer. Geology, rainfall, temperature, topography, all figure in how soils develop and provide nutritional support for microbes, plants, animals and man. Rainfall is under 10 inches east of the coastal ranges in the west. As the westerlies move moisture down track, rainfall increases. In central Kansas it ranges between 20 and 30 inches, with 30 to 40 inches of rainfall common in eastern Kansas and much of Missouri, the Ozark region excepted. There the rain falls harder and more often, leaching out soils, making them heavier in texture. Depending on the swing between cycles— hot-dry (as in the 1930s), warm-wet (as in the late 40s, 50s and 60s), cool-dry (starting in the 70s), and cool-wet (toward the end of this century)— here are the generalized rainfall patterns in the continental United States.

Rainfall weathers rocks, and those nutrients remain or depart, depending on whether evaporation exceeds precipitation. Scant vegetation and high winds account for scattered and balanced nutrients in the semi-arid west, where seasons are often too dry for perennial vegetation and tracts of forest land, yet these same conditions permit grass, "the forgiveness of nature," to alternate between growth and dormancy in a single season.

Mineral rich plants provide microbes with a diet that has a narrow carbon-nitrogen ratio, the ratio one finds in rapidly spoiling meat. This means rapid decay and elimination of a contest between microbes and the next

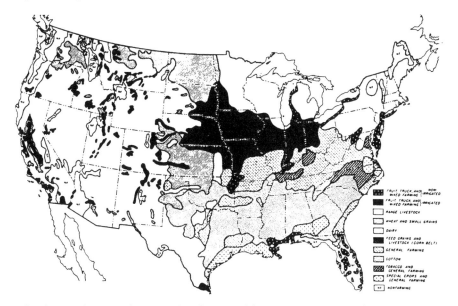

USDA map showing the generalized types of farming in the United States.

round of plant life. One of the grasses that grows best in the midcontinental part of the United States is wheat. The endosperm is high in protein, the source of its hardness.

It is true, there is a shortage of sheer bulk in this area of moderate rainfall, a consequence of the short water supply. But it is precisely this shortage of rainfall that keeps for future crops inorganic nutrients and nitrogenous organic fertility.

Rainfall will produce bulk, and this has always been a confusing factor when one considers biotic geography. The soft wheat belt will produce less bushels of a low protein starchy grain than western Kansas, but there will likely be more bulk, more beautiful foliage for the color photos in farm magazines.

Soils of the east certainly produce more bulk vegetation, more forest litter for the soil's surface with a wide carbon-nitrogen ratio. Microbes find such a diet short of growth essentials because of the great carbon excess. This means the microbes will pounce on inorganic and organic fertility units that a farmer elects to make available. Soils of this territory are apt to give plants great exercise in the photosynthetic process of making carbohydrate bulk.

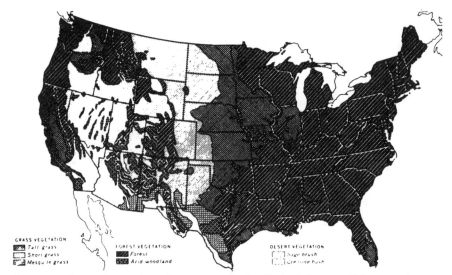

This map shows the native plant cover in the United States. This map does not show the small areas of tall-grass prairie in Indiana, Michigan, Ohio, and Wisconsin.

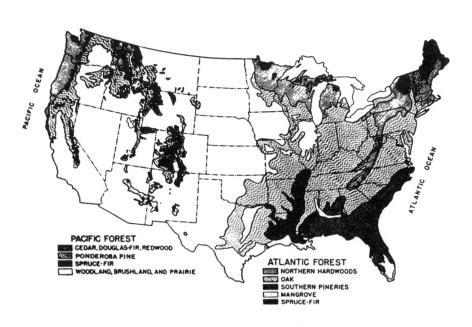

The two forest belts of the United States, the Atlantic forest and the Pacific forest, and their major subdivisions.

3. special problems

Special problems present themselves to geographical location. The monovalent alkaline element sodium is often a problem in the west. Not so as waters unseat this element from its weak bond and move it out of the soil system and off to the sea. This gives greater prominence for the more securely held divalent calcium, magnesium and so on. Still calcium is removed more readily than potassium. And, of course, potassium figures prominently in synthesis and storage of carbohydrates.

Plants still grow on exhausted soils—either bulk plants, or weeds. As fertility fades, the array of plants favors those that reproduce vegetatively. The instructors in botany talk about such plants growing naturally in the tropics. The farmer is more apt to say faltering soils "grow hay crops but not seed crops." Plants that have plenty of air, water and sunshine proceed with the business of delivering vegetative bulk—no problem! But without some two dozen nutrient elements from the soil to make proteins—and reserve carbohydrates—in the seeds, it is not likely that this plant growth will serve as good cow feed. This what is meant by the adage that health comes from the soil and "we are what we eat."

The soil profiles of the nation are further complicated by increasing temperatures from north to south, with a rainbelt taking command over acres from eastern Texas to the Atlantic, and from the Mason-Dixon line to the Gulf of Mexico. Here soils are destitute of the nutrient minerals so common in Kansas, and clay has little adsorbing and exchange capacity. Fertilizers are leached away with fantastic rapidity. The two alkaline earth elements—calcium and magnesium—are usually missing, making it difficult to grow proteinaceous crops. And yet active potassium hangs on, enough it would seem to grow cellulosic fiber crops and coniferous forests. Heavy use of fertilizers has enabled such soils to continue in the production of cotton, sugar cane, even a few other crops, bulk production being the hallmark of success. But because such forages have proved deficient, the south gained a reputation as a good bull market and a good mule market, simply because animals can't be bred to tolerate starvation.

4. enclaves of perfectly balanced soil

It was no accident that champion race horses were produced in certain areas of Tennessee and Kentucky, where biotic geography had endowed enclaves of perfectly balanced soils. There were valleys in Pennsylvania where the buffalo once roamed. But as civilization pushed the buffalo into higher, less fertile ground and homesteads and farms required mountain goats and deer to live off rocky highlands, disease became the legacy, and extinction of species the likely outcome. Indeed, Missouri researchers found that even the cottontail rabbit lived in greater number, had larger bodies, stronger bones—all according to the fertility based on high protein production. Missouri rabbits decreased in the breaking strength of the femur bones from less well developed fertile soils of the northwest to the cotton soils of the southeast, really not unlike patterns discovered in evaluating draft age men during WWII.

5. from field to field

Soils vary not only from county to county, but from field to field, and sometimes within a field. There are maps that define areas in which there is a copper deficiency, where nature has blessed the farmer with plenty of phosphorus, potassium or calcium, and where there is an overload of selenium or sodium or iron. Here is a typical phosphorus map.

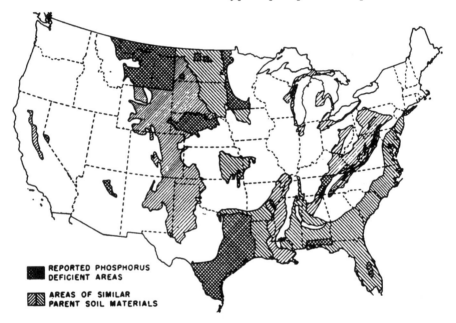

REPORTED PHOSPHORUS DEFICIENT AREAS

AREAS OF SIMILAR PARENT SOIL MATERIALS

We are not reproducing many of these maps simply because they may be of marginal help. In the wake of a chemical binge, soils in the U.S. are frequently complexed and confused, and this imposes many problems in eco-soil management. The answers can't be found in commercial formulations styled N, P and K.

Extension people will tell you lime is important, but an eco-farmer has to know the character of that lime, how many pounds per acre must be put on each area within a field to bring it into equilibrium.

As you drive through the countryside or walk the acres of your own farm, you'll note the different shades of green. If it is early fall, you'll note how some beans have died back early because they lacked the right nutritional support to keep them living until there was a mature bean in every pod. Within a few square rods, you'll see shriveled up beans and some that are incompletely developed. In such cases the amino acids are not built into complete protein. When animals or human beings eat such a crop, the amino acids will not be assimilated by the body as well as might be the case when a crop has been managed so as to build a complete mature protein. Some parts of management have to do with knowing what nature gifted an acre in the first place. More often than not the kind and quality of farming in the past will determine crop potential for the coming year.

6. not all farmers understand

People running the farms of this country are different. Some want to know little more than what the labels say. Accordingly they have come to favor cropping systems that rely on toxic rescue chemistry, drying equipment and embalming potions of several sorts so that insect and fungal predators cannot enjoy the meal that mismanagement has provided. Not all farmers understand that insects respond to areas in a field that are literally sick, nor do they accept the scientific intelligence that sick plants invite insects or plant diseases. Yet this is nature's way of regulating her own balance.

When a farmer grows a plant on sick soil, nature will destroy that plant so that the lower species of plants can take its place. A later lesson on weeds will amply illustrate this point. All of the insect and health problems in plants are caused by inefficiencies or complexed problems within a soil system. All relate to the availability of nutrients which are either in excess supply or in short supply, or locked up in a bank account in the wrong form.

7. capital programs are different.

Capital programs are different with each farmer. Some farmers buy space, and some buy land. We have seen young farmers pay more for paper-thin topsoil in the Ozarks, possibly because they liked the scenery, than good deep topsoil would have cost them in the Little Dixie area of northern Missouri. In addition to commercial farmers who took up business after the wars, the nation now has people interested in homestead land, *leben straum* in which to grow a garden and raise a family.

Most farmers buy land because it is available in a contiguous area, and because they want to expand. Yet land isn't really land. One acre isn't as good as another. It takes a good soil audit to reveal whether one acre is worth as much as another, or whether it has the potential to be as good. Or, it takes a farmer well in command of nature's art to understand soil value. Weeds provide a fair index, if you know how to read them. Take cockleburs. Cockleburs grow on rich soil. All the old timers knew that if you buy a farm, buy one in cockleburs because it has rich soil and will grow good crops. The cocklebur grows in an environment with a high level of phosphorus that is available, and where there is a good pH system. Obviously, phosphate can only be available in ample supply when pH is in line. At the same time a high level of phosphate tends to complex zinc. And it is in this kind of environment that the cocklebur finds its hormone system activated, or nature telling it that the cocklebur has its turn to grow this year. Farmers who know that everything relates to everything else instantly know that you don't need phosphate when you have cockleburs.

8. long before the american experience

Soil is a product of agencies that gather, grind, mix and deposit it in place. Long before the American experience, other civilizations rose and fell. Along the Tigris and Euphrates—rivers that flooded frequently—civilizations survived for a long time, then went into decline. In North Africa, civilization fell as soil fertility became exhausted. Only the farmers of 40 centuries, the Chinese, the Korean and the Japanese, seem to have maintained their fertility by living with a rule of return, "and its meaning carries a world of new-old values upon its back." The quoted words are from *England and the Farmer*, a short book published during WWII, one that ventured to head off the thrust of a synthetic era. "In the most primal sense it implies the return of all waste products to the soil, wherein myriads of microorganisms convert them into the one indispensable substance for every type of plant growth: humus," wrote editor H. C. Massingham.

Subtract humus from a soil system, and you have a corpse, as is the Gobi or the Sahara.

Remember what farmer Gene Poirot said in the opening pages of his book: nature had to invent death. And an eco-farmer has to recruit death to serve the balance of life dislocated by the modern fantasy called conquering nature. Only by making proper use of the resources of waste can we cut the heart out of another kind of waste—the waste of soil's fertility.

Only during the last few seconds of man's history has he understood the fundamentals of soil fertility. And during some of those seconds he's forgotten what he knew. Mountain-building, erosion, sedimentation, and geomorphological processes have caused land surfaces to exist in great complexity. Vegetation has adjusted in different regions, and animals have done the same. Climate has always exercised disciplinary control over development and distribution of plants. As a consequence, all plants have differing moisture and heat requirements. Weathering and organic decay and leaching are all influenced by the same climatic elements. At the same time, climate is in turn affected by the nature of vegetative inhabitants which profoundly affect wind, humidity and the distribution of animals.

In biotic geography these several phenomena are called the biotic complex or the eco-system. Either term should do. Either describes nature's long dedicated task of shattering rocks via frost action, the splitting, tumbling, drying, wetting, freezing of argillaceous rocks, mechanical actions all. Chemical weathering also splits fragments into smaller fragments.

As might be reasoned, mechanical weathering predominates in cold and dry regions of the world. Chemical weathering takes on front burner status in hot and wet regions.

The atmosphere contains small amounts of carbon dioxide. Any raindrop dissolves a little of this as it falls to the ground. This mild acid is able to detach potassium, magnesium and calcium from rock minerals. It carries them into the soil solution as carbonates. For instance, feldspar loses its potassium in the form of potassium carbonate, leaving a quite simple clay residue called kaolinite. Biotic mica breaks down to the form potassium and magnesium carbonates, sesquioxides of iron and aluminum, and some clay. Quartz, the prince of inert substances, is merely released when mildly acid raindrops take away those nutrients and hand them over to a soil system for plant growth.

There are other forces that fit into the soil building picture—wind blown loess, for instance, the superb grinding of gravel by slow moving glaciers, nature's steady assault on igneous and sedimentary rocks. Whatever the legacy of this action, the plants and soil profit accordingly, the soil forever treating life much like a churchgoer, allowing no more to be taken out than is put in.

Plants that adapt to an environment which is very poor ultimately die and contribute their organic matter to the soil. Having lived on soil poor in minerals, they have little to return.

9. a moot point

It is a moot point whether it is more difficult to discuss anything connected with eco-agriculture in a closed airtight compartment, or as an open, rambling whole. Sooner or later the farmer has to bring it all together so that his mind can handle the whole of it all. The level of organic matter has to do with pH management, and pH management cannot be accomplished properly without calcium, magnesium, potassium and sodium in equilibrium in terms of organic matter. This not-so-closed compartment leads to the cation exchange capacity, the trace element keys to enzyme system, and all the other considerations of the several facets in eco-farming. To know where you're going, you have to know where you are. To this end, the soil audit is a valid tool.

BORON DEFICIENCY AREAS

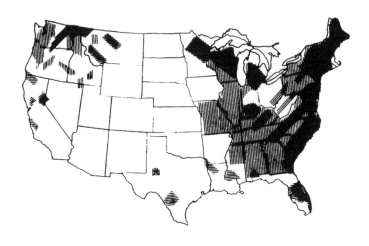

▦ *Field Crops*
clover, crimson, white
alfalfa
cotton
tobacco

▤ *Truck or Vegetable Crops*
beets
carrots
celery
general

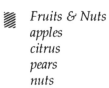 *Fruits & Nuts*
apples
citrus
pears
nuts

The chemical era has confused and complexed soils nationwide, and for this reason man-made problems often outweigh the shortfalls that nature has bestowed.

SECOND SEMESTER

LESSON 10

Soil Audit

The steps that must be taken to repair a soil system depend entirely on the condition of the soil system "as is." To really know, you have to read what the ground has to say. Some few farmers, well schooled in nature's signs and signals, get a fair idea by reading the weeds, the symptoms on plants, water intake capillarity, the quality of the end product as measured by animal metabolism, albeit always after the fact.

1. the in-depth test

It is a mistake to believe a soil test is a soil test, and therefore any test will do. So don't confuse the in-depth test called for by eco-agriculture with the usual quick test reports made by many universities, fertilizer companies and jacks-of-all-trades that roam across the farming scene. Quick tests simply comply with the conventional soil science of the hour. This means continued mining of the soil and spoon-feeding of crops with fertilizers that stimulate growth and yield.

To restore soil to its optimum nutrient release capacity, it is basic and absolutely necessary to start with a reliable and precise laboratory analysis of the soil. This analysis must define the real causes of limitations.

Limitations, after all, are merely an index of mismanagement, weather effects, nutrient disequilibrium, and chemical mistreatment—all factors that restrict the viability of the life systems in the soil. Limitations prevent the natural release and energy exchange systems from working, and these systems are vital if a productive soil is to be perpetuated.

To put it in another way, we might say that a wholesome soil environment, in good pH and nutritional balance, is actually the first priority. It naturally follows that the nitrogen-carbon system can begin to work to fuel the whole soil energy system for perpetual release of nutrition. The soil now becomes self-sustaining, but only if we cultivate the art and exercise common sense judgment as we farm the soils. It is then that protein production results and synthetic chemical and herbicidal systems are really not required. In the final analysis, it is protein agriculture that makes it possible for an acre to be farmed decade after decade, century after century.

When a farmer gets into his car for a drive to Timbuktu or Skiatook, he has to take along a map—unless, of course, he knows the roads the way an artful physician knows the symptoms and signs. Reading the soil isn't all that different. You have to take samples, but to sample properly you need a map. Start with fence rows, then put in the details the way a painter fills in features and colors on a good fine painting. If a gully divides a field, put that in. Characterize each area that has had good, bad or indifferent production. Note the weed patterns, eroded spots and areas where crops have either field ripened or failed to do so. A soil map soon takes on the personality of your farm and tells you that not all soil is the same even in one field, certainly not on one farm. Soil Conservation Service soil survey maps can be used to guide a farmer in taking proper samples and locating them accurately to scale. If you have an artistic bent, it might be well to draw up the farm, taking a helicopter view. Here is a sample.

An overview is not absolutely necessary. A map is. Here is a map of the same farm.

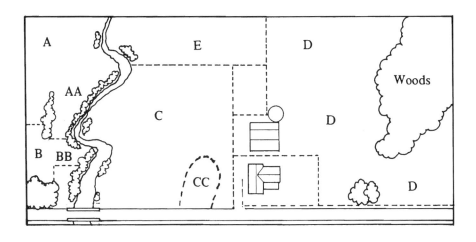

Areas obviously labeled differently require special probes. Several rules rate consideration in effecting proper sampling.

1. After making a soil map, divide fields according to soil types and conditions, past fertilizer treatments, and take samples from each area large enough to permit separate management. Composite samples should not include more than 20 acres under most circumstances.

2. Avoid sampling unusual areas or spots such as old fence lines, dead furrows, eroded spots, fertilizer spill spots, field depressions. These areas will not be representative of the whole field. Such areas should be sampled separately if tests are desired.

3. Individual samples can be taken with a clean spade, a sampling tube or an auger. A sampling tube will provide the most accurate composite since it tubes a uniform core of soil the full depth of the aerated topsoil. If a spade is used, be sure the sample is uniform in profile to represent the soil from the surface to the depth being sampled. The proper depth should include the major feeder root zone for the crop being grown (the top 6 to 8 inches for most crops and soils). Subsoil samples and other special samples desired should be taken according to special needs.

4. Collect an adequate number of individual samples of uniform depth to represent each area and mix them thoroughly in a clean pail to make a composite sample for testing.

5. Record sample locations on a map to help you or your consultant interpret the tests and make proper recommendations.

6. Package and label the samples correctly to avoid breakage and mixups. Include adequate information such as name, field number, sample identity, depth of sample, date, crops being grown, etc., to insure complete and reliable results.

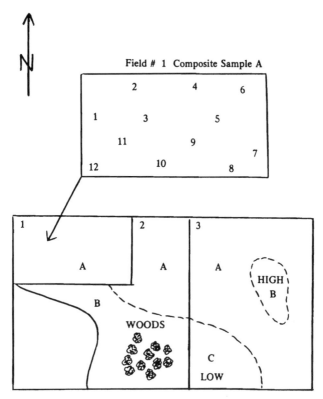

This diagram illustrates the points being made. One composite soil sample usually covers a 5 to 10 acre area. Note how 12 probes were made in the section designated as 1-A. Certain obvious conditions made this area distinct. The soil seemed consistent. Its history required it to be treated as a unit—on and on. It is well to build a history of each field the way real estate companies build an abstract of real property.

2. lots of laboratories and lots of tests

There are lots of laboratories and lots of tests, and farmers sometimes complain that different labs will return different test results based on the same sample. At *Acres U.S.A.* this problem was confirmed when the same samples were sent to four different laboratories. When recommendations came back, drastic disagreement surfaced on at last one sample, and general inconsistencies laced the entire set of readouts. One lab recommended 5.5 tons of lime, and another recommended 1 ton, both basing their recommendations on the same sample. If the 5.5 ton people were right, then 1 ton would be a waste of time and money. If one ton is right, then 5.5 tons meant wasted money and a soil system over-saturated with calcium. Neither were specific as to the quality of the lime in any case. A single ton of fine $CaCO_3$ would be of more value than 5 tons of coarse-ground 20% $CaCO_3$ as lime. Laboratories often do not know the calcium or calcium/magnesium content of ag lime, nor do they know the coarseness or its availability.

Ag courses use a manual of approved procedures for testing soils, and not all labs have the same equipment or use the same procedure. The results should nevertheless be close, but sometimes they aren't. Some laboratories avoid this dilemma by reporting in terms of high, low, medium for certain cations, and expressing pH, organic matter and humus in the same approximate terms.

Also, a good part of the game called eco-agriculture has to do with making nutrients available. Eco-farmers are interested in just what the soil contains in all its forms, and not just what's laboratory available immersed in citric acid. These data are devoutly to be wished, but are not generally a part of the readout scene.

3. the form

Since laboratory forms have different styles and sequences, we elect here to print one of the first forms drafted by William A. Albrecht for his associates, when Louis Bromfield's Malabar Farm was eco-agriculture's Mecca, and Friends of the Land were the organized disciples. Take a good look at this form, then follow it with us as we explain what each item means.

Shortly before William A. Albrecht passed from the scene at age 85, he admitted he'd learned a bit so far, but there was so much he didn't know. Yet the great scientist was certain of a few observations even children could make.

Plants are not mobile, and they do not have a stomach. Indeed, the soil system has to provide the anchor and the digestive apparatus for the entire crop, and earth nutrient overload as well. The soil system is the stomach. And into it are fed nutrients—whether as canal mud or manure or factory fertilizer—and these must be digested and made ready for plant use. It is the conceptualization of laboratory scientists that these nutrients must become ions, cations if they are charged positively, anions if they are negatively charged acid elements. They are perceived to enter plant roots the way electricity flows, so to speak, in chemical equivalents. This has been compared to the action one gets when positive and negative poles of a battery are connected. Interactions of cations and anions dovetail with the sun's energy to sustain plant life.

SOIL AUDIT AND INVENTORY REPORT

Account of_____City_____State_____

Service   Representative_____Date_____

	1	2	3	
House, Plot or Field No.	*1*	*2*	*3*	
Soils Sample No.	Untilled	in use	Compost	
Total Exchange Capacity (M.E.)	24	30	37.5	
pH of Soil Sample	7.5	7.9	7.8	
Organic Matter, Percent	2.40	2.30	8.30	
Anions Nitrogen: lbs/acre	62	59	111	
Sulfates: lbs/acre	1714	2171	3345	
Phosphates: Desired Value	253	275	315	
as (P$_2$O$_5$) Value Found	524	617	403	
lbs/acre Deficit				
Exchangeable Cations Calcium: Desired Value	6048	7200	9000	
lbs/acre Value Found	6240	8224	6768	
Deficit			2232	
Magnesium: Desired Value	980	1440	1800	
lbs/acre Value Found	1040	1480	2880	
Deficit				
Potassium: Desired Value	749	796	850	
lbs/acre Value Found	2336	1648	4800	
Deficit				
Sodium: lbs/acre	84	62	470	
Base Saturation Percent				
Calcium (60 to 70%)) 80%	64.97	68.60	45.34	
Magnesium (10 to 20%)	17.83	20.32	31.75	
Potassium (2 to 5%)	12.45	7.04	16.48	
Sodium (.5 to 3%)	.75	.43	2.73	
Other Bases (Variable)	4.00	3.61	3.71	
Exchangeable Hydrogen (10 to 15%)				
Salt Concentration (p.p.m.)	1530	600	6600	
Chlorides (p.p.m.)	100	None	1000	
Boron (p.p.m.)	17.2	18.5	9.6	
Iron (p.p.m.)	9.9	6.6	9.9	
Manganese (p.p.m.)	59	25.3	45.7	
Copper (p.p.m.)	28.2	28.2	176	
Zinc (p.p.m.)	74	61	61	

Remarks and/or suggestions:

4. cation exchange capacity

The first order of business for the soil colloid, then, is to hold nutrients—nutrients that can be traded off as the roots of a plant demand them. Thus the first index from the laboratory—the energy in the clay and the humus. Let's consider this entry on the soil audit sheet.

Soils Sample No.	ꬱᴛᴜᴇᴀ	ᴍ ᴜᴏᴇ	ᴄᴏⁿⁱʳⁿⁱ	
Total Exchange Capacity (M.E.)	24	30	37.5	
pH of Soil Sample	7.5	7.6	7.0	

Almost all laboratories report cation exchange capacity, and they do this in terms of milliequivalents, or ME. If it helps, you can think of an electrician measuring in terms of volts and amperes, or a physicist measuring magnetic energy in terms of ergs and joules. The soil laboratory has its own lexicon. It measures colloidal energy in terms of milliequivalents of a total exchange capacity, since soil colloids—composed of clay and organic matter—are negatively charged particles. Negative attracts positive. Cation nutrients are attracted and held on the soil colloids. Since anions are not attracted by the negative soil colloids, they remain free to move in the soil solution or water.

A milliequivalent is the exchange capacity of 100 grams of over-dry soil involved with 1/1000 of a gram of hydrogen (hence the prefix *milli*). This is laboratory talk. What this means is that the ME figure on a soil audit report will tell the farmer the pedigree of his soil, just the way horsepower rating of a tractor spells out the machine's power, or the way the cup, pint or quart defines the volume capacity of a vessel. Some soils have no great capacity for holding plant nutrients, and accordingly they can do very little to deliver the goodies when a hungry root arrives. Cation exchange capacity of a soil depends on the type of clay and the amount of humus. These can vary widely. There are CEC figures as low as 1 and as high as 80. The clays themselves vary, kaolinite in the south often measuring 10 to 20, montmorillonite in the west measuring 40 to 80. CEC values for organic matter are higher yet—100 to 200 ME being common. Obviously, pure sand gravel would have a CEC near zero. But the moment sand or gravel starts accumulating colloidal clay and humus, its CEC can go up dramatically.

Let's continue: ME represents the amount of colloidal energy needed to absorb and hold to the soil's colloid in the top seven inches of one acre of soil 400 pounds of calcium, or 240 pounds of magnesium, or 780 pounds of potassium, or simply 20 pounds of exchangeable hydrogen.

The upper portion of the earth's solid crust, to a depth of a few miles, has been estimated to consist of only eight chemical elements to the tune of 98.3%. They are in the order of their decreasing abundance, oxygen, silicon, aluminum and iron; and calcium, magnesium, potassium and sodium.

Of the 98.3%, oxygen comprises 47%; silicon 27%; aluminum 8% and iron 5%. These first four elements are either acidic or weakly alkaline in character. The last four—calcium, magnesium, potassium and sodium—comprise 11.3% of this 98.3%, and they are all four strongly alkaline in character. They are also positively charged elements, meaning their atoms bear positive rather than negative electrical charges. The remaining 84 known elements are contained in the remaining 1.7% of the earth's crust. Remember *calcium, magnesium, potassium* and *sodium*. Those four nutrient elements have a lot to do with pH management, maintenance of balanced hormone and enzyme systems, healthy plants, protection against insect, bacterial and fungal attack, weed control, and the business called eco-agriculture.

In any sample audit readout, it will be noted that sandy soil has a lower CEC in terms of ME than silty loam, clay loam or rich muck. The CEC then is an index of what the soil complex can take or will require if a farmer tries to effect balance. Let's move on.

5. pH

...ange Capacity, (M.E.,	...	30	...	
pH of Soil Sample	7.5	7.9	7.8	
Organic Matter, Percent	? ./?	? ??	8.30	

Soil laboratories deal with pH values, often on the basis of premises eco-farmers find wanting. Formally stated, the pH figure is the logarithm of the concentration of the active hydrogen or ions, not molecules. Distilled water has one millionth of a gram (.000001 or 10 to the minus 7 grams) of active hydrogen per liter, said to be pH 7. Such water tastes flat. With a bit more acidity or ionized hydrogen (.000001 or 10 to the minus 6 grams) or more acidity at a pH of 6.0, taste takes on character. Thus as the pH value is smaller, the degree of acidity is higher because there are more active hydrogen ions per unit volume as the ion concentration is higher. In eco-agriculture, an acid soil is viewed as a deficient soil. Calcium may be in short supply, but this may also be true of magnesium, sodium and potassium. The acid condition of the soil means very little if not related to the availability or absence of these major cation nutrients. In fact we will not deal with pH as such. By bringing calcium, magnesium, sodium and potassium into equilibrium, we will automatically adjust pH in a soil system suitable for plant growth. In the meantime pH provides a clue, albeit one that can be easily misread if fundamentals of soil balance are not kept in mind.

The clay-humus complex can exist in two fairly distinct physical states. Everyone is familiar with the colloid called gelatin. This substance can exist as a mobile fluid or a flexible solid, the agency for determining the difference being temperature. Much the same is true in a soil system, although acidity, not temperature determines the state. In a neutral medium (high pH), soil colloids tend to be stable (gel). Under conditions of acidity (low pH), a more mobile (801) state is achieved. Obviously when molecules of clay and humus become mobile in response to acidity, there is more chemical action in the soil. More nutrients are etched from the parent material. It is not possible, for instance, to farm without factory acidulated fertilizers if pH is kept too high, or neutral. Natural farming in any form requires mild acidity in the soil, more specifically, ideal acidity of the colloidal portions within the soil.

6. organic matter

Much the same is true of organic matter. Almost all soil audits state organic matter in terms of percent.

Organic Matter, Percent	2.40	2.30	8.30	

The road from organic matter in the soil to humus, punctuated with bacteria, dressed with earthworm castings at times, is long and delicate, but in the end it is the balancewheel of the clay-humus complex. From the chair of soil fertility and structure, clay particles resemble humus molecules more than they do the fragments of silt and sand, which are considered inert chemically. The clay-humus complex is chemically active. Molecules and humus not only link to each other, they also have spare valences with which to trap mineral elements and hold them available for plant use. When this active complex is not operative in a soil system, a rain shower washes the soluble salts of potassium, magnesium and ammonium down and away. In other words, it takes the clay-humus complex to make salt fertilizers work, and yet when the clay-humus complex is maintained at peak performance, salt forms of fertilizer are no longer necessary.

Much as a good physician might read signs and symptoms, and treat a patient accordingly, some few farmers and agronomists read what the plants have to say and take the saving action in time. Most farmers, however, tooled into compliance with theory-period instruction, would do best to take inventory, to draw appropriate conclusions, and to act accordingly.

7. anions

The next section of the soil audit has to do with anions, the negatively charged nutrients. Since the soil colloid—composed of humus and clay—is negative in electrical charge, anions cannot attach themselves. Negative repels negative just as positive repels positive. That is why anion elements are held in the soil solution. Plants absorb them through root hairs.

Anions	Nitrogen: lbs/acre	62	59	111
	Sulfates: lbs/acre	1714	2171	3345
	Phosphates: Desired Value	253	275	315
	as (P_2O_5) Value Found	524	617	403
	lbs/acre Deficit			

The conceptualization of chemistry puts it this way. Phosphorus, sulfur, chloride, boron and molybdenum as plant nutrients must be converted by the soil's refining process to each element's anion phase. Microorganisms handle this chore. Using their bodies as a chelating apparatus, microorganisms convert nitrogen to a nitrate form, phosphorus to a phosphate, sulfur to sulfate, chlorine to chloride, boron to borate, molybdenum to molybdate. Of the anions, phosphorus is the most difficult to comprehend. The country is full of farms with several thousands of pounds P_2O_5 per acre, and crops on those farms still suffering a phosphorus deficiency. Usually release is the limiting factor, and release is governed by the physical and biological condition of the soil. The soil audit simply states P_2O_5 in pounds per acre. But this is a laboratory chemical judgment and must relate to other plant food elements, how they are related to the values found in the exchange capacity, conditions of temperature and moisture. By itself a laboratory reading means very little, then, unless it is understood by the farmer in terms of the grand mosaic of the whole.

Much the same is true of sulfur in a sulfate form. The soil audit states values found by chemical evaluation in terms of pounds per acre. We will discuss this nutrient in a later lesson.

Ditto for nitrogen. Even though the atmosphere is 78% nitrogen, it has to become fixed with one or more elements in the soil before it can be of use to plant life. Rain does a little fixing, but not much, perhaps only 5 pounds per acre per year. Soil bacteria and microorganisms fix nitrogen in their bodies. Some of these little fellows are symbiotic and some are non-symbiotic, the former being associated with legumes, the latter living independently. A natural nitrogen cycle depends on healthy living soil. This is the reason nitrogen costs become a farmer's ball and chain as soils go dead, In any case, this is what figures say, but what they really mean is so interdependent that we must defer explanation until a few more facets of the soil profile can be brought into the arena.

8. the linchpin in achieving balance

Remember the figure at the top of the soil readout exhibited in this lesson? It said CEC (cation exchange capacity). This is the part of the soil readout that catches most attention. It is that part scientific eco-farmers use to adjust pH as a starter. It is the part that is the linchpin in achieving balance and bringing the anion nutrients into the ballpark as well.

		lbs/acre — Deficit				
Exchangeable Cations	Calcium: lbs/acre	Desired Value	6048	7200	9000	
		Value Found	6240	8224	6768	
		Deficit			2232	
	Magnesium: lbs/acre	Desired Value	980	1440	1800	
		Value Found	1040	1480	2880	
		Deficit				
	Potassium: lbs/acre	Desired Value	749	796	850	
		Value Found	2336	1648	4800	
		Deficit				
	Sodium:	lbs/acre	84	62	470	

As explained earlier, soil colloids, both clay and humus, have a holding capacity for plant nutrients. They operate more like a catchpen. They determine the pedigree of the soil, its potential, its possibilities in crop management. Some soils have no great capacity for holding plant nutrients and accordingly they can do very little to deliver the goodies when hungry roots arrive. CEC qualifies what you have and what you can expect if you take certain actions.

These concepts are not as difficult to handle as first inspection might suggest. Let's forget the laboratory for a minute and think in terms of something we all "think" we understand. Your pickup truck, tractor or automobile all have batteries, each with a positive pole and a negative pole. A flick of a switch will cause movement from one pole to the other—electrical energy! This energy starts the vehicle, operates the lights, permits you to broadcast "Breaker, Breaker" on the CB radio. The flow of current or energy is always positive to negative.

Nature is inclined to like electrical balance. Therefore a natural movement is an example. Lightning, a product of thunderstorms, not only balances energy in the heavens, it serves as a great producer of nitrogen. An activity that runs counterclockwise to natural balanced energy is the explosion of an atomic bomb.

Farmers of course deal with soil systems, not laboratory experiments or nuclear theory. They look to soil structure, tilth, biotic activity and nutrient uptake for healthy plant construction.

Soil colloids are negatively charged. That is why there is an attraction between the soil colloid for base or cation elements—the old business of positive to negative. This attraction is called colloid adsorption. The capacity to exchange base elements from soil colloids is called cation exchange capacity, or CEC. When negative colloids are neutralized by an accumulation of positive base elements, the system is said to have achieved 100% base saturation.

In your soil audit, the base elements are calcium, magnesium, potassium, sodium and the base trace elements. A soil should be saturated with usable cation nutrients up to 85 or 90% of its capacity. This leaves 10 to 15% for hydrogen, a non-fertilizer.

This is an oversimplification, but hopefully it will provide a fair insight into the scope of soil chemistry. This base element adsorption is called chemical binding. Cation exchange capacity—to put it plainly—means the ability of a soil to act as a source of supply for the exchange base so it can give up nutrient elements to plants in exchange for hydrogen, another base. This means that loading nutrients over to the soil colloid, not feeding the plant, is the name of the soil fertilization game, and in this sense the term plant food is a misnomer. Once an eco-farmer gets this straight he's well on the way to making it happen with nature.

A paragraph or two ago we related that it would take 400 pounds of calcium to fully load the top 7 inches of an acre of ME soil. Therefore it would take 4,000 pounds per acre for a 10 ME tested soil. Needless to say, no farmer would want to go whole hog on one nutrient and neglect the rest. Obviously an audit tells only about cation inventory, not the objective. There are differing theories, and there are sub-schools of thought within each group as to what target figures ought to be. Among eco-farmers, the Albrecht formula is both gaining strength and rolling up millions of acres of proof as well. It works.

It was Professor William A. Albrecht, working at the University of Missouri, who found that for best crop production, the soil's colloid had to be loaded with 65% calcium, only 15% magnesium, and that potassium had to be in the 2 to 4% range. Sodium was never much of a problem since it was normally satisfied under the other bases. These values, Albrecht held, were the saturation figures when nature was at her finest balance and capable of producing healthy crops.

The setup in this old Albrecht design (see page 120) for a soil audit reflects the schoolmaster's touch. A desired value is stated in terms of pounds per acre for calcium, magnesium and potassium. Then the value actually found is stated. Simple addition or subtraction yields a deficit or surplus.

Take calcium for a 12 CEC soil. If such a soil were fully loaded with calcium, the equation would read 400 x 12 = 4,800, (400 pounds for each ME in the top 7 inches of an acre). Calcium should take up only 65% of the

CEC. Thus the computation should read:

$$12 \times 400 = 4{,}800 \text{ pounds}$$
$$\underline{\times\ 65\%}$$
$$3{,}120$$

Since this is the value found in the soil audit for this CEC soil, the calcium part would appear to be in perfect harmony.

Computations can be made this way for each of the cations. In the case of magnesium, multiplication would read:

$$12 \times 240 = 2{,}880$$
$$\underline{\times\ 15\%}$$
$$432$$

Here, again, if this is the value actually found, magnesium should be no problem in this soil. But let's follow the arithmetic all the way on CEC 12, taking potassium.

$$12 \times 780 = 9{,}360$$
$$\underline{\times\ 03\%}$$
$$280$$
$$\text{less hypothetical value found } \underline{-312}$$
$$-32$$

The negative 32 is not far off target, and probably this soil requires no treatment in terms of potassium.

9. wrapup

The next section of a sample soil audit ought to come clear at this point.

Sodium:	lbs/acre			
Base Saturation Percent				
Calcium (60 to 70%)) 80%	64.97	68.60	45.34	
Magnesium (10 to 20%)	17.83	20.32	31.75	
Potassium (2 to 5%)	12.45	7.04	16.48	
Sodium (.5 to 3%)	.75	.43	2.73	
Other Bases (Variable)	4.00	3.61	3.71	

These data merely represent the extensions simple arithmetic would duly account for. Each of the soil types has a saturation percentage once computations are completed. Note how the first soil is relatively in cation balance. The second soil is fair also. The third needs repair.

As each of the equations for each of the chief cation elements fall into place, a principle of eco-farming becomes transparently clear. pH becomes self-adjusting when calcium, magnesium, potassium and sodium are in proper equilibrium. To answer low pH with lime regardless of the character of that lime rates attention as frustration agronomy.

pH means acidity to some people, but to eco-farmers it means a shortage of fertility elements, the same fertility elements named in the paragraph above. Professor C. E. Marshall of the University of Missouri once designed electrodes and membranes to measure ionic activity of calcium and potassium in the same way hydrogen ions are measured to determine pH. Marshall's pK and pCa illustrated clearly that ionic activities of a mixture of elements were not independent of each other. Taking calcium and potassium in combination, he discovered that potassium gained ascendancy in relative activity as cations become narrower. Magnesium, pound for pound, could raise pH up to 1.4 times higher than calcium. A soil high in magnesium and low in calcium could test pH 6.5 and still be entirely inadequate for the growth of alfalfa. Any of the major cations—calcium, magnesium, potassium and sodium—in excess can push pH up, and any one of them in lower amounts can take pH down. They have to be in balance, or the pH reading is likely to be meaningless. An equilibrium of pH at 6.2 or 6.3 for farm crops (based on these four elements in balance) will prompt plants to grow well and produce bins and bushels in tune with both the pedigree of the seed and the character of the soil.

10. colloidal position

As the reader might suspect, the equation presented here represents pure theory—theory that has been proved out to a great extent—yet one missing on a few cylinders. Advanced Ag Associates—West for years found that the plants were saying something not evident in those fine-tuned soil reports. The laboratory measured *availability*, but it failed to tell whether nutrients were in *colloidal position*, or floating around somewhere else, feeding the readout, albeit not the plant. Plant growth seemed to say nutrients were remaining water soluble. Some obviously hadn't been loaded over to the colloid—the humus/clay complex. Yet usual laboratory readouts handled them as one and the same.

This prompted co-author C. J. Fenzau to design a new soil readout form, one that accounted for this discrepancy. A good laboratory has the capability of giving the colloidal position answer, but before Advanced Ag Associates—West came along, no one had ever asked the question.

Looking at almost any of the hundreds of research efforts on Advanced Ag plots, the thesis and its proof becomes at once apparent. The usual procedure has been to divide a plot, using a full NPK fertilizer program on one section, pH modifiers plus a ton of compost on the other. pH modifiers are blends of calcium, magnesium and potassium (or sulfur) deionized and prepared for field application in solution form. Rate of application is usually 10 to 20 pounds per acre, each spray treatment being worth 3 to 4 pounds calcium in perfect form. Next the crop is planted. A sample is

Advanced Ag Associates - West

SOIL ANALYSIS REPORT

NAME: AA Ranch Exp. Plots
ACRES: Plots 41 & 42 Last Crop: Spuds
DATE SAMPLED: 11/26
Purpose: Fertilizer vs Full Program

DESIRED LEVELS	LAB NO. / FIELD or SAMPLE NO.		7-24 (Program + Compost) 41	C6094 41	3-15 4AB	C6095 42 (Full Fertilizer Only)	7-24 42
%	EXCHANGE CAPACITY		12.9	14.0	9.9	12.9	12.9
4	% ORGANIC MATERIAL		5.4	5.4	4.8	5.2	5.4
	% HUMUS		2.8	2.9	1.7	2.5	3.0
.5-7.0	pH		7.1	6.3	7.5	6.0	6.7
5-10	% BASE SAT. HYDROGEN		-	.0	-	.0	4.5
50-100	SODIUM - lbs per Acre		70	70.0	70	50.0	60
	CALCIUM	ACTUAL lbs per Acre	4100	3900	4116	3500	4000
104		% of Desired Level	122.0	106.4	116.5	103.5	118.6
95		% Desired Level in Colloidal Positions	113.1	92.7	110.8	85.8	109.7
68		% BASE SATURATION	79.3	69.1	77.9	67.3	77.1
	MAGNESIUM	ACTUAL lbs per Acre	400	410	363	320	350
103		% of Desired Level	86	80.8	98.2	68.3	74.7
95		% Desired Level in Colloidal Positions	71	72.9	64.0	55.5	59.7
15-16		% BASE SATURATION	12.9	12.1	14.6	10.2	11.2
	POTASSIUM	ACTUAL lbs per Acre	220	210	170	110	130
60-100		% of Desired Level	60	54.5	44.8	31.0	34.3
95		% Desired Level in Colloidal Positions	57.3	44.1	37.6	22.5	31.7
3%		% BASE SATURATION	2.1	1.9	1.8	1.0	1.2
35	P_1 lbs per Acre		73.3	35	23	9	50.4
300	P_2O_5 lbs per Acre		132	238	130	123	141
600-900	SO_4 lbs per Acre		270	495	495	420	255
80	Manganese ppm		128	98	82	81	134
4	Copper ppm		2	2	3	2	2
12-14	Zinc ppm		13	13	13	11	12
4-5	Boron ppm		15.4	3.8	8.8	4.6	11.4
200-300	Iron ppm		310	210	290	260	320

taken July 24, another in November, one during a growth season, one during a dormant season. The procedure is followed year after year. One can only marvel at the readout changes that take place according to plot.

In each case CEC is adjusted upward on the eco-plot. Organic matter improves dramatically. pH moves in the desired direction toward 6.0 and 6.2, often in as little as 7 months. As though they were trained circus performers, calcium, magnesium, potassium and sodium move toward and

THE HAUGHLEY EXPERIMENT

In the Haughley Experiment, started in 1939 in England, soils and crops were tested by a biochemist, R. F. Milton. They tested soils in each field every month for ten years running. They discovered something unsuspected by the fossil fuel crowd, namely that levels of available minerals in the soil fluctuate according to the season, maximum levels coinciding with the time of maximum plant demand. Moreover, fluctuations were more pronounced in the organic section than in the mixed or stockless section. In the field with the highest organic matter content, ten times more available phosphorus was recorded in the growing season than in the dormant period. Potash and nitrogen followed the same general pattern. The upshot was that a single soil test told very little in high organic matter soils—those with 4 to 5% organic matter. In soils with less than 3% organic matter, one test was sufficient—if taken at the right time.

Fertilizer salesmen sometimes know this. They are perfectly safe in guaranteeing fertilizers will triple available nutrients between March and June. They know all they have to do is wait for warm weather and nature to unlock the vault and break out some of those nutrients.

reach their equilibrium.

If, say, 17% of 500 pounds of calcium is water soluble, it will still measure available via the laboratory test, but there will be a shortage on the colloid. Ditto for the other cations. If, on the other hand, a system of humus is working properly etching out nutrients from locked-up positions that aren't even measured, the working colloidal system remains close to constant, a dinner table set for full meals as plants require them.

Complexed supplies mean very little to a farming system that relies on bagged N, P and K year after year because the organic matter complex, the humus supply will be low and poorly functional. In any soil there is a tremendous amount of calcium, aluminum, magnesium, silica, iron and zinc, all a part of the permanent base. At any given point they will not show on the colloidal position or on the soluble position. But they are there. If a good humus system is working, they can be etched out every day they are needed, leaving the level unchanged.

Unfortunately there is no commercial test that will give a farmer a readout of nutrients in a soil in all their forms. This would be too costly. Yet it is known that many soils carry from 40,000 to 80,000 pounds of P_2O_5 per acre in the top 7 inches. The farmer does not want that much for a crop year. But if he can figure out how to etch it out—take it out of the bank account, so to speak—then it makes poor economic sense to buy the bagged product and pour it on the soil only to have it join more phosphorus already in the lockup.

As for sampling, the optimum date should make little difference if governing principles are kept in mind, and if enough records exist to explain the variation. Warm soils release more phosphate. A sample taken in February will be different from the one taken in May. This does not mean either are wrong. If a sample is taken in the middle of winter, it is only logical to expect the potassium to be lower than in July.

By managing organic matter, water, air, and temperature properly, the farmer can diminish the amount of N, P and K he has to use. Indeed, the actual yield potential is determined by these other principles to a greater extent than by N, P and K. Nearly 95% of the nutritional factors are related to carbon dioxide, sunlight and water—photosynthesis, in short—together with temperature.

The soil audit, then, can hint at how well and how poorly the chores of farm soil management are being accomplished.

SOIL ANALYSIS REPORT   COMPOST VARIABLE

NAME: AA Ranch Exp. Plots
ACRES: Plots 51 & 53   Last Crop: Spuds & Pop Corn

DESIRED LEVELS	FIELD or SAMPLE NO.	7-24 / 51	C6096 11/26 3-15 (Full Program) / 51	5B	11/26 C6097 (Program – NO Manure Compost) / 53	7-24 / 52
	EXCHANGE CAPACITY	14.5	15.0	10.2	12.8	16.7
8	% ORGANIC MATERIAL	5.6	5.6	5.0	5.5	5.7
4	% HUMUS	2.9	2.4	2.2	2.5	2.6
6.5-7.0	pH	6.5	6.2	7.0	6.0	6.5
5-10	% BASE SAT. HYDROGEN	7.5	.0	—	.0	7.5
50-100	SODIUM - lbs per Acre	160	55.0	70	45.0	60
	CALCIUM ACTUAL lbs per Acre	4100	3900	4837	3400	4900
104	% of Desired Level	108.0	99.8	118	101.8	112.6
95	% Desired Level in Colloidal Positions	102.7	84.5	112	80.8	101.1
68	% BASE SATURATION	70.2	64.9	80.5	66.2	73.2
	MAGNESIUM ACTUAL lbs per Acre	420	560	392	340	570
103	% of Desired Level	79.3	103.5	82.8	73.5	84.7
95	% Desired Level in Colloidal Positions	68.0	92.4	88.4	58.4	71.3
15-16	% BASE SATURATION	11.9	15.5	12.2	11.0	12.7
	POTASSIUM ACTUAL lbs per Acre	330	170	160	150	100
60-100	% of Desired Level	80	41.4	30.1	42.7	20
95	% Desired Level in Colloidal Positions	75.2	34.1	23.0	34.2	18
3%	% BASE SATURATION	2.8	1.4	1.2	1.4	.7
35	P_1 lbs per Acre	77.9	.29	18	12.	25.2
300	P_2O_5 lbs per Acre	183	224	107	148	119
600-900	SO_4 lbs per Acre	420	450	555	435	165
80	Manganese ppm	130	89	96	93	147
4	Copper ppm	2	2	3	3	2
12-14	Zinc ppm	12	12	12	12	13
4-5	Boron ppm	17.4	4.2	13.2	3.2	19.4
200-300	Iron ppm	300	230	340	250	340

The evolution of one soil audit form is depicted in this Advanced Ag Associates-West update. It embodies William A. Albrecht's concepts, the early Brookside refinement thereof, and finally the requirement that calcium, magnesium and potassium be stated in terms of desired level in colloidal position. Percentages are used. With hand calculators, translation from percent to pounds, and pounds to percent, is no problem.

LESSON 11

The N, P and K Concept

So far this primer has a little more than hinted at the vast complexities involved in crop production. Air, sunshine, water, temperature, soil structure, tilth are all topics we've touched base on. We'll touch those bases again. But first, let's take a passing look at what the plants have to say in terms of a chemist's ability to read. This lesson represents the conventional wisdom, only a fragment of which is ever passed on to farmers by the industrially owned farm journals. We need to understand what the land grant colleges and Extension people have to say before we can move on to the higher plane of an updated technology.

1. carbon

The plants tell us via the laboratory that they need carbon. Between 45 and 56% of a plant's compounds are structured with the help of carbon. As we have seen, carbon is taken from the carbon dioxide in the air. A single human being gives off enough carbon dioxide in 24 hours to fill the photosynthesis requirement of a single tree. It has been computed that it takes 20 trees to handle the CO_2 given off by every five gallons of gasoline used by an internal combustion engine. In the theory of the chemist there is always plenty of carbon, thus this element is seldom thought of as a limiting factor in plant growth. It's not that simple, of course. Carbon is governed by the character of the nitrogen source. The many ramifications involved are covered in lessons 7 and 12.

2. hydrogen

The laboratory also tells us that 6% of a plant's compounds involve hydrogen. Water supplies hydrogen, and water is certainly a limiting factor, hence irrigation in drier climates.

3. oxygen

Of all the plant compounds measured by laboratory techniques, oxygen is second only to carbon in the numbers game. Approximately 43% of all the compounds in a plant are composed of oxygen. This comes from both air and water. Obviously plants will not grow unless there is a sufficient amount of oxygen.

4. nitrogen

"You can't farm without nitrogen" is more than a farmer's old saw. Nitrogen accounts for 16 to 18% of plant proteins and up to 3% of all plant compounds. Nitrogen puts the paint in plants regardless of its source. However, when synthetic nitrogen is used it can delay maturity of many plants, cause lodging, and there are many questions as to whether synthetic nitrogen can indeed substitute for protein nitrogen. Again, see lesson 12.

5. phosphorus

Phosphorus is a "go" food for plants. It tends to concentrate itself in seeds and fruits—but it also migrates to many of the organic compounds in plants. Laboratory examinations have revealed that up to 1% of all plant tissues contain phosphorus. The plants will tell you when phosphorus intake is deficient by growing red and purple leaf colors and by exhibiting stunted root and tops growth.

6. potassium

In some plants potassium as part of tissues is small—no more than .3%. In others it reaches 6%. Styled as K on the Mendeleyeff chart, it rides along in solution over the plant's vascular highway, and figures mightily in translocation of vital sugars in plant structures. Potassium strengthens plant stalks and helps undo the stress accounted for by excess nitrogen in the wrong form. When a plant grows first white then brown spots, it is screaming for potassium. This symptom is commonlyseen on ladino clover. Corn leaves tell you about a potassium shortage with brown edges on the bottom leaves.

7. calcium

Calcium is often called the prince of nutrients because the soil colloid has to have a great saturation for plant uptake. Yet it accounts for no more than 4% of plant tissue at the upper range of laboratory expression, .1% at the lower range. Calcium is used by nature to make a material called calcium pectate, a sturdy building material component of cell walls. We will discuss the complexities that attend calcium's role in plant chemistry in a later lesson. For now it is enough to note that missing calcium means stunted roots and stress symptoms on newer leaves. Discoloration and distortion of growth are common symptoms.

8. magnesium

Each chlorophyll molecule is built around a single atom of magnesium. As a consequence, magnesium accounts for as little as .05% and as much as 1% of plant tissue. Obviously poor photosynthesis affects crop production. If plants could talk, the moan would be loud and horrible. In terms of the silent language called signs and symptoms, the deficiency message comes through just as clear. Corn farmers know them as the light and dark stripes running the length of a plant leaf. An eco-farmer, however, can just sense the right green color that is a good photosynthesis green—not an over-fed nitrogen green.

9. sulfur

Sulfur is an anion, a constituent part of proteins, and it accounts for .05 to 1.5% of a plant's tissue. Legumes demand it for good nodule development, and when it isn't offered to plant life pale green leaves develop. It is easy to mistake sulfur deficiency for nitrogen deficiency as well as magnesium, iron and available P_2O_5 deficiencies. All are essential ingredients of the sunlight energy production system.

10. iron

Iron is what scientists call a micro-element. Although it figures in the formation of chlorophyll, it is not a part of the chlorophyll molecule. It is a part of the plant, however, and is expressed in terms of parts per million. Plants ashed out will test with as little as 10 p.p.m. and sometimes will have 1,000 or 2,000 p.p.m. When iron is missing in the nutrient uptake, chlorophyll production for greater photosynthesis is also punished. Whitened young leaves and the white and dark stripes of a magnesium lack may also bespeak an iron problem.

11. *manganese*

Manganese is also measured in terms of parts per million. A plant may test out with as little as 5 ppm, and it may have 500 ppm or more. It has the same deficiency fingerprint as iron.

12. *zinc*

Zinc is only now joining the crowd of elements "required" for healthy plant production in the conventional lexicon. Up to very recently agronomists couldn't agree on its function except to note that certain weed patterns appeared when zinc disappeared. Plants do require it in the 3 to 100 ppm range.

13. *copper*

Copper is essential, but the ramifications of that essentiality are seldom explored to the fullest in conventional agronomy texts. As a part of Bordeaux mixture in grape arbors, it functions as a nutrient and not as an insecticide as is often believed.

14. *boron*

Plants must have boron, again in the trace range. Texts quote 2 to 75 ppm as being essential, but note that plants vary in their required amounts according to species. Boron is quite lethal to seeds when used in the salt form. Classic among plants requiring boron is alfalfa. Absent it has the earmarks of leafhopper injury, and general rosetting of leaves due to stunted stem growth. Celery, tobacco, cauliflower and many other plants are affected by unavailability of the boron nutrient.

15. *molybdenum*

Plants that are ashed out reveal that between .01 and 10 ppm molybdenum are required. Beyond that there is much mystery. Certain beneficial microorganisms need molybdenum or they won't be around to set the stage for anion nutrient uptake, and for this reason it is difficult to discuss this nutrient—or any of the other nutrients—in a closed airtight compartment. They all relate to each other and are complexed by each other.

16. *chlorine*

Chlorine stimulates crops. It is seldom deficient.

17. *the folklore of fossil fuel chemistry*

It is of course not possible to tell how many elements are essential or exactly how they affect plant life. Knowledge is still very primitive in this

---

COMPOSITION OF SOILS

For 1 Plow Acre 6-2/3 Inches in Depth
Approximately 2,000,000 lbs., or 1,000 tons

Elements	Sandy Loam lbs. per acre	Silt Loam lbs. per acre	Clay Loam lbs. per acre
Organic Matter	20,000 =	54,000=	96,000 =
Lbs. of Nitrogen	1,340	3,618	6,432
Live Portion	1,000	3,600	4,000
(Earthworms, Bacteria)			
Silicon Dioxide	1,905,000	1,570,000	1,440,000
Aluminum Oxide	22,600	190,000	240,000
Iron Oxide	17,000	60,000	80,000
Calcium Oxide	5,400	6,800	26,000
Magnesium Oxide	4,000	10,400	17,000
Potash	2,600	35,000	40,000
Phosphate	400	5,200	10,000
Sodium Oxide	4,600	26,000	24,000
Titanium Oxide	13,600	18,000	14,400
Sulfur Trioxide	600	8,500	6,000
Manganese	2,500	2,000	2,000
Zinc	100	220	320
Copper	120	60	60
Molybdenum	40	40	40
Boron	90	130	130
Cobalt	50	50	50
Chlorine	50	200	200

Compiled by J. L. Halbeisen and W. R. Franklin.

From rock to availability covers a multitude of sin. In the laboratory the scientist takes a sample and subjects it to extracting solutions as decreed by official manuals. Unfortunately, plants cannot go out and buy those solutions for the purpose of extracting nutrients. A living, dynamic complex known as the living soil must do this. Here is what that life system has to work with.

Crop	Approximate Acre Yields	Nitrogen (N)	Phosphate (P$_2$O$_5$)	Potash (K$_2$O)
		pounds per acre		
Legume Seed				
Soybeans	40 bu. (2,400 lb.)	150	35	55
Corn (grain)	100 bu. (5,600 lb.)	90	36	26
Small Grains (Seed)				
Barley	40 bu. (1,920 lb.)	35	15	10
Oats	80 bu. (2,560 lb.)	50	20	15
Rye	30 bu. (1,680 lb.)	35	10	10
Wheat	40 bu. (2,400 lb.)	50	25	15
Straw and Stover				
Corn stover	3 tons	67	24	96
Barley straw	1 ton	15	5	30
Oats straw	2 tons	25	15	80
Rye straw	1.5 tons	15	8	25
Wheat straw	1.5 tons	20	5	35
Legume Hay				
Alfalfa	4 tons	180	40	180
Red clover	2.5 tons	100	25	100
Soybean	2 tons	90	20	50
Non-Legume Hay				
Bluegrass	2 tons	60	20	60
Orchardgrass	3 tons	90	40	140
Timothy	2.5 tons	60	25	95
Fruits and Vegetables				
Apples (fruit)	500 bu. (22,000 lb.)	30	10	45
Beans (seed)	30 bu. (1,800 lb.)	75	25	25
Cabbage (heads)	20 tons	130	35	130
Onions (bulbs)	7.5 tons	45	20	40
Peaches (fruit)	600 bu. (30,000 lb.)	35	20	65
Potatoes (tubers)	400 bu. (2,400 lb.)	80	30	150
Tomatoes (fruit)	15 tons	90	30	120
Turnips (roots)	10 tons	45	20	90
Tobacco (leaves)	1,500 lb.	55	10	80
Tobacco (stalks)	1,500 lb.	25	10	35
Animal Products				
1,000 lb. of milk takes off the farm		6	2	2
1,000 lb. of beef (live weight) takes off the farm	27	17	2	2

Bushel weights are computed in terms of the following average weights: soybeans, wheat, beans, potatoes, 60 pounds per bushel; barley, 48 pounds per bushel; oats, 32 pounds per bushel; rye, 56 pounds per bushel.

A WORD OF CAUTION

The implication in tables such as the foregoing is that to remove the nutrient it must first be there in available form. Yet the term shortage must be understood in terms of the technology available to the laboratory. Nature marches to a different drummer. She decides availability according to rules all her own. So far agriculture does not have an economic test that will give the soil's nutrients in all their forms. Agronomists thus seem content to follow the chemist's pronouncements as though they were words from Olympus.

These words have recast both soil management and fertilizer consumption in the United States since the hard-sell onset of the salt fertilizer and toxic chemical era. Before 1951, most salt fertilizer sales were made in worn out farm territory, where hard chemistry had upset soil life, where feed the plant rather than feed the soil had become copybook maxim. This map amply illustrates the point.

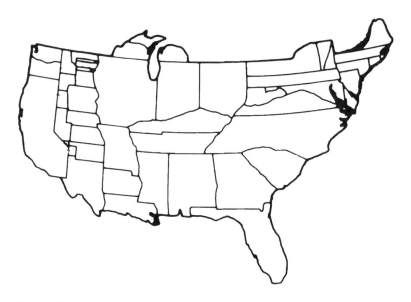

1951 - 1952
The United States of Fertilizer
A dealer's view of the nation as shaped by plant food consumption.

area, but it is developing rapidly—and it will continue to develop as quickly as the folklore of fossil fuel chemistry can be outgrown.

The "ash mentality" has given agriculture data that still governs most farming operations. While these equations are interesting, they are not as definitive as fertilizer purveyors would have clients believe. On page 138, for instance, is a standard table on "major" nutrients removed with each harvest.

18. soluble salt arithmetic

It was largely to answer this arithmetic that soluble salt fertilizers were developed. Still it must be remembered that the soil chemistry we call modern didn't really figure until the start of the 19th century. Theodore de Saussure of Switzerland led the way, riding high on the shoulders of van Helmont, Joseph Priestley, Henry Cavendish and Antonine Lavoisier, some of the individuals discussed in our opener on photosynthesis. Theodore de Saussure went a bit beyond separating air and water into component gases and concluding that gases made up most of the plant. He noted that the ash of a plant was derived from the soil. He challenged the old humus theory of Albrecht Thaer of Germany—a point modern symposiums seem fond of quoting in an effort to put down organic gardeners.

There was that "discordant note," to quote Firman Bear, that note called *Elements of Agricultural Chemistry* by Humphrey Davey. Davey wrote in 1813, and what he wrote was widely accepted. He had credentials and standing in the scientific community, and this gave weight to his concept that decomposed animal and vegetable matter in the soil was the primary food for plants. It was Davey's opinion that inorganic elements in a soil system acted merely as stimulants. The real food for plants, he held, was the same as that of animals, namely organic matter.

A few years later Justus von Liebig's famous lectures before the British Association for the Advancement of Science exploded across Europe. von Liebig argued orally and in print that "the primary source whence man and animals derive the means of their growth and support is the vegetable kingdom. Plants, on the other hand, find new nutritive materials only in inorganic substances." von Liebig proved so persuasive that he conquered the academic community.

Sir Albert Howard has commented on the impact of von Liebig's *Chemistry in Its Application to Agriculture.* "Inquiries," he wrote, "into general organic chemistry were so vast and so illuminating that scientists and farmers alike naturally yielded to the influence." Sir Albert Howard further commented that von Liebig's views "remained those of a chemist who understood nothing about humus because he was enchanted with laboratory chemistry and dealing with only a fraction of the facts." The idea that plants could be fed back the same three nutrients that they had drawn seemed dazzling in its purity, too good to let go. Again, let us quote Sir Albert Howard. "There was a kind of superb arrogance in the idea that we had only to put the ashes of a few plants in a test tube, analyze them, and scatter back into the soil equivalent quantities of dead minerals. It is true that the plants are the supreme, the only agents capable of converting the inorganic materials of nature into the organic; that is their great function, their justification, if we like to use the word. But it was expecting altogether too much of the vegetable kingdom that it should work only in this crude, brutal way; as we shall see, the apparent submission of nature has turned out to be only a great refusal to have so childish a manipulation imposed upon her."

The world has turned over several times since von Liebig, and science has come full circle, a point that knowledge and application thereof seems to deny at times. Plants in fact do absorb and translocate complex organic molecules of systemic insecticides. Thus there is little ground to hold that plants cannot also absorb soluble forms of humus.

In any case, there was more to von Liebig's concept about the cation potassium (or potash) and the anion phosphorous than simplistic followers would have you believe. And there was a great deal the experts didn't know 150 years ago.

Even in von Liebig's day, the source of nitrogen of plants was controversial. von Liebig figured it came from the ammonia of the atmosphere. One of von Liebig's lecture students turned out to be John Lawes of Rothamsted, who established the world's best known experiment stations. At Rothamsted, he caught the rain and determined that the rain wash of air accounted for only 3 to 5 pounds of nitrogen per acre per year. That's not enough, not when plants contain 10 to 20 times these quantities. This led Lawes to use mineral forms of nitrogen on his wheat plots. The results were spectacular. And thus was born the end leg of N, P and K fertilization.

It was known, of course, that nitrogen from organic matter was released in the form of ammonia, and subsequently changed to nitrate. It was not known until 1878 that changes in soil nitrogen were the handiwork of biological agencies. Later, bacteria in the nodules of legumes got full recognition, and finally Russia's Serge N. Winogradsky and Holland's M. W. Beijerinck proved that nitrogen fixation took place in the soils without any legumes whatsoever. The essentials of rapid fixation were found to be absence of readily available nitrogen and the presence of carbohydrates, phosphate and lime, and the bio-starters of molybdate and borate.

Almost from that moment on, scientific agriculture became knowledgeable mining, or killer agriculture—something less than working with the cycles of nature. Business got into the act. As if by magic potash salts were discovered in Germany. Phosphate deposits filled railcars heading out of Florida. And nitrate beds in Chile swamped boats and spawned local wars even before there were bigger wars. Chemistry was perceived to have made a breakthrough, and business said it was much cheaper to supply missing nutrients than to deal with nature on nature's hard terms.

19. nitrates NO3

It didn't take chemists long to make sodium nitrate synthetically from sodium carbonate and nitric acid. Styled NO_3, it contains 16% nitrogen, and is used for top and side dressing small grains and grass, and for side dressing corn. The chemical rationale has it that it is instantly available because it is soluble, and that it does not increase soil acidity as do other nitrogen forms. As a soluble, it leaches.

Another form of nitrogen is ammonium nitrate. This one came to the farm scene when war plants were pressed into civilian use hard on the heels of the wars. Ammonium nitrate is made by passing ammonia gas into nitric acid. The end product contains 33.5% nitrogen. This is a 50-50 fertilizer. Fully 50% of the nitrogen is in the nitrate form, the rest in an ammonia form. It is of course highly soluble and has a reputation for being quick acting. Ammonium nitrate is flammable and it can go up with a bang if stored in a closed warehouse. It is deliquescent. This means it takes on water, which is exactly what happens when it is left in open storage. Most makers of ammonium nitrate prill the material and condition it with a water repellent. It is used as a top dressing and side dressing approximately the same as sodium nitrate.

Also NO_3 forms are potassium nitrate and calcium nitrate, 13 and 15% nitrogen respectively.

20. the NH3 forms

There are five entries in the NH_3 forms of nitrogen—ammonium nitrate, ammonium sulfate, ammonium phosphate, ammonia solutions and anhydrous ammonia (82%). Ammonium nitrate NH_3 is sold at the same 33.5 strength as the NO_3 form.

Ammonium sulfate is also a synthetic nitrogen form, available as a white to grayish green material, usually containing 20.5% nitrogen. In terms of water solubility, ammonium sulfate is a lot like sodium nitrate. In the soil system, it loads over to the clay component of the soil colloid quite readily and is not as subject to leaching as many salt fertilizers. Moreover, microbiologists say that ammonium sulfate gets along well with most microorganisms in the soil. These properties make it one of the better forms of the salt fertilizers. Rice in particular gets along better with the ammonium forms of nitrogen, whereas nitrates under damp and poorly aerated conditions form toxic substances neither a plant nor a soil system can cope with.

21. cal-nitro

Cal-nitro designates ammonium nitrate-lime. This one is no more than a mixture of ammonium nitrate and lime jacked up to the usual 20.5% nitrogen. Basic to this process has been the flammable character of ammonium nitrate, which is thus reduced.

Ammonium phosphate is another nitrogen form with 11 to 16% nitrogen.

21. the nitrogen solutions

In addition to the other NH_3 forms, the market now offers to the farmer nitrogen solutions. These usually contain urea, ammonia and ammonium nitrate, or various combinations, and sometimes they are simply ammonia solutions. The nitrogen strength is variable—sometimes up to 49%. Since no vapor pressure is involved there is no free ammonia. Obviously if there is vapor pressure, nitrogen has to be knifed into the soil and covered before it all escapes into the atmosphere.

The latter refers to anhydrous ammonia, a particularly lethal form of nitrogen available at 82%. As pressure is released, anhydrous changes from a liquid to a gas form. In theory this gas has a quick reaction with soil moisture and loads over to the soil colloid for plant use. When applied to soils low in humus, calcium and those that are poorly drained, over two-thirds of the material can be lost into the atmosphere, sometimes before a farmer can move from one end of the field to the other.

22. the synthetic organic nitrogen forms

In addition to the above, there are the synthetic organic nitrogen forms—Calcium Cyanamide and urea. Calcium Cyanamide has several aliases—aerocyanamide, lime nitrogen, or nitro-lime, all of them containing approximately 21 to 22% nitrogen. Exposed to air it locks up the nitrogen. It is black because of the carbon content, and is often a constituent part of mixed fertilizers. In the soil this material moves pH up, sterilizes the soil to some extent and forms a material that is toxic to plants. A lot of chemical farmers use it as a part of their weed killer program. The theory has it that it can be used well before planting with impunity to deliver both calcium and nitrogen.

Urea, the product of the animal kidney, is now made synthetically by reacting ammonia and carbon dioxide under high pressure and temperature. As prilled, it drills into the soil. It dissolves rapidly and suffers leaching losses. Again the conceptualization is that in reaction with enzymes, urea forms into ammonium carbonate, then breaks down to form the ammonium ion. Then the ion loads over to the soil colloid and better resists leaching. Soil bacteria convert the ammonium ion to nitrate form. Urea has 45% nitrogen, is fairly versatile in terms of this form of farming, can be plowed down, used as top dressing and side dressing and even be laid on via airplane or irrigation pipes.

23. the phosphate materials

There are several phosphate materials available under the NPK system—nitric phosphates, ammonium phosphates, Rhenania-type phosphates, calcium metaphosphate, liquid phosphoric acid, superphosphate, triple superphosphate and diammonium phosphate.

Superphosphate is made by treating phosphate with sulfuric acid. In this way the tricalcium phosphate form of the rock phosphate is converted to the water soluble monocalcium phosphate form. This means that 20% superphosphate contains about 45% monocalcium phosphate and about 55% calcium sulfate, or gypsum. Thus a superphospate will contain 12% sulfur, and 20% P_2O_5—or 8.7% P.

In the case of triple superphosphate, rock is treated with phosphoric acid. This eliminates the calcium sulfate. Such a material hits the market with very little sulfur, possibly only 1.4%. It, however, runs 44 to 52% P_2O_5, or 19 to 22% P. Ammonium phosphate such as 8-32-0 or 11-48-0 also involves phosphoric acid treatment in processing.

In the case of diammonium phosphate, ammonia is reacted with phosphoric acid. Diammonium phosphate is used directly and as a constituent part of the blended NPK fertilizers.

24. the potassium forms

More than 75% of all potassium moving into trade channels is sold as muriate of potash, or potassium chloride. Here one atom of potassium is combined with one atom of chlorine, an acid element. This red or white salt contains from 60 to 63% potash as K_2O with a 50 to 52% K equivalent.

Potassium sulfate is sulfate of potash. It has 50 to 53% potash, or 42 to 44% potassium, only 2.5% chlorine.

Sul-po-mag is low in chlorine content. As the name implies, it contains magnesium, 11% in fact, and 22% potash—approximately 18% K.

25. mixed fertilizers

When we talk of fertilizers, we usually use the nomenclature of the Association of Official Agricultural Chemists. It has been a long standing rule of the trade—and the law—that fertilizers be tagged in terms of nitrogen, phosphorus and potassium—NPK. The mixed fertilizers will be styled 5-10-10, 17-17-17, or whatever. Even when a fertilizer is designated by its call number, it must be remembered that phosphorus is never simply P_2O_5 and potassium is never K₂O. There are carriers. Even when a combination fertilizer has one or more elements missing, the official designation is written in terms of N, P and K. We've mentioned superphosphate 0-20-0. We've mentioned triple superphosphate—0-45-0. In legal terminology, a fertilizer is complete if it has all three elements. It is incomplete if it has one or more elements missing. But these are simply legal definitions and have all too little to do with complete nutrition of plants.

William Albrecht (see *The Albrecht Papers*) gave his critique on all this in a paper entitled *Insoluble Yet Available*. "In our attempts to comprehend how the insoluble rock can make soil which, under proper moisture, will nourish plants, we have too long believed that the nutrient inorganic elements from the soil must be water soluble. That has seemed necessary, at least for the first step of that service, namely the entrance of them as ions into the plant root, or for their movement through the outer wall or membrane of it. The state's inspections, and their license of commercial fertilizers, use water solubility and the solubility in citrate solutions, as the criteria of the fertilizers' services. The inspection assesses fees on the tonnages of fertilizers distributed according to those solubility requirements for the nitrogen and potassium units and for those of phosphorus, respectively."

Albrecht's point on solubility is well taken. Laws and schoolman advice tell farmers that rock phosphate is not effective, and water soluble phosphorus fertilizers is. A point in fact is that the P in NPK comes from rock phosphate in the first place. The formulas on the bags have an esoteric meaning most farmers and most elevator operators do not understand. For instance, 1,400 pounds of a 33% rock phosphate material treated with 1,200 pounds of sulfuric acid will yield a 20% superphosphate. The tricalcium form of phosphorus in the rock is thereby converted to water soluble monocalcium phosphate. The end product in a 20% superphosphate is about 45% phosphate and 55% calcium sulfate, or gypsum. Thus 0-20-0 really means that two separate materials will be delivered in the 100 pound package—in this case, 45 pounds of monocalcium phosphate, which is probably what the soil audit indicates and the farmer wants, and 55 pounds of calcium sulfate. which the farmer may or may not want or need. It might also be said that in 100 pounds of 0-20-0 there are 20 pounds of P₂O₅, but it is in the 80 pounds of filler (not regulated, licensed or measured) where the real goodies are. Most cornbelt soils will respond well to the use of 0-20-0, the real value coming from the calcium and sulfur content of the filler, each meaning more than the P₂O₅.

The ball game changes entirely when a farmer orders triple superphosphate (0-45-0). Here acid treatment is supplied via phosphoric acid. This eliminates the calcium sulfate in the fertilizer material. In 8-23-0, ammonium phosphate, phosphoric acid also replaces sulfuric acid. Triple-super is quite different from super. Yet too many farmers think of NPK as low grade vs. high grade.

Not emphasized in the official manuals is the fact that unstable monocalcium phosphate reverts back to the stable tricalcium phosphate form quite readily. The amount of humus and organic matter in the soil, the free calcium, the pH—all figure in the rate of reversion. Many agronomists estimate that 75% of the monocalcium phosphate reverts to the stable tri-calcium form in no more than 90 days in even the best soil systems. In some of the chemically shocked soils, reversion takes place in a matter of hours.

26. how they compute the formula

All of the confusion expressed above is further complicated by the nature of the weighted NPK formula system. A bag might say 0-0-60. Does this mean that in a 100 pound bag there are 60 pounds of K, the rest being inert filler? Not exactly. The fact is that farmers get a lot of things they may not even want when they buy 0-0-60, 18-46-0, 17-17-17, or whatever.

As an example, take ammonium nitrate—or 33.5-0-0. Here's how they compute the formula. Ammonium nitrate of course is NH_4NO_3. Each of the elements in this affair has a Mendeleyeff atomic weight. N has an atomic weight of 14. O has an atomic weight of 16. There are two N's in the formula, or 2 N. There are four H's, or 4 H. There are three O's, or 3 O. Count them. This means—

$$2 \text{ N} \times 14 = 28$$
$$4 \text{ H} \times 1 = 4$$
$$3 \text{ O} \times 16 = \underline{48}$$
$$\text{Total } 80$$

The entire compound has a weight of 80. This must be divided into 28 (because we're presumably interested in the legislated formula for N), and the figure is 35%. There is a bit of tolerance here because of binding agents used. As a result the bag is 33.5 (the fertilizer grade of the product). By law the H and the O are defined into the designation N.

Potassium can be computed the same way, whatever the designation, but the compound involved must be known. KCl means K x 39 and Cl x 35, total 74, with 39 divided by 74, or 52.7% K in 0-0-60.

The general idea holds for all the NPKs. Take the common mix, 18-46-0—diammonium phosphate, or $(NH_4)_2HPO_4$. Here N and P are combined with no K in tow.

$$2 \text{ N} \times 14 = 28$$
$$9 \text{ H} \times 1 = 9$$
$$1 \text{ P} \times 31 = 31$$
$$4 \text{ O} \times 16 = \underline{64}$$
$$\text{Total } 132$$

The total for N divided by the total for the compound—28 + 132—will equal 21.2%. The P end of the affair is arrived at by dividing 31 by 132 to equal 23.5%. Since the law says P is really P_2O_5, a further conversion must be made to yield the bag designation. In this case the factor is 2.29 x 23.5, and this equals 53.8%. That's close enough to 18-46-0 as far as the law is concerned—thus diammonium phosphate is bagged as 18-46-0 instead of the pure stuff, 21-53-0, and the complicated affair styled NP is really $(NH_4)_2HPO_4$.

A lot of water has gone over the dam since von Liebig and the fossil fuel crowd got into the fertilizer game, but those old laws abide as though farming were an industrial procedure rather than a biological procedure. New knowledge about humates, trace elements, enzymes—all are proscribed here and there because they don't fit the formulas and the laws. The pundits talk about the best science, not remembering that the best science once prescribed holes in the skull to let out the demons, bleeding to dispose of bad humors, laudable pus to exit toxins and the intelligence that the world was flat.

Chemical inspection of factory fertilizers and state regulatory control gained a toehold in agriculture because—in the words of Albrecht—of "the early beliefs that water solubility of them was an index of their absorption by the plant roots."

It was Albrecht's argument that soil behavior in nature did not conform to the laws of the laboratory. Some substances tagged as soluble in the laboratory became insoluble when placed into the soil. He pointed out that plain old land plaster—known as the chemical compound calcium sulfate, or the mineral gypsum—is soluble in the laboratory. Yet it can break down lung tissue by a process suggesting calcification. Some of the water soluble fertilizers are promptly lost because they are water soluble. Some become immediately insoluble and difficult for plants to take in as nutrients.

Management of the cation exchange capacity we visited about in the section on soil audits is not best accomplished with soluble salts. Indeed, calcium, magnesium, nitrogen, phosphorus, potassium and other essential nutrients are not swept into plants because they are flooded into the soil in a water soluble form. Plants force-fed with solubles alone may seem well-fed, but remain completely undernourished. Synthetic and hydroponic systems simply are not basic to life essentials.

27. nature's management

Hopefully we can learn from nature's management. Biotic geography has told us how nature has conserved both inorganic and organic fertility. It warns that there is little of the water soluble in a living soil. It cautions that man's takeover of climax soil hardly permits natural conservation, and it points with great finality to the disparity between man's conceptualizations and nature's revelations.

The soil audit and the college classroom both emphasize nutritive elements as active ions, yet the creative atoms of life are not in an ionic condition either in the rock, the soil microbes, plants, animals or man. Albrecht's lines on the subject are so pregnant they bear italicized repetition. *"Those elements need to be viewed in their compounds, in their larger molecules and as adsorbed ions to give the lowest possible ionic activity when held by colloids requiring energies, or work, to exchange them and make them more ionic, or separately active. Living soils are stocked with organic compounds, which control the chemical activities of the essential inorganic elements emphasized for their water solubilities but are unknown, or forgotten, for their unique combinations into unionized and insoluble parts of organic reactions, and which supply nutrient directly in the form of those inorganic compounds."*

Nature deals in insoluble nutrients, serves them up on signal, and adapts plants in well-suited climatic setting. It is here that the crop grows best, has the most self-protection against insect, bacterial and fungal attack, and best reproduces its own kind.

The mandate for rethinking soil science can best be stated by ten cautions first framed by Albrecht.

1. Insects and disease are the symptoms of a failing crop, not the cause of it.

2. The use of sprays is an act of desperation in a dying agriculture. It is not the overpowering invader we must fear, but the weakened condition of the victim.

3. The excessive use of chemical salts in fertilizers is upsetting plant nutrition.

4. Manure forms an organic shield around the salts. It is a buffer against salt injury. As soils become lower in organic matter we will not be able to use salts so directly.

5. Fertilizer placement is the art of putting the salts in the ground so the plant roots can dodge it.

6. To help them maintain their soils, farmers should be given a depletion allowance on their income tax the same as owners of mines, oil wells and timber tracts.

7. Most of the south will always be a good bull market, just as it was a good mule market. The soils of the south do not have the virgin fertility to produce a good breeding stock.

8. We are exhausting the quality of our soils. As we do so the quality of our plants goes down. And we are accepting this.

9. Don't lime to fight soil acidity. Use lime to feed the plant.

10. Study nature more than you study books.

28. the ancients

The ancients saw the elements as earth, air, fire and water. Since they equated fire with heat, and heat with sunshine, they in fact named the equation for food power from nature—the secret of life, sunshine, air and water (95%) and earth minerals (5%). Even at the primer level we know that all is not answered by N, P and K fertilization. A few points suggest themselves.

1. The conversion and availability of mineral elements are related to and regulated by the system of decay in the soil. The proper decay processes are initiated and determined by usually full levels of calcium and reasonable levels of magnesium.

2. A balanced equilibrium of calcium and magnesium creates a soil environment for bacteria and fungus activity for the proper decay of organic residues into carbon dioxide, carbonic acid and a host of many weak and mild organic acids, all so necessary to convert and release mineral elements in the soil system.

3. An imbalance of calcium and magnesium permits organic residues to decay into alcohol, a sterilant to bacteria, and into formaldehyde, a preservative of cell tissue. The symptoms of this improper decay system can be observed when previous year's stalks are plowed back up just as shiny and fresh as they were when turned down.

4. Soils degenerating as the result of a farmer's blissful ignorance require larger and increasing amounts of calcium and humus as well as proper colloidally processed trace minerals to maintain crop yields. A dying soil is not capable of releasing the minerals or other soil essentials needed for optimum growth and wholesome maturity of plants, grain and food products. A dead plant system cannot mature or ripen its fruit or seed and a dying synthetic soil nutrient system cannot sustain itself.

5. Large applications of nitrogen consume extra amounts of calcium as well as "burn up" crop residues and humus. This synthetic form of agriculture will permit increased yields for a few years—via mining of stored up humus—but eventually diminished returns will set in.

6. Without an active organic matter system in the soil a farmer cannot produce a crop at all, no matter how much N, P and K is used. The soil is a living complex system that not only holds necessary minerals needed for plant life, it is also the factory that produces carbon dioxide, digests lignin into humus, provides nutrition and energy for desirable bacterial and soil animal life, and at the same time contains air and water.

7. In the absence of a system of organic matter management and calcium, synthetic farming will have to account for carbon dioxide (perhaps by inserting dry ice into the soil), and carbon (perhaps by knifing propane gas into the soil). As with improper reliance on N, P and K fertilizers, such synthetic farming stands as a sad excuse for ignorance and mismanage-

ment of calcium and magnesium, and continually extracts a dear price for continued disregard of vital soil equilibrium.

8. A soil program with managed levels of calcium and magnesium will allow the nature of the soil to function complementary to the process of life. It will account for more effective photosynthesis, maximum use of heat-degree days—a natural time clock of the life systems of plants. It will create and maintain root and stem capacities for optimum use of sunlight energy and thermal efficiency by the leaves of the plant. Consequently plants can use water, carbon dioxide, nitrogen and mineral nutrients with greater efficiency.

9. A healthy and normal functioning plant can maintain an adequate hormone and enzyme system so vital to resistance to insect and disease hazards.

10. A balanced soil equilibrium will regulate and manage the quality and availability of all mineral elements needed by growing plants. Excesses of minerals during the early growth stages often plug up the vessels of the stem and are the frequent cause of early death. Indeed, an excess of magnesium as well as nitrogen in the soil initiates these very processes which prevent the crop from growing dry and nutritionally ripe.

29. chemical, biological, physical

One thing ought to be evident at this point. The naturalist John Muir expressed it well when he said everything is connected to everything else. In the lessons to follow, we will make an attempt at dicing off a section of knowledge and discussing it in isolation. It will prove to be helpful to the student, yet inadequate for on-scene decision, and for this reason we will seem unaccountably to be darting back and forth, recalling a bit of information from lesson 1, interlacing it with an item under some other lesson, and relating all to some seemingly unrelated topic. In short, we can't separate chemical, biological and physical, try as we might. Here are a few for instances.

When a farmer plows up corn stover or wheat stubble that has been underground for half a dozen years, it can be assumed he has an improper decay system going. This much should be simple. We're talking about the biological equation, right?

Wrong. Organic matter can decay in two different directions. One system produces alcohol. Alcohol is a sterilant and the alcohol present in the soil from this initial process tends to sterilize most of the desirable forms of bacteria. As the decay process ferments the organic juices into alcohol, it next forms formaldehyde, and formaldehyde is a preservative. And this protects the residue from biological activity in the soil for years. Chemical, right?

Wrong. Farmers who answer this problem with heavy applications of nitrogen set in motion a system that oxidizes or burns organic material and reduces this living complex to an ash heap. Fast acting nitrogen never returns the remains of a decay system into humus. Rather, the extra nitrogen burns even deeper into the soil humus system and continues to deplete the humus buffer potential of the soil.

Bacterial systems in a soil are determined by the right equilibrium of the pH factors of calcium, magnesium, potassium and sodium as explained above. And these bacterial and fungal systems, not deep-till plows and tiled fields, account for tilth. Physical, right?

Wrong. To warm a cubic foot of dark soil one degree with 80 degree ambient temperature from the sun during six hours takes four times as much energy as it does to warm lighter soil that is well drained. Good tilth will do more to drain soil than spillways, tile systems, and chug-chug pumps. The warmer the soil, the faster the bacterial action.

Well, can an eco-farmer use the salt fertilizers with impunity? Rather than answer this question right away, let's pause a moment and listen to what the microbiologists have to say, soil microbes concurring.

They do not seem to be concerned about urea, diammonium phosphate, monoammonium phosphate, or ammonium sulfate. They are more concerned about straight forms of nitrogen—anhydrous ammonia, for instance—and muriate of potash. They're concerned about the survival and activity of little livestock in the soil system. These little workers can live with the forms endorsed above.

If a farmer is not going to pay out the price for maintenance of his organic matter, he will not be able to farm without using some of the synthetic tools mentioned here. If he does not keep his soil's colloidal pH a little on the acid side, he will not be able to account for acidulation in the soil, fetching nutrients from so-called unavailable mineral forms. Yes, there are indications that some of the synthetic tools can be fit into an ecologically sound agriculture, one geared to permanence and living with nature's harsh requirements. But we must walk lightly.

IS NITROGEN NITROGEN?

"So far as the plant is concerned, nitrogen is nitrogen," writes a researcher while seated behind his desk at the Library of Congress. Is this true?

Barry Commoner of Washington University, St. Louis, has shown that nitrogen made synthetically has a different structure from natural nitrogen. This synthetic nitrogen, because it has a different structure, was shown to have polluted the water supply for Decatur, Illinois. The key word here is isotope. Natural nitrogen exists as an isotope quite different from the synthetic product. This difference was flagged in the Commoner studies. Nature also flags that difference, and lets it all hang out when synthetic nitrogen fed crops fail to field ripen, and when immature proteins lure in insects for a feast.

LESSON 12

pH

pH measures soil acidity, and the pH figure merely expresses negative logarithms of active hydrogen concentration. All soil audits hand out a pH figure and designate the soil as acid, neutral or alkaline. Early agronomists discovered that low pH soils produced poorly, and thus was born the idea of liming to the neutral point. Soil acidity was blamed for poor crop performance.

The old triple A program of the New Deal seized upon this half-truth and inaugurated a liming program nationwide. "Lime to the neutral point," ran the advice. "You can't overdo it. And we'll pay you to do it."

This position was unscientific simply because it ignored the services of soil acidity in mobilizing nutrients in the rocks and minerals of the soil.

1. fertility depletion

Excess acidity is nothing more than the reciprocal of fertility depletion. Nature simply puts acidity into the soil from plant roots so that this acid action can break nutrients out of locked-up position and feed the several life forms. In liming an acid soil to correct pH, the farmer is merely providing rock that can react with the acid clay of the soil system. Acid goes from the clay to the lime, and lime being calcium breaks down to give carbonic acid while the calcium is absorbed by the clay. Held on the clay it is available for plant use, and it corrects or adjusts the pH in the colloidal domain where plant roots expect their energy exchange. In the meantime the carbonic acid decomposes into water and carbon dioxide. Carbon dioxide escapes from the soil, taking acidity with it.

There is a benefit here, but it has all too little to do with making the soil neutral. The benefit is derived by trading calcium off for hydrogen, and hydrogen is of no direct nutritional service to plants.

Adjusting pH, then, has to do with loading nutrients back into the soil system, and not removing acidity. In fact it is bad business to remove all the acidity. Since time began, soil acidity has been breaking potassium out of potash feldspars. It has been taking magnesium from dolomitic limestone much the same as it has taken calcium. Acid soil uses the same mechanism to make phosphate rock available for plant uptake as factories do to convert mineral rocks into soluble fertilizer.

In other words, acid soil is beneficial. It is also a free source of virgin plant nutrients. The pH measure is of great value—if we pause to understand what it means. Yet it should stand to reason that adjusting pH with one nutrient alone has got to result in shortages or marked imbalances of some fertility elements. As a matter of fact, it would be well to consider *pCa, pMg, pNa* and *pK* in addition to *pH*.

Lime to the neutral point is a breath-taking error. It may well be a self-serving error proposed and furbished by those interested in selling water soluble salt fertilizers. A farmer with neutral soil cannot possibly grow crops without having a factory acidulate the nutrients indicated for crop production. Without an acid soil, or a good colloidal humus energy exchange system, it is no longer possible to rely on phosphate in the rock, potassium from green sand, magnesium from dolomite, or any of the trace nutrients available in the gravels that nature has provided. Soil without a measure of acidity condemns crops to simplistic fertilization, and when essential nutrients are restricted, plants exhibit their stress in the customary way: lodging, fungal diseases, bacterial debilitation, and insect attack.

2. the demands of most commercial crops

The conventional wisdom has it that pH of 6.2 either in the soil or in its colloidal pantry meets the demands of most commercial crops. This wisdom permits exceptions, the reasons for these exceptions often being subtle. In legend there are the lime-loving legumes. The foundations of thought expressed by the simple statistical array of pH variables shown in the chart on page 157, must now be challenged and corrected. We can no longer permit the usual lip service that chemistry agronomists use to ignore away or skip over the responsibilities that molecular biologists or biological agronomists are tending to. Moreover, continued ignorance by the chemical agronomists is quietly sustained and perpetuated by the chart that expresses the tolerance of our usual farm crops. Inversely it covers up and gives integrity to those who wish to ignore the power of the soil and the factors that permit degeneration of the biological potential of soil nutrition. This chart could just as well have indicated the phylums of life that can grow at the various levels of degenerated soil systems. It could also have associated itself to types of weeds, soil insects, soil diseases and bacterial and fungal habitats and then to plant limitations as well—the type of insight covered in their primer. (See the lessons on weeds and insects.) We may ask this simple question: why do Wisconsin potatoes require an acid soil whereas the famous Idaho potatoes do so well on virgin, low humus, alkaline soil? A pH of 5.8 in Wisconsin surely cannot be the equivalent of pH 8.5 in Idaho. So perhaps we no longer can accept the fact that there are lime sensitive potatoes, melons and berries, nor can we continue to depend on a pH of 6.2 to 6.5 for most crops. It is necessary that we recognize the biological and colloidal influence involved as well.

Legumes take in pure nitrogen from the air and hold it in the soil as their contribution to the natural nitrogen cycle. They like calcium to match this natural intake (pCa). Lettuce, sugar beets, field peas—on the scale of 4 to 7—move off that alkaline position in their likes and dislikes (either pCa or pNa). At the far end of the acidity spectrum is the rhododendron plant, said to be acid loving. Actually it isn't that the rhododendron loves acid as much as it is a case of this flowering plant loving magnesium (pMg). This pattern holds for so-called lime sensitive plants, plants that require the availability of iron, manganese, copper, boron, all elements locked up in high pH (generally overlimed) soils or in soils without a humus buffer system.

3. the pH process

Any pH tolerance chart must be read with these points in mind. In a manner of speaking, soil is pH dependent, but so is everything that lives. Animals are pH dependent. Human beings are pH dependent. Bacteria are pH dependent. Each has a pH range in which it exists the best. Unfortunately the more favorable bacteria we want working for us are the least able to cope with some of the environmental conditions modern farming has manufactured for them. The nitrogen fixing bacteria like the lower pH range levels. Only soils with some acid activity will unlock the trace nutrients required for bacteria life.

Nematodes are regulated by the pH process in the soil. The byproducts of a decay system—and the presence of certain antibiotics of the whole system of nature working in the soil—tends to cleanse. The decay system produces antibiotics. It regulates the hormone availability of each of these life systems. All cells are regulated by hormones, and there is ample evidence that they are affected by the light of both the sun and the moon, by magnetic influence, and by atmospheric pressure, temperature and moisture.

Thus the pH index hints at the final answer to insect and fungal attack. When a plant is grown under ideal conditions—pH adjusted with calcium, magnesium, sodium, and potassium in equilibrium—it will have a specific direct influence on the balance of hormone and enzyme systems, and the root cell membranes will be structured so that insects cannot attack or endure around it. Needless to say, the trace metals also figure here.

This is what the pH figure on a soil audit says, but too few manage to read the ramifications.

Over the years research workers have worked out the most favored pH for certain plants. We have compiled this information and present it here. Certain cautions must be kept in mind. All march hand in hand with earlier observations on desirability of acidity together with a well stocked nutrient catchpen.

The chart (page 157) innocently suggests that at a pH of 4 up to 5.5, phosphates become complexed with iron and aluminum. Manganese, aluminum, iron, copper, zinc, cobalt and boron become increasingly soluble, especially if calcium or humus is in low supply. Acidity acting on soil minerals in the rock may even account for manganese and aluminum toxicity. As a consequence, harmful fungi thrive. Friendly bacteria cease to proliferate. This can mean that nitrogen cannot be fixed as part of the natural nitrogen cycle, now that a reasonable balanced meal of soil nutrients can be supplied for good plant growth.

Between pH 5.5 and 6.5, friendly bacteria and fungi activity takes off. This is the arena in which most plants thrive best because nitrogen is freely fixed, traces of manganese, iron, copper, zinc, boron, cobalt, et al, become available for plant uptake. Phosphates also become available to plants between pH 5.5 and 6.8 (approximately). Between pH 6.8 and a neutral 7 and higher, phosphates become complexed with calcium.

Indeed, manganese, iron, copper, zinc, boron—all become increasingly complexed in the soil system when pH moves above pH 6.5. Friendly fungi languish, although many bacteria continue to thrive.

TOLERANCE CHART FOR USUAL FARM CROPS

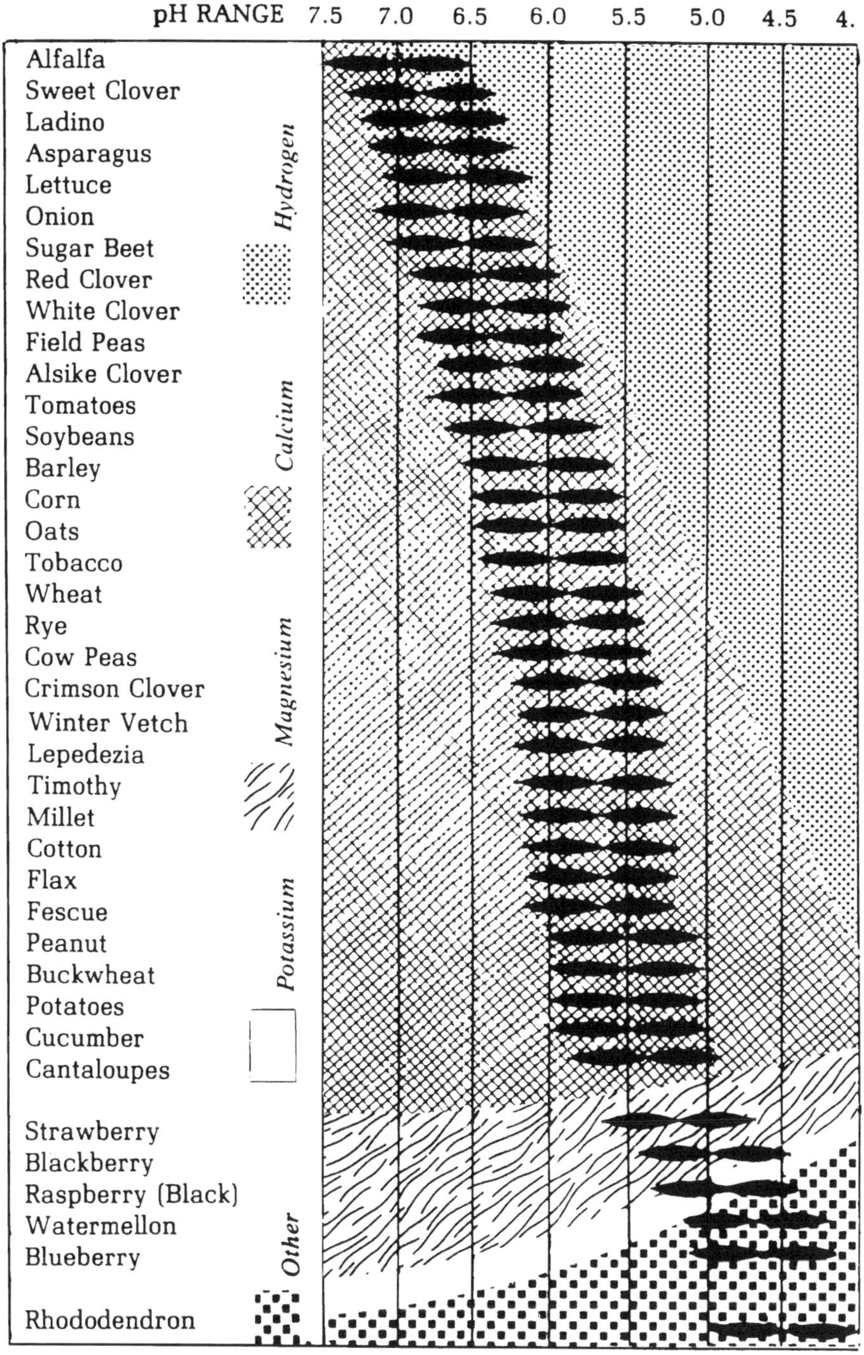

pH RANGE 7.5 7.0 6.5 6.0 5.5 5.0 4.5 4.

Alfalfa
Sweet Clover
Ladino
Asparagus
Lettuce
Onion
Sugar Beet
Red Clover
White Clover
Field Peas
Alsike Clover
Tomatoes
Soybeans
Barley
Corn
Oats
Tobacco
Wheat
Rye
Cow Peas
Crimson Clover
Winter Vetch
Lepedezia
Timothy
Millet
Cotton
Flax
Fescue
Peanut
Buckwheat
Potatoes
Cucumber
Cantaloupes

Strawberry
Blackberry
Raspberry (Black)
Watermellon
Blueberry

Rhododendron

Hydrogen
Calcium
Magnesium
Potassium
Other

	Optimum pH Range
Common Names	
Cereals	
Barley	6.5-8.0
Buckwheat	5.5-7.0
Corn, Indian	5.5-7.0
Oats	5.5-7.0
Rice	5.5-6.5
Rye	5.0-7.0
Wheat	6.0-8.0
Legumes, Field & Garden	
Alfalfa	6.5-8.0
Bean, White Navy	6.0-7.5
Bean, Lima	6.0-7.0
Bean, Snap & Wax	6.0-7.5
Bean, Velvet	6.0-7.0
Clover, Alsike	5.5-7.5
Clover, Bur	5.0-6.5
Clover, Crimson	5.5-7.0
Clover, Japan	5.5-6.5
Clover, Ladino	6.0-7.5
Clover, Mam. Red	6.0-7.5
Clover, Red	6.0-7.5
Clover, Sweet White	6.5-8.0
Clover, Sweet Yellow	6.5-8.0
Clover, White	6.0-7.5
Cowpea	5.5-7.0
Lespedeza, Jap.	5.5-7.0
Lespedeza, Kor.	5.5-7.0
Lupine, Blue	5.0-6.5
Lupine, White	5.5-7.0
Lupine, Yellow	5.0-6.0
Pea, Canning & garden	6.0-7.5
Peanut	6.0-7.0
Serradella	6.0-7.0
Soybean	6.0-7.0
Vetch, Hairy	5.5-7.0
Hay Forage & Pasture Grasses	
Bermuda Grass	6.0-7.0
Bluegrass, Canada	6.0-7.5

Common Names	Optimum pH Range
Bluegrass, Kentucky	6.0-7.5
Brome Grass, Smooth	6.0-8.0
Fescue, Chewing's	5.5-6.5
Fescue Fine leaves	6.5-7.5
Fescue Meadow	5.0-7.0
Fescue, Sheep's	5.0-6.5
Italian Rye Grass	6.0-7.0
Johnson Grass	5.0-6.0
Meadow Foxtail	6.0-7.5
Orchard Grass	6.0-7.0
Kafir Corn	6.0-7.5
Millet	5.5-7.0
Milo, Dwarf Yellow	5.5-7.5
Redtop	6.0-7.0
Reed Canarygrass	6.0-7.0
Sudan Grass	5.5-7.0
Timothy	5-5-7-5
Miscellaneous Field Crops	
Beet Mangel Wursel	6.0-7.5
Beet Sugar	6.5-8.0
Broom Corn	5.0-6.5
Cotton, Upland	5.5-7.0
Flax	6.0-7.5
Hemp	6.0-7.0
Potato	5.0-6.5
Potato, Sweet	5.5-6.5
Sugar Cane	6.0-8.0
Sunflower	6.0-7.5
Tobacco	5.5-7.5
Turnip	6.0-7.0
Rape	6.0-7.5
Garden Crops	
Asparagus	6.0-8.0
Beet (Table)	6.0-7.5
Broccoli	6.0-7.0
Brussels Sprouts	6.0-7.5
Cabbage	6.0-7.5
Cantaloupe	6.0-7.5

Common Names	Optimum pH Range
Carrot, Garden	5.5-7.0
Cauliflower	6.0-7.5
Celery	6.0-7.5
Cucumber	5.5-7.0
Eggplant	5 5-6 0
Horseradish	5.5-7.0
Kale	6.0-7.5
Kohl-Rabi	6.0-7.5
Lettuce	6.0-7.5
Mushroom	6.5-7.5
Onion	6.0-7.5
Paprika, Red Pepper	6.5-8.0
Parsley	5-5-7-5
Parsnip	5.5-7.5
Pepper, Garden	5.5-7.0
Pumpkin	5.5-7.5
Radish	6.0-7.5
Rhubarb	6.0-7.0
Spinach	6.0-7.5
Squash, Hubbard	5.5-7.5
Tomato	5.5-7.5
Watermelon	5.5-6.5
Small Fruits	
Blackberry	5.5-7.0
Blueberry	4.5-5.8
Cranberry, Large	4.3-5.3
Currant, Black & Red	5.5-7.5
Dewberry	4.5-6.0
Gooseberry, American	5.5-7.0
Grapes	6.0-7.5
Raspberry, Black & Red	5.5-7.0
Strawberry	5.0-6.5
Tree & Tropical Fruits, Nuts & Grapes	
Apple	5.5-7.0
Apricot	6.0-7.0
Cherry, Sour—Sweet	6.0-7.5
Coffee	4.0-5.0
Lemon	6.0-7.5

Common Names	Optimum pH Range
Grapefruit	6.0-7.5
Orange, Sweet	6.0-7.5
Peach	6.0-7.5
Pear	6.0-7.5
Pecan	6.4-8.0
Pineapple	5.0-6.0
Plum, American	6.0-8.0
Trees in General	
Ash, White	6.0-7.5
Aspen, American	5.0-6.5
Beech	5.0-6.7
Birch, Paper	5.0-6.5
Birch, Yellow	5.0-5.5
Catalpa, Western	6.0-8.0
Cedar, Red	6.0-8.0
Cedar, White	5.0-8.0
Cottonwood	6.0-7.5
Chestnut, American	5.0-6.5
Elm, American White	5.5-8.0
Fir, Balsam	5.0-6.0
Fir, Douglas	6.0-7.0
Hemlock	5.0-5.5
Hickory, Shagbark	6.0-7.5
Larch, European	6.0-8.0
Locust, Black	6.0-7.5
Maple, Sugar	5.5-7.0
Mulberry	6.5-7.5
Oak, Black	5.5-6.5
Oak, Bur	5.5-7.5
Oak, Pin	5.0-6.5
Oak, Red	5.5-6.5
Oak, White	6.0-7.5
Pine, Jack	5.0-6.5
Pine, Long leaf	4.0-6.0
Pine, Red	5.0-6.0
Pine, Western yellow	5.5-7.5
Pine, White	5.0-6.5
Pine, Short Leaf	5.0-6.0

Common Names	Optimum pH Range
Poplar, Silver	6.0-8.0
Spruce, Black	4.5-5.0
Spruce, White	5.0-6.0
Sycamore	6.0-7.5
Tamarack	4.5-5.5
Tung-oil Tree	5.0-6.0
Walnut, Black	6.0-7.5
Willow, Weeping	5.5-7.0

Ornamental Plants

Flowers, Shrubs & Vines

Arborvitae, American	6.5-8.0
Chrysanthemum	5.5-7.0
Cinerarias	5.0-6.5
Clematis, Curly	5.5 7.0
Columbines	6.0-7.0
Cyclamens	5.0-6.5
Dahlias	6.0-7.5
Delphinium	6.0-7.5
Eucalyptus	6.0-8.0
Fern, Boston	5.0-6.5
Fern, Xmas	6.5-8.0
Fern, Cliff	6.5-8.0
Fuchsias	5.5-6.5
Gardenia	5.0-6.0
Geranium	6.5-8.0
Gladiolus	6.0-7.5
Holly, American	5.0-6.0
Hydrangea	4.5-6.5
Iris	5.0-6.5
Ivy, English	6.0-8.0
Iris, Japanese	5.0-6.5
Iris, Vernal	4.0-5.0
Laurel, Great	4.5-6.0
Lilac, common	6.0-7.5
Lily, Easter	6.0-7.0
Lily, Tiger	6.0-7.0
Magnolia, Umbrella	5.0-6.0
Mayflower	6.0-7.0

Common Names	Optimum pH Range
Morning Glory	6.0-7.5
Narcissus, Polyanthus	6.0-7.5
Orchids	4.5-5.5
Pansy, Heart's Ease	5.5-6.5
Peony	6.0-7.5
Phlox	5.0-6.0
Pitcher plants	4.5-5.5
Poinsettia	6.0-7.5
Primrose	6.0-8.0
Rhododendron	4.5-6.0
Roses	6.0-7.5
Spirca	6.0-7.5
Trailing Arbutus	4.5-6.0
Tulip, common	6.0-7.0
Weeds	
Bindweed, Field	6.0-7.5
Dandelion	5.5-7.0
Dodder, Field	5.5-7.0
Ivy, Poison	5.0-6.5
Lamb's Quarters	5.0-7.5
Mustard, Wild	6.0-8.0
Plantain, Common	6.0-7.5
Quack Grass	5.5-6.5
Sorrel, Sheep	5.5-7.0
Thistle, Canada	5.0-7.5
Thistle, Russian	6.5-8.0

Having provided tables and charts on pH, it is only fair to warn the reader that these data are conventional wisdom speaking. When pH is properly constructed—with calcium, magnesium, sodium and potassium in appropriate ratio on the soil colloid—then the ranges expressed by this conventional wisdom fade away and become more or less obsolete, like the battle ax and the blunderbuss. A properly constructed pH governs crop production because it governs the welfare of the soil's microbial workers. Indeed—as the rest of this primer will illustrate, the character of the pH as well as its level is judge and jury in giving weeds and insects permission for life.

CALCIUM, MAGNESIUM & POTASSIUM REQUIRED ACCORDING TO CEC

If your CEC is	Total Saturation by calcium would be (x 400)	70% of total saturation by calcium would be	Total saturation by magnesium would be (x 240)	12% of total saturation by magnesium would be	Total saturation by potassium would be (x 780)	3% of total saturation by potassium would be
1	400 lb.	280 lb.	240 lb.	28.8 lb.	780 lb.	23.4 lb.
2	800 lb.	560 lb.	480 lb.	57.6 lb.	1,560 lb.	46.8 lb.
3	1,200 lb.	840 lb.	720 lb.	86.4 lb.	2,340 lb.	70.2 lb.
4	1,600 lb.	1,120 lb.	960 lb.	115.2 lb.	3,120 lb.	93.6 lb.
5	2,000 lb.	1,400 lb.	1,200 lb.	144. lb.	3,900 lb.	117. lb.
6	2,400 lb.	1,680 lb.	1,440 lb.	172.8 lb.	4,680 lb.	140.4 lb.
7	2,800 lb.	1,960 lb.	1,680 lb.	201.6 lb.	5,460 lb.	163.8 lb.
8	3,200 lb.	2,240 lb.	1,920 lb.	230.4 lb.	6,240 lb.	187.2 lb.
9	3,600 lb.	2,520 lb.	2,160 lb.	259.2 lb.	7,020 lb.	210.6 lb.
10	4,000 lb.	2,800 lb.	2,400 lb.	288. lb.	7,800 lb.	234. lb.
15	6,000 lb.	4,200 lb.	3,600 lb.	432. lb.	11,700 lb.	351. lb.
20	8,000 lb.	5,600 lb.	4,800 lb.	576. lb.	15,600 lb.	468. lb.
25	10,000 lb.	7,000 lb.	6,000 lb.	720. lb.	19,500 lb.	585. lb.
30	12,000 lb.	8,400 lb.	7,200 lb.	864. lb.	23,400 lb.	702. lb.

4. character of the lime

Lime is not simply lime. The character of the lime must be known to adjust pH in tune with the cation exchange capacity of the soil according to the proper ratio. As pointed out in the lesson dealing with soil audits, lime in excess of a soil's capacity to fix calcium is expensively foolish, and also a one-way ticket to upsetting the physiological balance of the soil with reference to other plant nutrients.

A pH of 6 to 6.2 will require some acidity, or hydrogen between 10 and 20%—average 12%. Calcium should range between 65 and 75%—a halfway house average being 70%. Magnesium: 10 to 15%—12% being a good average. Potassium: 2 to 5%, average 3%. Other cation elements, sodium, manganese, iron, aluminum and some two dozen others are never computed by soil audit analysis, but these should account for from 1 to 10% of the cation sites on the soil colloid. An average here would be 3 to 5%.

One can easily compute the calcium, magnesium, and potassium required according to CEC.

Limestone materials can range from 98% calcium carbonate stone to 54% calcium carbonate associated with 45% magnesium carbonate. Obviously the right material must be obtained, one with a calcium/magnesium very close to the deficit revealed by soil audit. It isn't always possible to find such limestone, and it isn't always possible to fine-tune everything the way the laboratory suggests it ought to be done. The objective here is simply to put the soil system into the ball park.

Even if the right limestone is obtained, it has to be screened properly. Limestone particles account for nutrient activity hardly 1/16th of an inch from the particle's outside perimeter. Limestone the size of a garden pea will do little to adjust pH or nourish a plant. Moreover, limestone the size of a garden pea would last for years and have little effect in a soil system. It takes finely screened material to do the job, material that can pass through a 100 mesh screen. Whenever the mesh is reduced, payload must be increased accordingly. A 20 mesh screen, for instance, would require a farmer to use another 50% as much of the right limestone to get the same effect. Some eco-farmers make at least one more adjustment in their liming practices when alfalfa and sweet clover are to be grown. Accordingly, they use 1,000 pounds per acre just prior to seeding when pH is 6.0; 900 at pH 6.1; 800 pounds at pH 6.2; 500 pounds at pH 6.5, and so on.

A rule of thumb has it that 1,000 pounds of pure calcium carbonate will provide 400 pounds calcium in the top 7 inches of soil (replacing 1 ME hydrogen). Going the other way, 335 pounds of flowers of sulfur will create 1 ME of hydrogen, thereby offsetting 1 ME of base materials, magnesium, potassium, or calcium.

The central lesson in the pH story ought by now to be at once apparent.

5. the task of acidulation

Soils held at pH 7 cannot handle the task of acidulation in the ground. This is why so many organic growers fail. They use rock phosphate and granite dust and other materials and go broke. Their crops look awful and production of bins and bushels is well under par. Having allowed organic matter and proper pH management to go out of whack, and having obeyed the schoolman's injunctions to lime to the neutral point, they must use water soluble salts or deprive plants of certain nutrients. Having made the mistake of touting neutral soil and wasting organic matter, it became much simpler for the colleges and the Extension people to counsel fossil fuel fertilization for low fertility than to handle soils out of balance because of previous over-applications of certain types of materials.

GEOLOGICAL SURVEY OFFICES

All states have geological survey offices. Often these offices produce bulletins, booklets and reports that set out an inventory of mineral deposits in the state. For instance, Bulletin 199, Part 5, is a State Geological Survey of Kansas publication styled Inventory of Industrial, Metallic and Solid-Fuel Minerals in Kansas. It shows where the bentonite, limestone, gypsum, diatomaceous marl, magnesium, and phosphate are located. It tells something about their quality, designates "proven," "probable," "possible," etc. When a soil system calls for gypsum or high calcium lime, a farmer and his friends might do well to look in their own back yard. Most major library reference departments can provide addresses for geological survey offices.

Carbon/Nitrogen Ratio

Carbon dioxide availability is more important to high yield potential than nitrogen or any mineral element. The supply in the atmosphere could sustain life for not much more than 30 days. Plants depend on the soil's decay system to replenish the supply. Without the beneficial effects of calcium in the soil system and its effect on soil structure and tilth, the decay processes are retarded and inefficient.

As organic matter, the wealth of the nation is wasted away, the carbon-nitrogen ratio finds it difficult to function. Organic matter is the source of additional carbon dioxide, and farmers who fail to manage their soil systems properly suddenly discover a deficit that isn't even in the NPK lexicon.

1. the carbon problem

Nevertheless, the carbon problem rates attention. A few years ago there was research afloat involving the use of dry ice in the cornfield. It was calculated that dry ice could account for an extra 15 to 20 bushels an acre. Sure enough, some business firms were rolling their eyes like Harpo Marx chasing blondes over the prospect of selling carbon to the farmer.

A carbon problem often has something to do with the form of nitrogen being used. Excess amounts of nitrogen reduce the humus content of the soil, and with that reduction develops a sudden need for carbon. Propane gas has been suggested to make up this shortfall. And there have been tests using propane gas. Research people made neat tabulations, replicated their findings in worn out plots of ground, and announced a new scientific finding to the world. Needless to say, if such tests had been conducted on soils managed by organic gardeners or good eco-farmers, they'd have gotten no response at all. In this regard, Rodale and the organic gardening people have been right all along. Presence or absence of organic matter in the soil determines how these college experiments come out. The accident of organic matter in research plots enables one university to show benefits, whereas another shows deficits. Those who do not deal with field conditions frequently resort to single factor analysis. As they used to say at Iowa State, "For $100,000 we can prove anything."

We're not interested in scientific gangsterism or in sophistry, thus it is necessary to forge each of these sentences in the teeth of irreducible fact. Approximately 58% of soil organic matter exists as organic carbon, the universal element that belongs almost everywhere and figures in almost everything from the fragrance molecule of a blossom to being the brick in cell wall tissue. For many decades running it was believed that plants simply took their carbon from carbon dioxide in the air, and that soil substances, ergo decaying organic matter, were unimportant in this realm. Some of the behind-the-times texts on the subject still echo that claim. In the 1920s, a Swede named Lundegardh discovered that carbon dioxide released from decaying organic matter in aerated soil was more vital than CO_2 in the atmosphere.

2. four tons of carbon dioxide

An acre of 100 bushel corn needs something like four tons of carbon dioxide. Yet on a quiet summer day, the carbon dioxide level over a cornfield reaches the near vanishing point. Soundly fertilized corn operating under bright sunlight could use several times the normal available level. It is true that some 20 tons of carbon dioxide tower to the sky above each acre of farm ground. As with any essential, it has to be immediately available, and the corn plant can't suck it down that easily.

When carbon dioxide is present in soil solutions as bicarbonates, carbon is absorbed by plant roots as CO_3 along with other essential elements, according to research completed as early as 1927. But the chief source of carbon remains the soil itself. When organic matter is digested by soil organisms, it is the carbon cycle speaking. Energy from the sun is used in plant transpiration. It acts on the green sacs of chlorophyll in the cells of leaves, and pulls carbon dioxide out of the air and builds them into the many structures in plants. Should a food shortage appear, or should imbalance occur, the construction site closes down, and farmers stand off and observe symptoms.

The symptom least likely to be observed is the decay system in the soil. And the decay system is dependent on there being something to decay. As soils become infertile, the CO_2 level over them diminishes accordingly simply because the microbes of decay cannot function when the cycle is broken.

Here is an artist's conception of the rotation of carbon in nature.

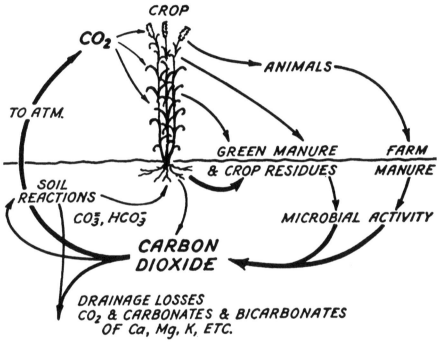

Soil audits do not deal in carbon or carbon-nitrogen ratios, and yet the mandate for understanding this key to crop production is clear. A field crop such as corn will require perhaps 160 pounds of nitrogen per acre, and the reasoning has it that this can be knifed in as gas, applied as a salt, or hammered into the ground as a combination of some NPK formula. And it is the thinking of the fossil fuel salesman that this procedure sidesteps the organic matter requirement. Forgotten or at least mislaid in the synthetic nitrogen scheme is the requirement of carbon and the carbon-nitrogen ratio.

It takes an organic system to supply the carbon, and when heavy doses of salts and anhydrous gas have taken out the organic matter and the carbon, the fault of this technology at last becomes apparent.

Greenhouses often supply carbon to their plants, drawing on carbon dioxide obtained from coal, coke and charcoal. Indeed, it was an attempt to buy a little more time for a faltering technology that caused investigators to follow the greenhouse lead and test dry ice for carbon and to use propane in the field the way some chemical amateurs use anhydrous ammonia.

3. limiting factors

Dead soil accounts for many limiting factors, the most overlooked being carbon. Carbon fixed by plants is an energy source. In a living organism, as in ordinary combustion, it is the union of oxygen and carbon that generates energy. The cycle thus completes itself. Liberated carbon arrives at the leaf surface, is taken in, given room and board, fixed or converted into food—perpetual motion, if ineptness does not intervene.

As usual, the most dramatic cases illustrate the point. In black alkali soils, there is an inability to absorb phosphorus and nitrate nitrogen due to the absence of carbon dioxide. At pH 8 or 9, wheat seedlings fail to absorb and assimilate phosphorus very well because carbon dioxide exudation is faltering, and this does not change until the pH is lowered to at least a little above neutral, possibly pH 7.6. Alkaline soils in general benefit with the addition of organic matter, compost, green manure (because green manures help the carbon dioxide load).

Even before the 1930s it was known that how CO_2 evolved varied a great deal, depending on the temperature and the absorption of oxygen in the living soil. As carbonic acid, it helps dissolve soil minerals. Soil respiration put the carbon dioxide at ground level, just where the plant wants it. How much is entirely dependent on the quality of life it manages to deliver to the anaerobes, not only by way of food, but in terms of a toxic free environment.

When on the job, soil organisms go about their work with great alacrity.

Clay soils high in magnesium and low in calcium content cement together tightly, and are subject to compaction and clodding. They crust over easily and this prevents insoak of rainwater and recovery of capillary water during dry periods of the season. Even more important, pancaked soils prevent a continual interchange of oxygen with liberated carbon dioxide.

Some few other factors figure—temperature, moisture, barometric pressure, even wind sweeping the field. But in the main interdiffusion handles the lion's share of the chore. In the top six or seven inches of a soil system with good tilth, the exchange of carbon dioxide into the atmosphere and the entry of oxygen into the soil is rapid.

It has to be rapid, or toxicity to roots could become an unwanted accomplishment. Important as it is, carbon dioxide can't be allowed to linger.

It has to move out, reach for that cycle of life. With other elements it can be used to build thousands of molecules. Without it, life for plants, animals and man comes to a screeching halt. Its link with nitrogen is so profound that it becomes impossible to discuss one without leaning on the other, and it is impossible to deal with one without dealing with the other. This is what we were trying to say when organic matter as such was under discussion on pages 89 to 92.

4. "go" foods and "grow" foods

Microbes do need both "go" foods and "grow" foods. They need carbohydrates just as they need nitrogen. Without nitrogen, decay of crop trash heavy in wood stalk residue is slow and incomplete. Sawdust. for instance, can be added to the soil to fluff it up, but this is a high carbon, or energy, food and low body-building food. Since microbes eat first, they readily take ammonia, nitrogen, potassium, phosphorus, calcium and other growth nutrients from the clay to balance the sawdust. While the microorganisms are taking these nutrients, they are denying them to the plants. When the soil is too low in nutrients to feed both microbes and crops, the crops will suffer.

This continues until excess carbon dioxide escapes to the atmosphere. Nitrogen and inorganic nutrient elements are kept in the soil. As the process continues, the carbon-nitrogen ratio narrows. Indeed, ratios come very near to that of microbial body composition, or that of protein.

The process can start with a carbon-nitrogen ratio of 80 or 100 to 1 and end up as humus, or very near a ratio of 12 to 1. As carbonaceous organic matter is consumed, microorganism continue to grow by consuming their predecessors.

Thus the equation changes in the opposite direction, and this is of benefit to crops above the soil. This is what happens when green manures are plowed under. There will be a high content of nitrogen together with a high content of calcium, phosphorus, magnesium, potassium and the several dozen other inorganic nutrients, and this means the carbon-nitrogen ratio is narrow. Energy foods will be low in supply, but this isn't bad because nitrogen and minerals will be in surplus. Since this surplus is not built into microbial bodies, it is available in simpler forms—in a word, it will be there for the use by production crops.

It is the fact that microbes eat first that makes nitrogen so important when stover is turned under. Soils that have been pushed to the brink come to house microbes that work against crop production, not for it.

5. the procedure has to be fine-tuned

Does this not add up the proposition that when you put nitrogen in the soil you burn the carbon out? Does this not mean that the name of the game is to use nitrogen without imposing a limiting dimension on the carbon load, and doesn't it add up to the proposition that nitrogen has to be used sparingly if at all?

Obviously a farmer still has to take care of the phosphorus requirement, the cation balances of calcium, magnesium, potassium, et al. But the procedure has to be fine-tuned so that there will be no shortage of carbon with enough nitrogen being used this way.

Microbes are the agency that prevent nitrogen from escaping. And microbes have their own requirements. Azotobacter, a bacteria that fixes nitrogen from the air without living in symbiosis with plants, has to have its zinc and molybdenum. If there is no zinc and no molybdenum in the soil, there simply won't be an azotobacter population.

Peas, beans, lentils and other legumes have achieved something of a reputation as nitrogen fixers. They are higher in proteins than other plants, meaning they take from the soil elements essential for amino protein construction. Clovers, alfalfa, soybeans and some few others are also high protein field crops. These plants can take gaseous nitrogen from the air via the agency of symbiotic bacteria and combine with other nutrients to form a valued food constituent, the proteins. As Albrecht used to say, legumes are feeds for growing animals. Corn is feed for fattening animals—usually castrated males that are not burdened with the load of reproduction or growing fetal young.

6. ashing plants

Scientists tried to put a handle on the nitrogen requirements of plants in the days when ashing plants represented the final solution to almost everything. It was determined that a chemical structure called amino nitrogen took up about 16% of compounds. By measuring the nitrogen by ashing foods and feeds in sulfuric acid it was determined that 16% of the compound was amino nitrogen. This yielded a factor of 6.25. Thus the measured value nowadays is multiplied by 6.25, and the results are called protein. Cereal chemists used a factor of 5.73 to measure protein in grain. That's how nitrogen became synonymous with proteins, and that's how many dangerous confusions became even more dangerous with the touted application of nitrogen to soil systems in highly soluble and active forms. Plants often fail to take up the nitrogen, or they may take it up and fail to convert it into protein, and yet the simplistic chemical analysis will spit out a figure that is more fiction than fact. How well plants convert nitrogen into proteins is governed by calcium, magnesium, potassium, boron, manganese, copper, zinc, and trace mineral levels in a manner not completely understood even now. Albrecht handled the general equation with these trenchant lines.

"Only after legumes have a balanced soil fertility in terms of all the essential elements except nitrogen, and those in good supply first, will those plants add the nitrogen of the atmosphere to that stock and carry out this process of nitrogen fixation by which they grow, protect themselves and reproduce. The nitrogen is not taken up in that natural process by legumes except as it moves via the protein forms of it."

Deception is not uncommon. The corn plant is a monocot, a shirttail relative of the grass family. During dry seasons it is notorious in failing to make the nitrogen-to-protein conversion, as victims of silo fillers disease readily illustrate. Poor soil tilth, nutrient shortage or imbalance, dry weather, all conspire to destroy enzyme systems necessary to convert nitrates coming into the plant from the soil into other plant compounds. As a consequence, nitrate nitrogen accumulates excessively in the corn stalks. Since nitrates and nitrites are deadly compounds for both animal and man, death in the field and death in the silo have resulted. And yet chemical analysis would call this deadly toxin protein.

Even a cow knows better. The cow, after all, has been distributing nitrogen since day number one. Some of this nitrogen has been deposited on fields in organic form, some has been delivered in soluble salt form, namely urea. Each time droppings and urine hit the turf, that's nitrogen fertilizer! Whenever nitrogen hits, grass jumps skyward, and it is painted very green. These extra lush growths of grass, bitter and overly green, are the reflection of poor or incomplete digestion of nitrogenous feedstuffs. Undigested portions of the feed are usually excreted by the animal while still containing the indoles and skatoles—nature's reminder again, of things in dis-equilibrium.

Farmers growing that kind of a crop uniformly over a field would be proud indeed, but the cow knows better. She avoids and refuses to eat the tall, green grass. She's after complete protein, and her assay is more accurate than anything that can come out of a laboratory on the basis of nitrogen times 6.25 equals crude protein.

There is a natural nitrogen, and there is a natural nitrogen cycle.

On pages 174 and 175, facing diagrams illustrate the difference between natural nitrogen and nitrogen in salt form. This conceptualization was first drawn up as a pen and ink sketch by Ehrenfried Pfeiffer and styled *The Bio-Dynamic Versus the Mineral Concept of Availability*. The basic idea in any natural system is to feed the soil life. Soil life in turn will feed the plant.

The basic idea behind water soluble salt forms of fertilizer is to feed the plant directly and use the soil as a vehicle to carry the nutrient solution and to anchor the plant.

NITROGEN
THE BIO-DYNAMIC CONCEPT

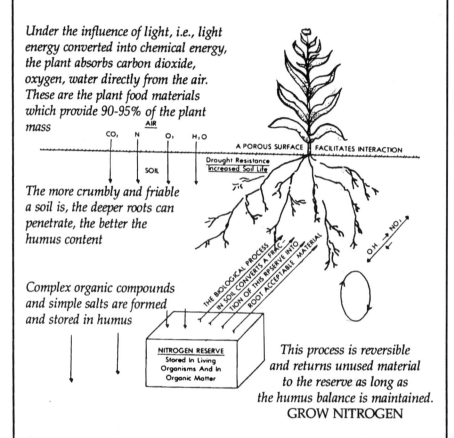

Under the influence of light, i.e., light energy converted into chemical energy, the plant absorbs carbon dioxide, oxygen, water directly from the air. These are the plant food materials which provide 90-95% of the plant mass

AIR

CO_2 N O_2 H_2O

A POROUS SURFACE | FACILITATES INTERACTION

SOIL

Drought Resistance
Increased Soil Life

The more crumbly and friable a soil is, the deeper roots can penetrate, the better the humus content

Complex organic compounds and simple salts are formed and stored in humus

THE BIOLOGICAL PROCESS
IN SOIL CONVERTS A FRAC-
TION OF THIS RPSERVE INTO
ROOT ACCEPTABLE MATERIAL

O_4H NO_3

NITROGEN RESERVE
Stored In Living
Organisms And In
Organic Matter

This process is reversible and returns unused material to the reserve as long as the humus balance is maintained.
GROW NITROGEN

2% organic matter—the organic nitrogen reserve can be 2,000 lbs. / acre
3% organic matter—the organic nitrogen reserve can be 3,000 lbs. / acre
4% organic matter—the organic nitrogen reserve can be 4,000 lbs. / acre

In biodynamic soils there are often two and three times as much. This is possible because of the activation of the living process

Basic idea: Feed the soil life
and soil will feed the plant

E.E. PFEIFFER

NITROGEN
THE MINERAL CONCEPT

Excess NO3 in plant tissue toxic to animals. Sometimes more protein (%), but less nutritious value because of unfavorable amino acid balance

Nitrogen in the air is inert and has to pass through the soil and nitrogen-fixing organisms in the soil to be activated and converted into acceptable material, or tremendous electrical energies are employed to convert nitrogen into useful forms. These energies exist in lightning or artificially in factories

EXCESS FORCED ABSORPTION CAUSES LUSH GREEN

LOSSES INTO AIR

SCATTERED AND/OR UNTREATED MANURE

NO₃

NO₃   NH₃

CAKED CRUSTED   MINERAL SOILS:

BUT MORE WATER IN TISSUE

DECREASED SOIL LIFE

Drought Damage   NO₃ & NO₃

Prevent Absorption and Aeration

N

Ammonia
Ammonium Sulfate
Ammonium Nitrates
INCREASES ACID
REACTION OF SOIL

WASHED OUT. PARTIAL LOSS OF AMMONIA AND AMMONIUM SALTS

quick release no lasting effect each year you need more

roots are flooded with easily available soluble ammonia and nitrates

Compaction and unfavorable reactions in mineralized soil cause horizons and hardpan. Much of the excess fertilizer can be found here—inaccessible to roots

excess not utilized and lost
BUY NITROGEN

Contamination of groundwater with NH3 from manure sludge or NO3 from excess in deep well water

These unfavorable conditions prevail below
1. 5% organic matter in mineralized soils

Virgin soil: 4 to 6% organic matter
Today's soil: U.S. average 1.5%

Basic idea: Feed the plant directly and use the soil as a vehicle only to carry nutrient solution

E.E. PFEIFFER

UREA

Urea was discovered as a white crystalline compound in urine as early as 1773. It represented a waste product of the body's breakdown of proteins, and was therefore considered organic for half a century or better. Early in the 19th century Frederick Wohler succeeded in making urea from potassium cyanate and ammonium sulfate. As of that date it has been possible to call urea organic or inorganic according to one's wish.

In terms of chemistry, urea is organic. It is composed of carbon, hydrogen, oxygen and nitrogen, and it leaves no ash on burning. In this case nitrogen has two hydrogens connected with it. This is attached to a carbon.

This is the chemical structure of nitrogen in proteins, the living compounds of carbon, hydrogen, and nitrogen (sometimes sulfur and phosphorus). These proteins can grow, protect themselves and reproduce. In urea this nitrogen arrangement is called the "amide" of carbonic acid.

When urea is used in the compost pile or the soil system, it behaves the same whether it originates in the kidney of an animal or in the manufacturing plant.

There is a difference, however. Factory urea is chemically pure. Urea via urine carries organic and inorganic compounds not present in the factory product.

Otherwise, urea changes into ammonia, water and carbon dioxide under the influence of microbes or enzymes from plant products. This ammonia can be absorbed on the soil colloid, be taken as nutrition by microbes, or converted into nitrites and then nitrates for nutritional service.

7. synthetic nitrogen worsens the problem

Every cell in a plant needs nitrogen in the right form. It is probably correct to note that it governs the use made of potassium, phosphorus—on and on—but it would be incorrect to suggest that nitrogen can govern without subjects. Like a king on the mountain, it can operate only as long as the peasants do their part. Cell division and cell enlargement account for plant growth. Growth as a process is dependent on new protoplasm being formed, protoplasm being what bolts cells together. Proteinaceous material is made up of carbon, hydrogen, oxygen, nitrogen, sulfur and water. Sulfur and nitrogen arrive from the soil. The other elements are contained in the sugars of photosynthesis. Soils in which organic matter is not maintained soon falter in their carbon-nitrogen relationship, and attempts to add back synthetic nitrogen further worsens the problem.

Nearly 4,000 thunderstorms have been lashing the earth 24 hours a day, year in and year out, through at least the most recent eons of time. These storms are necessary. They not only settle pollution, smoke and dust, they set up the electrical exchange known as lightning. Lightning is the world's greatest producer of nitrogen compounds. Scientists have concluded that perhaps 5% of the nitrogen supply is carried right into the soil with the showers themselves.

Some 75 million pounds of nitrogen exist over every acre of soil, gaseous and useless, as is seawater to a thirsting sailor. That's the reason for being of organic matter, soil microorganisms, and the soil system itself—to store and hold and to deliver to plants the needed nitrogen on a free choice basis. Nitrogen is always acted upon. Microorganisms chew up and spit out ammonium and leave a residue in a nitrate form. Each step along the way finds nitrogen being acted upon, used up and recycled—vaporized, in short, pushed back into the towering column from whence it came.

Nature likes to keep her cafeteria in order. No plant is expected to eat for an entire season, then live off body stores. That's why organic matter stores nitrogen on a slowly available basis, giving up what the plant needs as it needs it.

8. the primacy of organic matter

That's what the primacy of organic matter is all about. In an earlier lesson it was noted that *about 5% of the total organic matter in a soil is present as nitrogen in various compounds. In terms of a 2% organic matter soil, this means about 40,000 pounds of organic matter to the acre. Approximately 5% of this is nitrogen held fast in the bodies of microorganisms themselves. A soil with 2% organic matter therefore has a reserve of 2,000 pounds of nitrogen. A soil with 3% organic matter has a reserve of 3,000 pounds.* We have tried not to repeat things in this *Primer*, but here is an exception—one a farmer would do well to memorize.

Nitrifying or ammonifying microorganisms transform a small fraction of this storage material to nitrate of ammonia—enough in a living soil to sustain plant growth if all the other nutrients are in the right place in the right quantity at the right time. If we feed these organisms, if we maintain a good carbon level and otherwise care for proper balance, the biological process in soil and compost will make these resources available.

But if easily available ammonia or nitrate is offered, nitrogen fixing bacteria become spoiled. They become consumers and feeders, rather than fixers. This is why the simplistic salt fertilizers mean double losses. When nitrogen was only 3 to 4 cents per pound in the early 1950s, it seemed appropriate for the nitrogen salesman to state his position. "You can buy nitrogen cheaper than it costs you to manage your own organic matter," was the usual statement. How could any struggling farmer refuse to accept this basic economic fact? Now, after 25 years of nitrogen economics, and after costs of nitrogen have risen as high as 30 cents per unit, and the farmer is struggling even more, the farmer must surely reevaluate the true costs of cheaper nitrogen—plus its whole effect on soil humus and the soil's ability to convert the expensive nitrogen. Surely we can now begin to realize the true value of raw material that can be turned into composted manure.

Failure of a farmer to manage his soil system gives him a nitrogen problem, and his attempt to buy nitrogen gives him other fertility problems. Anhydrous ammonia, it has been noted, burns out the humus supply in a soil system, just as surely as a match touched to a stubble field. Yet it is a fact that as soon as organic matter content exceeds 2 to 2.5% in a soil system, these soils build up a reserve and draw from it. There is no washing out of potash or loss of nitrogen, and all three of the so-called major nutritive elements—N, P and K—are always available and root acceptable under eco-conditions. The problem with eco-farming is that it asks a farmer to do something for himself that he cannot buy—to fine-tune, taking many facets into consideration and bringing them all together for interrelated decisions.

So far it has proved all too simple for college men and Extension agents to counsel hotdog fertilization in the form of N, P and K—"Try 17-17-17 this year," runs the advice, depending on what inventory the elevator or the fertilizer dealer has in the back storage room. Mickey Mouse soil tests have been used to prescribe whatever the inventory can provide, and slogans from schoolmen have been hung on the wall to refurbish this brand of misfertilization. "There is no substitute for N, P and K," run the arresting streamers, and we agree. But N, P and K in salt form can't be made to substitute for the catalyst keys, for carbon, calcium, boron, copper, tilth, water retention and capillary return, and pH management with basic cations in equilibrium.

Of all the hotdog approaches ever used, applying nitrogen has been the leader. Many times farmers succeed in painting their crops green for neighbors to admire, but they don't get the healthy yields, the insect and fungus free crops, or the protein that nitrogen intake ought to deliver.

It is true, nitrogen is used in large amounts by plants. Agronomy manuals reveal that it takes 5 pounds of nitrogen to nurse a corn plant along for one month. By the end of the season a 100 bushel corn crop will have used 145 pounds of N, 71 pounds of P_2O_5, and 141 pounds of K_2O. Where does it go? Some is recycled as trash. Some is carried from the field into trade channels.

LESSON 14

Calcium and Magnesium

There are significant differences between greens of the mustard family—kale, mustard greens, turnip tops—and those of the goosefoot family, namely spinach, Swiss chard, beet greens, and New Zealand spinach. The mustard group accounts for nutritional superiority because of higher concentrations of calcium, vitamin C (ascorbic acid). Calcium is quite essential to body building, and vitamin C is a protective food. Greens of the goosefoot family do not concentrate as much calcium as do the mustard family. This shortage is magnified by the fact that calcium in the goosefoot family cannot be digestively utilized in the diet because of oxalic acid formed and present in these plants.

Oxalic acid combines with the plant's calcium and magnesium to convert these nutrients into indigestible oxalates. These indigestible oxalates are most pronounced when soils are near neutral, when so-called sour soils have been corrected with calcium to a neutral point and not balanced as discussed earlier in this primer.

Calcium has the leadership role among the nutrient ions that enter plant life. As protein concentrations rise, calcium concentration rises also. And with an increase in proteins there is an increase in vitamins.

1. the magnet for insect and fungal attack

All this calls into question the advice farmers have been getting since AAA came into being in the early 1930s. In those years the advice ran *lime and lime some more to fight soil acidity*. It was believed by many that acid soils were failing and that they needed to be sweetened. As we have pointed out earlier, these soils were failing not because they were sour, but because they were nutrient deficient and contained toxic loads of undesirable minerals.

The Albrecht equation says that calcium should occupy no more than 65 to 70% of the positions on the soil colloid in terms of cation exchange capacity. Magnesium should take 15 to 20% of the capacity, potassium around 5%—leaving room for a touch of acidity as expressed in terms of hydrogen—possibly 10%. Too much calcium or too much magnesium represent imbalance even if they adjust pH to the neutral point bureau people so ardently wish. The lip service statement on all this is, *calcareous soils are high in calcium and magnesium.* Unfortunately, this lip service merely allows fertilizer people to dodge full responsibility. Many high pH soils are built primarily by high levels of sodium and potassium, and are still very low in calcium and magnesium. We cannot lump high pH soils into an erroneous conclusion that it must be calcareous if the pH is high. True, certain soils cannot handle soil moisture properly. Hardpan soils are created by heavy mineralization in areas of field leaching and improper drainage. Excessive calcium will cause magnesium, phosphate and minor element deficiency. This assuredly means vegetables without *digestive calcium*. In general it means plants with imbalanced hormone and enzyme systems, ergo, poor health—the magnet for bacterial, fungal and insect attack.

Excessive magnesium will cause phosphate, potash and N_2 deficiency. High magnesium and low calcium permit organic residue to decay into alcohol, a sterilant to bacteria. It may also prevent normal dry down and nutritional ripening of any growing crop. It may cement clay soils tightly together, thereby providing a crust that can easily exclude oxygen and water retention and proper insoak or capillary return during dry spells. It is not by accident that such soils produce abundant weed crops, particularly the kinds of weeds that experience best seed germination under anaerobic conditions—foxtail, fall panicum, and many others.

Yet calcium levels within the tolerance range expressed above improve soil texture, make phosphorus and other micronutrients more available, then improve the environment for microorganisms and aid the growth of both symbiotic and non-symbiotic nitrogen fixing bacteria. Proper calcium levels help plants form better root systems, stems and leaves for efficient use of sunlight energy, water, carbon dioxide, nitrogen and mineral nutrients. They reduce the toxicity of several soil constituents and combinations.

2. proper magnesium level

At the same time a proper magnesium level is important to the plant's photosynthesis process, the formation of proteins and the movement of carbohydrates from the leaves to the stems.

3. a good horrible example

Today the whole country is trying to learn something about making a transition to eco-agriculture. The process has to start with analysis, seedbed and rootbed preparation, seed selection, and then it has to move into tillage, crop maintenance, foliar support nutrition, harvesting and storage. When the foundation of this sequence is short-circuited, rescue chemistry might pull a farmer's chestnuts out of the fire this year, the next and even the next after that, but ultimately destruction's payday must arrive.

A good horrible example of what can happen when soil imbalance is allowed to develop was handed to us by an Iowa consultant, Thorpe Friar. Friar had gone to the Loren Burkle farm near Monticello, Iowa. Some seven soil samples had been pulled on that farm. Five revealed magnesium levels in excess of 25%. The others turned in a 20% reading. The highest reading on magnesium was 31%. This last sample had a calcium level of 67.4%. Obviously between calcium and magnesium, there was 98.4% saturation of the cation exchange capacity. The pH for the soil was 6.9 (or above). Farmer Burkle had a fine looking herd of dairy cattle. He also had problems.

The calves were dying of pneumonia. There was too much magnesium in the soil and this was combining with the soil's aluminum to create a toxic condition. Magnesium and aluminum had chemically bonded to form a toxic substance right on the farm. Yet not one state ag college agronomist, not one Extension worker, not one fertilizer dealer had warned of this possibility before Friar came along. No one had identified for this farmer the fact that all limestone materials are not the same.

The usual simple *organic matter, phosphate, potassium* soil test made by a typical college, with pH and liming requirements provided, has no scientific basis for evaluation of the base elements. Still most prescriptions for soils are drawn from that source by fertilizer dealers. This service is worth approximately what it costs, which is nothing. In other words if it is important to define the percentages of nitrogen, potassium and phosphorus on hand and needed, it is just as important to state the percentage of calcium, magnesium and sulfur on hand and needed.

4. the limestone requirement

Here is a chart illustrating the calcium, magnesium, sulfur and potassium of materials applied for either their calcium, magnesium, sulfur or

Analyses of "Liming" Materials
(Average Percentage Content)

Material	Calcium	Magnesium	Sulfur	Potassium
High Calcium Limestone	38			
Magnesium Limestone (Dolomite)	22	14		
Magnesium Sulfate (Epsom Salt)		10	14	
Marl	Variable			
Calcium Sulfate (Gypsum)	23		18	
Sul-Po-Mag		11	9	21

potassium content.

The country is full of lime pits. Some of this lime is dolomite. Some is high calcium lime. Since calcium should represent from 65 to 75% of the saturation of the exchange capacity, and magnesium only 10 to 15%, it is not possible to reach balance for the cation bases with the wrong type of limestone.

To further drive this point home, let's consider the following. If the average dolomite is 22 pounds calcium per hundredweight, and 14 pounds magnesium per hundredweight, this would seem to mean a ratio of 22:14. Unfortunately this is an error. Recalling our information in the lesson on soil audits, it was revealed that computations for calcium and magnesium are different because each has a different balance. In other words, 200 parts per million, or 400 pounds of calcium per acre, equals 1 ME of calcium per 100 grams of soil, and 120 parts per million, or 240 pounds of magnesium per acre equals 1 ME of magnesium per 100 grams of soil.

Magnesium has 1.6666 times more exchange capacity than an equal amount of calcium. Thus the ratio becomes 22:23.33, or over 1 to 1.

This is one of the least understood problems in agriculture. Not only do limestone quarries contain different amounts of base elements, they also have ledges or layers of different material within the quarry itself. A quarry would have to shoulder the expense of continuing analysis in order to tell a farmer exactly what type of limestone he's getting. This is costly, but it is not as costly as using the wrong materials.

And yet the information is available. Highway Departments of the several states have the information on calcium quality in limestone deposits. Unfortunately this is not exchanged with Departments of Agriculture, nor is exchange sought by the Limestone Institute, a well funded and strong voice. Apparently the Institute is happy with things the way they are and is not interested in rocking the boat. Much the same is true of the ag colleges.

5. a poison to the calcium

The calcium-magnesium balance profoundly affects plant life, its health, its production record, its progeny. Oscar Loew, a German researcher, was perhaps the first to point out that excess magnesium is a poison to the calcium in the nucleus of the plant cell. This may seem odd since magnesium is a close associate in natural chemodynamics with calcium, and the latter is the major inorganic element in the human body.

Magnesium and calcium are alkaline earths, calcium being the dominant element in terms of sheer bulk in almost any soil system. In the human body the ratio is 32:1. Calcium is part of the construction material. Magnesium is more of a tool or gear in the biochemical performance of the body.

If, indeed, magnesium is a poison to the performance of calcium, it stands to reason that balance is of utmost importance. That balance requirement is further sharpened when it is realized that magnesium and calcium represent fully 75% of the soil system's available essential cations. This alone places calcium and magnesium at the top of the list of fertility elements, and cements into place the rather specific ratios they must occupy in diets offered plants by the soil.

It is ironic, then, that these two alkaline earths are not even listed on the fertilizer bag. State gumshoes checking freight cars or bag skids in warehouses don't look at calcium and magnesium: they look at N, P and K. When it is realized that potassium on the clay-humus colloid is ample at 2%—at most 5%—of that soil's exchange capacity, then the code calling for such and such a percentage of N, P and K becomes ludicrous.

6. one of the great myths

It is one of the great myths of agriculture that the rhododendron is an acid loving plant. Yet Scottish researchers were able to grow rhododendrons in soils near pH 8.0. Starting at pH 5, acidity was reduced to 7.0, then near 8.0 with the addition of magnesium carbonate. As soil acidity was reduced to the neutral point and beyond, the plants grew better. This fact alone tells us that the rhododendron really isn't an acid loving plant at all. It merely requires a soil highly loaded with magnesium. The rhododendron has a special appetite: it is a magnesiphile and a calciphobe. This means it loves magnesium and hates calcium, either of which will reduce acidity in terms of pH.

This lesson is critical to eco-farmers. High magnesium has an adverse effect on the movement of calcium, potassium and magnesium into plant cells. At the same time magnesium has a favorable effect on the movement of nitrogen and phosphorus into plants. The antagonistic effect by the magnesium on the calcium is almost directly inverse. As calcium concentrations go down, concentrations of magnesium go up approximately equally.

The antagonistic reduction of potassium in the plant by increased magnesium is equally dramatic.

Phosphorus, on the other hand, becomes a luxury consumption item when magnesium loads the colloid too much. Phosphorus was first connected with photosynthesis by a Nobel prize winner, Melvin Calvin. He demonstrated that, in the first stage, this process yields a compound of three carbons and one phosphorus. Two unite via the dropout of phosphorus to give the six carbon sugar, glucose.

What happens in a plant feeds its way into animals, finally into human beings. Professor William A. Albrecht believed that the poison to calcium became a poison to destroy plant tissue, and that therefore excess of magnesium on desert soils in the animal's circulation from eating herbage high in magnesium was destructive to liver tissues. Cirrhosis of the liver was thus explained—magnesium replacing calcium in the nucleus of cellular material. Not for nothing did Albrecht ask, "Have we been calling cirrhosis a disease when it is poisoning via attack on cellular structure?"

7. homeostasis

Homeostasis is one of those nice words we do not hear much nowadays. It means the body's ability to balance itself. This is why essential nutrients—such as magnesium—can be deadly poisons, the agency that makes operative that destructive power being imbalance. "May not an excess of magnesium along with alkalinity of potassium and sodium, all in excess, be a case of poisoning by desert soil alkalinity?" asked Albrecht.

Albrecht did more than ask questions. He stayed on for answers and then instructed farmers on how to take those answers into the field. Those who ignored the facts of nature did so in order to make a profit via the absurd business of trying to fool livestock kept on poisonous soils.

8. to summarize

To summarize so far, pH tells us nothing about excess of cation poisoning of forage feeds. Acidity can be reduced by any of several elements, thereby manufacturing the perfect pH with the wrong nutrient load. The calcium-magnesium ratio should be from 7 to 15 for calcium to 1 for magnesium. The calcium saturation to potassium at the widest should be 30:1 and at the narrowest 14:1. This is why gypsum drilled into desert soil proves so effective in improving the sugar in sugar beets. The sulfate in gypsum helps release the excess colloidal magnesium and replaces it with colloidal calcium for a more favorable balance of each. This is why streaking gypsum for later distribution by sprinkler irrigation to seedlings has proved so effective.

So there you have it. Calcium and magnesium are the less soluble of the alkaline earths. Potassium and sodium are highly soluble alkalis. The four behave quite differently in the soil, and it cannot be implied that even potassium and sodium act like blood brothers. Sodium has been carried to the sea so long it has turned sea water salty. Potassium has been held back by the clay in the soil and ashes higher in plants, even though calcium is first in the pecking order. Magnesium—by the ash test—appears hardly to be a trace element.

Yet these four elements take up most of the exchange capacity, save hydrogen, a non-nutrient. Movement of magnesium depends on the amount necessary to balance the concentrations of potassium, sodium and calcium and on which the withdrawal made by each species of herbage. These withdrawals vary with individual plant synthesis of carbohydrates or proteins.

Farmers don't think about this very often, except perhaps when grass tetany rears its ugly head. How is it that this magnesium shortage rears its ugly head in spring?

Magnesium is an activator of many enzyme systems. Magnesium is essential to photosynthesis since a single atom of magnesium serves as the inorganic core of chlorophyll.

Balance thus remains the one thing that can't be overdone. In a large measure, it is best administered by those who study nature, not books, although books can be of great help when we study nature.

Sodium and the Trace Elements

Sodium is number 12 among elements commonly found in plants. It is not regarded as essential to plant tissue, yet it plays an integral part in the life of all marine plants. Considering the primitive state of knowledge about nutrients, especially the trace elements, it is probably too early to dismiss sodium as being of no consequence except in its destructive power.

Maynard Murray, M.D. takes this view. After over two decades on research with sea water he has concluded that the aging process does not appear to occur in the sea. Chronic diseases are hardly to be found among fish and animal sea life unless introduced there by man. Moreover, there can be no shortage of essential nutrients in sea water.

There is reason for this. The sea is the center of gravity for the world of nutrients. For centuries nutrients have been leaching from soil and moving into the sea. It has been computed that within a single 24 hour day topsoil equivalent to seven inches off a 120 acre farm move into the sea at the mouth of the Mississippi. In terms of world agriculture, the computations tell us that 4 billion tons of dissolved materials are carried by rivers to the sea per annum. As nutrients leave farm acres, plants suffer.

Sea water has a density of 63.99 pounds per cubic foot, according to Smithsonian tables. Its composition has been determined in the laboratory.

Element	parts per million (mg/kg)
Chlorine	18980
Sodium	10561
Magnesium	1272
Sulfur	884
Calcium	400
Potassium	380
Bromine	65
Carbon	28
Strontium	13
Boron	4.6
Silicon	0.02-4.0
Fluorine	1.4
Nitrogen*	0.01-0. 7
Aluminum	0.5
Rubidium	0.2
Lithium	0. 1
Phosphorus	0.001-0.10
Barium	0-05
Iodine	0-05
Arsenic	0.01-0.02
Iron	0.002-0.02
Manganese	0.001-0.01
Copper	0.001-0.01
Zinc	0.005
Lead	0-004
Selenium	0.004
Cesium	0.002
Uranium	0.0015
Molybdenum	0.0005
Thorium	0.0005
Corium	0.0004
Silver	0.0003
Vanadium	0.0003
Lanthanum	0.0003
Yttrium	0.0003
Nickel	0.0001
Scandium	0.00004
Mercury	0-00003
Gold	0.000006
Radium	$0.2\text{-}3 \times 10^{10}$
Cadmium	—
Chromium	—

Element	parts per million (mg/kg)
Cobalt	—
Tin	—

*In dissolved compounds and not as dissolved atmospheric nitrogen.

According to Jacques Cousteau, scientists can take sea water apart and put it back together again, *except! Except* that ocean fish cannot live in the man-made version. There are four trace elements in sea water that exist only in the bodies of microorganisms. Cadmium, chromium, cobalt and tin have been found in the ash of marine organisms, but they have not been obtained directly from sea water. It is possible to synthesize sea water as far as it is understood, but it is not possible to synthesize life. Dead plus dead equals dead. It takes life to create life. It is in fact not possible to synthesize sea water because the microorganisms contribute a percentage of the trace elements, a lesson it might be well to remember when dealing with soil systems.

Here's how Murray puts it in his *Sea Energy Agriculture: Perfect Natural Nutrition.* "A plant can grow to maturity, and yet make dangerous substitutions of elements in its structure due to its chemical attempts to compensate for the dilution, or lack of elements, then they lose their resistance to disease." Not unlike plants, human bodies "are host to an enormous number of microbes that eagerly pounce when the slightest breakdown in cell function occurs."

College folklore notwithstanding, almost all commercial crops utilize some 40 earth minerals. These are never all supplied by N, P and K formulations. Indeed, even the most complex legislated fertilizers handle no more than 12 nutrients. Sodium is never one of them simply because it isn't considered essential.

Well, is it? We simply don't know. We do know that sodium and chloride are not toxic to plants in the presence of other elements as found in sea water, all this based on the work of the physician-scientist, Maynard Murray.

1. sodium

Likely as not, sodium is most important in physiological relation to plants in association with potassium. If this constitutes outlaw status, then look at chlorine. It isn't on the list of essential elements either, and yet some plants like tobacco find it necessary for best growth and quality. A decade or so ago there were only a dozen trace element laboratories in the United States. There are a few more than that now. Together they seem to find a definitive physiological role for a particular element in the Mendeleyeff table every decade. This means we may know a little more about trace elements in 200 or 300 years, and that it will take 600 years to run the gamut unless the pace of discovery is speeded up considerably.

Salt is very much a mystery even now.

Human beings and animals can use the several elements only in chelated form. If this complexing job isn't handled by plants, then it is up to bacteria in the gut to come to the rescue. Obviously this is impossible if there is a great overload, thus toxicity. When the senior author was in China, it was revealed that several tablespoons of salt was once considered an accepted form of suicide. Yet that same salt is used at mealtimes around the world by toddlers and grownups alike.

Physicians recognize this and prescribe salt-free diets for pregnant women and men with heart disease problems. Yet these diets never omit celery. Celery has as much sodium chloride in it as one usually distributes via shaker over potatoes or meat. There is a reason for this. Sodium chloride in celery is not toxic. Only sodium chloride in the inorganic state produces toxic effects. Withal, sodium remains pretty much a mystery.

As we move down the scale of nutrients, mystery widens. Aluminum is a good *for instance*. It seems to play an important role under some few conditions, and turns up prominently in green leaves. This has had scientists speculating that there may be an association between aluminum and chlorophyll.

Whether sodium could be substituted for potassium was investigated as early as WWII. The informed judgment became that it could be. The reasoning had it that sodium takes on the role of potassium in assimilation. This then allows more of the potassium to function in the seeds. It has been concluded that sodium makes potassium available, that it tends to conserve soil calcium, magnesium and potassium. It can assist in plant nutrition when soil potassium is not sufficient for the requirements of a crop. In 1929 it was found that a deficiency of sodium in some soils, coupled with application of none, because the probable reason why natural pastures in fields of Romney Marshes of England failed to fatten livestock.

It would take several books the size of this one to touch all the bases on trace minerals. Obviously farmers can't test for everything in "several books" simply because they aren't government and can't print their own money. Still the yield of a crop is controlled by and directly in proportion to the lowest growth factor, known or unknown. In a manner of speaking, there isn't such a thing as a major or minor nutrient. All are important and all are critical in their assigned roles. For this reason deficiencies in the soil must be corrected with the best available materials and these must be applied in the right amounts. Imbalanced fertilization—that is, fertilization to permit either a shortage or marked imbalance—disturbs the soil's living microflora—the algae, fungi, bacteria, protozoa. Using fertilizers beyond the corrective point will lead to over-fertilization, or massive disruption of the life system in the soil. In the soil as in the sea, life plus life equals life.

Dead plus dead equals dead.

All so-called major and minor nutrient elements are microflora in which efficiency is energetically coupled. Don't let the word frighten you. It simply means that overdosing with one growth factor will change the entire spectrum. An excess of nitrogen will cause potassium deficiency. In fact every excess disturbs the microflora's activity, chiefly through nitrification and fixation. Interrelations work their way all through the life chain. Here, for instance, are the mineral interrelationships in animals.

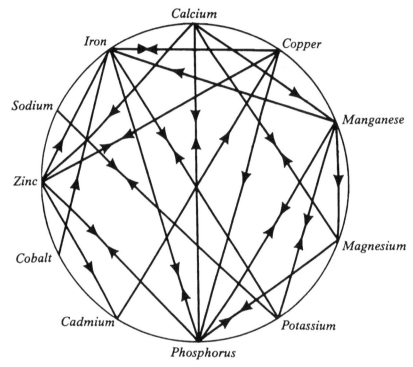

Based on the research of several investigators in animal testing, the above mineral interrelationships appear to be established. If a mineral has an arrow pointing to another mineral, it means a deficiency of that mineral or interference with its metabolism may be caused by excesses of the mineral from whence the arrow originates. This artwork and information was furnished to *Acres U.S.A. Primer* by Harvey Ashmead, Ph.D., of Albion Laboratories, Clearfield, Utah.

2. excesses and deficiencies

In the soil excessive magnesium will cause potassium, phosphorus and nitrogen deficiency. Excessive potassium, sodium and magnesium will cause calcium deficiency. Excessive calcium will cause magnesium, phosphorus and trace element deficiency. Excessive magnesium will cause effects similar to magnesium deficiency. Excessive boron will cause potassium and magnesium deficiency. Excessive sodium and/or chlorine will cause potassium deficiency.

These realizations need not be as troublesome as they sound. Most major imbalances can be repaired by the system illustrated in the soil audit depicted earlier in this text—that is, calcium, potassium, magnesium added or complexed—to achieve cation balance. Phosphorus and sulfates are also important.

Still it would cost a farmer too much money to run a comprehensive test on manganese, boron, zinc, copper, iron, and so on. Properly chelated, these can be applied safely without testing when a farmer has enough magnesium, potassium, phosphorus, calcium, and he has them in the right balance. Some of the so-called organic fertilizers fit the picture quite well for this purpose.

For instance, there are deposits from the Mississippi Valley that supply trace minerals in a well regulated form. The humates (of the right form) give a mighty assist to other problems, such as compaction and shortage of humus.

Everette Burdick once computed a formula for humate application: organic matter times 250, then subtract the quotient from 1,000. This yields the pounds of humates required in terms of a 30% humic acid content. (See humate in the glossary.)

O M	Lbs. Humates Required	O M	Lbs. Humates Required	O M	Lbs. Humates Required
4.0%	0	2.6	350	1.2	700
3.9	25	2.5	375	1.1	750
3.8	50	2.4	400	1.0	750
3.7	75	2.3	425	.9	775
3.6	100	2.2	450	.8	800
3.5	125	2.1	475	.7	825
3.4	150	2.0	500	.6	850
3.3	175	1.9	525	.5	875
3.2	200	1.8	550	.4	900
3.1	225	1.7	575	.3	925
3.0	250	1.6	600	.2	950
2.9	275	1.5	625	.1	975
2.8	300	1.4	650	.0	1000
2.7	325	1.3	675		

Many of the seed treatments, foliar solutions, soil conditioners help handle the trace nutrient shortfall in a broad-spectrum way. When the major elements are in relative balance, it is cheaper to apply trace minerals like a shotgun blast rather than to single out each of the numerous trace mineral requirements, all the necessary amounts of mineral elements such as zinc, iron, manganese, etc. These nutrients do not need to be totally scattered over the whole soil area of growth. Would it not follow that a "starter" supply of properly chelated minerals around the seed could guarantee the nutritional needs of such elements throughout the life of the plant? After all, why do we obtain as much as 10 to 20% more yield from the use of a mere 1 or 2 ounces of seed treatment material. Competent eco-farmers do obtain these yield improvements on acres too numerous to be considered a mere test plot.

Oftentimes it is not possible to balance a poor soil completely in one or two years. The use of seed treatments or seedzone starters of the right biological ingredients can make up for imbalanced limitations and grow increasingly better crops while soil adjustments are working.

Recent developments in planter applicator systems make it quite practical and economical to dribble a water solution of amino acid chelates, cultures and seed treatment materials directly on the seed. This virtually gives a nipple-bottle colostrum start to the germinating seed. The young root system is assured of a friendly supply of mineral nutrients.

Such chelated nutrients and seed-treated materials are quite stable, are quickly available, cannot be complexed by the hostile environment of imbalanced soils—and are always available to start the germinating plant with more vigor, vitality and a more complete natural absorption capacity. The cell wall structure can be more efficient for intake of nutrition and extra root hairs are stimulated for greater feeding capacity.

Economics being what it is, it will probably always be a requirement of the eco-farmer that he know something about the art of agriculture as well as the science. He will have to let the soil tell him what is right and what is wrong the way an old time physician looked to signs and symptoms rather than laboratory readouts.

The givens are clear enough. Healthy plants cannot grow in a sterile soil. Nor can they grow in a soil that is not balanced or buffered to synthesize balance. That is why American farmers grow a preponderance of sick plants and why they rely on toxic rescue chemicals to keep insects, bacteria and fungi from walking off with the crop. Healthy plants from a balanced soil system do not offend their reason for being, and therefore they do not have to be saved with high powered poisons.

3. the count

Not less than six trace elements comprise the prime trace element group: iron, manganese, zinc, copper, boron, and molybdenum. Without exception, plants require these elements for nutritional support. Without exception they are all indispensable in some physiological function—such as the buildup of enzymes or coenzymes. If the trace nutrient is not present, the specific function is not performed.

There are the so-called secondary trace elements, such as aluminum, chlorine, sodium, iodine, boron, silicon. These clearly influence growth in some plants, but not in all—at least as far as our primitive science understands it today. There are agronomists who are convinced all the trace elements figure in plant production and health, and that only our veil of ignorance keeps us from full knowledge of the subject. This may be. Certainly each decade sees nature revealing herself a bit more.

She has been doing so for a long time. Dmitri I. Mendeleyeff's periodic table of elements is one of several discovered at approximately the same time, each setting up the pecking order of the elements that make up the universe. In human medicine, Bertrand's Law and Weinberg's Principle are used to explain why some trace elements are necessary and yet become toxic. An explanation of homeostasis is reproduced below.

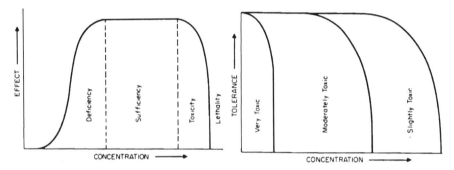

Left, a schematic of Bertrand 's Law. Right, a schematic of effects of toxic trace elements. All soluble elements can be toxic in large enough amounts, according to Henry A. Schroeder, M.D. These diagrams appear in The Trace Elements and Man, *published by the Devin-Adair Company.*

It will be noted that when there is no trace element present, there is no growth. As concentration increases, the deficiency area is diminished in size. At the plateau, function and nutrient availability are at a peak. The width of this plateau varies with the life form involved. In mammals it is quite wide because natural trace nutrients are under homeostatic control. The human body has a fantastically exact mechanism for maintaining balance of its inorganic elemental composition. Plants are not as lucky. For this reason, the plateau for plants is narrower. Beyond the plateau needed for proper growth, the same nutrient that might have been characterized as deficient at one time is now toxic and even lethal.

(Indeed, trace metals are under homeostatic control for human beings. Chlorinated hydrocarbons and man made toxic genetic chemicals are not and may not be for hundreds of centuries of evolution. With fossil fuel firms pouring man-made molecules into the market in excess of 1,000 new compounds a year, man is not likely to adjust to this pollution factor. Conversely, the organic elemental content of human food, carbon, hydrogen and oxygen are under poor homeostatic control, and this is why these elements build up and deposit themselves as fat. When people eat more than they burn, they might as well tack the new poundage directly to their hips.)

Trace nutrients blend into the farmer's art because almost all can be recognized in terms of hunger signs. When such plants are treated with the deficient element, physiology is often normalized. Here, then, are a few reasons for being of trace nutrients and the hunger signs by which absence can be recognized.

4. boron

Boron is required so that calcium can perform its metabolic chore. It is essential in several other metabolic processes. Present in the cafeteria of required nutrients, it prevents such abnormalities as cracked stem in celery, internal cork in apples, black heart in beets and turnips, yellowing of alfalfa leaves. When boron deficiency is a problem, death of the terminal bud is a common symptom. Lateral buds continue to produce side shoots, but terminal buds on these side shoots fade away. Rebranching may occur, but the multi-branched plant will take on the appearance of a rosette.

In cauliflower, heads fail to mature properly and remain small. Reddish-brown areas become evident. Terminal buds take on a light green color, graduation being from pale at the base, not so pale at the tips. Root crops are affected by brown heart, dark spots, or by splintering and cracking at the middle in, say, spuds, sweet potatoes, radishes, carrots.

Boron is required for translocation of sugar, and this means boron deficiency can be spotted as a sugar deficiency. Important as it is, a 100 bushel crop of corn requires only 4 ounces of boron. The big spender among farm crops is peaches, where 100 bushels contain fully 4 ounces of this valued nutrient. A ton of alfalfa requires only a single ounce. A ton of sugar beets needs 2.5 ounces. This gives you a picture in terms of fertilizer tonnage.

In terms of the biological procedure called farming, boron regulates flowering and fruiting, cell division, salt absorption, hormone movement and pollen germination, carbohydrate metabolism, water use, and nitrogen assimilation. Boron is more or less immobile in the plant unless caught up in the vascular system.

We've talked about cations before—positively charged base nutrients. Boron is a negatively charged acid element. In order to be available to plant life, it first must be converted into the element's anion phase, into borate, by the soil's refining process—namely the microorganism complex. In most soils boron is known as a highly insoluble tourmaline, the supply being somewhere between 20 and 200 pounds per acre. It takes a life in the soil to draw on this bank account, and the Creator has supplied this life in the form of microorganism species which simply have to have boron to live. By using the nutrient themselves and then contributing their bodies to the soil's fertility load, microorganisms change boron into an organic form.

When dry weather hits, microorganisms in soil without tilth and structure go dormant. This means the boron supply is cut off. Generally speaking there is more boron in the subsoil, and nature has wisely provided for roots that dig deeper, perhaps both for moisture and for this very essential nutrient. As microorganisms die, organic boron is oxidized into inorganic boron, and thus the cycle repeats itself. Each crop year takes very little boron from the bank account, yet there is a withdrawal. If the soil is alive and the organic matter is at its peak, the likelihood of a zero bank account in a 200 year old country is near nil. Yet it is possible in certain areas, and the requirement for this trace nutrient can be quite real.

Borax and sodium borates are the commonly available hot dog fertilizers in this case. There are, however, certain dangers in using this form. Most eco-farmers prefer to rely on nature's blend of trace minerals, either in the soil or on the leaf. Borox is one of the oldest weed killers known. Moreover, low rainfall areas of the country are frequently irrigated nowadays. In soils with high boron content, irrigation waters can carry enough boron to injure crops. So always remember that while a little can be good, it does not mean more is better.

Too much boron will kick plant vigor in the head, restrict growth, cause plants to exhibit that sickly pale green color sometimes mistaken for nitrogen deficiency, preside over root deterioration and poor yield. In short, either a shortage or marked imbalance of boron will set up a plant for insect and fungal attack. Unfortunately decline will be well underway by the time signs and symptoms rear their ugly heads.

5. cobalt

Cobalt is a constituent of vitamin B-12. In mammals it is essential for hemoglobin formation in the blood and to prevent degeneration of nerves. A deficiency of B-12 causes weakness, pain in the extremities, liverenlargement and skin problems. Cobalt and iron are two important necessities for legume nodule formation and nitrogen conversion as is demonstrated by the bean families, clovers, alfalfa, etc. Seeds started without cobalt and iron will not grow a viable good working plant.

6. iron

Iron is indispensable as a carrier of oxygen in the process of biologic oxidation. It is needed in the production of chlorophyll and aids in the prevention of chlorosis. When iron deficiency is serious, the entire leaf will turn yellow, leaving only the veins to stand out like roadmaps in a life process that has been shortchanged. The delicate flavor of nature's system is evident here. Chlorosis is possible even in the presence of iron, the interplay or imbalance taking its toll. Lime can complex iron, and yet in the human being calcium and copper must be present for iron to function properly. In order to free iron, the farmer must complex the calcium in this case, and this means using either iron sulfates or iron chelates, or substituting a proper foliar blend.

It is very difficult to supply iron to plants in any form that can be characterized as available. Also, there are times when plants starve for iron in the midst of plenty, much like sailors thirsting in an ocean of water. The top six inches of a farm acre can hold perhaps 20 tons of iron, most of it unavailable due to soil acidity levels and complexed soil nutrient arrangements. Recent developments in the chelation of iron to the amino acid form has produced a very reliable source for soil, plant or animal use.

Plants that are deficient in iron become food that is deficient in iron, a precursor to anemia in people. As all good TV ad watchers know, a woman requires three to four times more iron than a man. Computed for a city like New York, this would mean an iron requirement measured in tons per annum.

7. the micro-nutrients

There are the nutrients that one must deal with in terms of ounces rather than pounds per acre—molybdenum, for instance. These elements—trace and otherwise—are absolutely vital to the life process of microorganisms. Azotobacter, the nitrogen fixing non-symbiotic microorganism, has to have molybdenum and a quality supply of zinc, or it won't sustain its life process in the soil.

Copper, vitally important to root metabolism, helps form compounds and proteins, amino acids and a host of organic compounds. It acts as a catalyst or part of the enzyme systems. It helps produce dry matter through stimulation of growth, prevents development of chlorosis, rosetting and dieback. Manganese aids the oxidase enzyme in carrying oxygen, and it enters into the oxidation and reduction reactions needed in carbohydrate metabolism and in seed formation.

Zinc may act in the formation of chlorophyll. It too has a role in chlorosis prevention in some plants. It certainly stimulates plant growth and prevents the occurrence of mottled leaf in citrus, white bud in corn and other disorders.

It may perhaps be helpful to view trace nutrients not so much as constituents of mass production, but more as tools used in bringing on this growth. Trace nutrient support of microbes, plants, animals and man can be measured at any stopping point along the way. Perhaps it would be helpful to see what the several nutrients—macro and micro—have to say in terms of animal health, specifically via the health of the cow. In order to pick up this transition, see the lesson on animal health.

LESSON 16

Phosphorus

When farmers talk about phosphorus they usually talk about phosphorus deficiency. Indeed, phosphorus deficiency in humid soils has become legend. It is an anion, but it is united rapidly with divalent calcium, a cation, to become insoluble, and therein lies the whiff and whoof of phosphorus fertilization. The weak carbonic acid from a plant root can't seem to activate bound up phosphorus. Apatite is quite insoluble. Rock phosphate in any form becomes available reluctantly, and this is one of the reasons why soils should be at least slightly acid. But the fact is science still does not have a clear picture of what happens with phosphorus in the soil.

1. the phosphorus connection

We ran this problem up the flagpole with William A. Albrecht one day while a tape recorder was running. His answer exhibited the usual patience as he repeated what he had gone over a few dozen times before—only this time he made the phosphorus connection. "In soil management," he said, "we associate the inorganic nutrients calcium, magnesium, potassium, sodium, iron, aluminum, zinc, copper, cobalt, manganese and others with positive ionic characteristics with the soil's colloidal humus-clay fraction of the opposite, negative electrical character. The essential nutrient elements, namely nitrogen, sulfur, phosphorus, boron, chlorine, iodine, molybdenum, and others are empirically associated with the organic matter of the soil. So why not envision these anions or negative elements corrected with decay stages of organic matter when a wider carbon-nutrient ratio serves as a source of energy by the carbon first to fungi, then to bacteria in the soil. Why not visualize the negative elements as serving simultaneously in plant nutrition as chelated and larger molecular complexes of microbial origin."

Sing Chen Chang made the point in 1931 that considerable phosphorus, ranging from 0.1% to 0.5%, reached the soil in organic materials. Release of this phosphorus depends on decomposition of organic compounds by microorganisms. It is for this reason that manures—green and barnyard—or compost refinements thereof, should not be valued alone for their nitrogen.

All plant tissues contain phosphorus, with concentrations most pronounced in young plant parts—in seeds, flowers. The very life processes—photosynthesis, the construction and breakdown of carbohydrates, even energy transfer—all require phosphorus. Cell division is largely dependent on phosphorus, and this is the reason plants fail to grow, and stand spindly when phosphorus is limited. Agronomists like to compute the amount of phosphorus in terms of crop use—10 to 15 pounds for a 35 bushel wheat crop, 25 pounds for a 75 bushel corn crop, and so on, the ash mentality! Some soils are low in phosphorus the way the lab measures things, especially sandy soils, and some soils are too alkaline for ready use of non-water soluble phosphorus. Add a word to your vocabulary here: labile.

Superphosphate, triple-superphosphate, rock phosphate, or organic fraction phosphates—proteins, phospholipids, and so on—are labile. It means that while these phosphate forms are undergoing chemical or physical change, they are not necessarily available.

2. a somewhat acid soil

Phosphorus more clearly than any other plant nutrient explains why some farmers fail in organic farming, or any of several biological agriculture forms we like to report about under the head of eco-agriculture. It takes a somewhat acid soil to make rock minerals available for plant use. Yet acidity has been blamed when soils fail to deliver. In earlier times, farmers would taste the soil. If it tasted bitter it was acid, and that is how the symptom of nutrient deficiency came to be considered the cause. For some 40 years farmers have been liming to fight soil acidity when they should have been fertilizing to fight low fertility levels.

When the ag colleges decided to set the target for hydrogen at near neutral, they automatically made everyone have to use water soluble fertilizers. This may have been ignorance speaking, and it may have been thoroughly informed self-interest. Yet all types of soil need to have a pH of 6.5 or under in order to feel the acid conditions necessary to etch out the bound up forms of phosphorus, and to separate the calcium from the phosphate. A pH near 7 means no activity. Therefore farmers with near neutral pH cannot possibly make a success of organic agriculture. They cannot get the nutrients out of the rock phosphate without buffering with nitrogen, and procedure prohibited by the tenets of folklore organic farming.

Yet below pH 6 phosphorus tends to become complexed with iron, aluminum and manganese. At very low pH levels aluminum becomes soluble in toxic amounts. Above pH 7.0 phosphorus becomes complexed by calcium.

This is what makes organic matter so important. High organic matter soils are almost always populated by a variety of microorganisms. This microbial population serves as a phosphorus reservoir. Not unlike plants, these unpaid workers also require phosphorus, and they are more efficient than roots at the task of getting this vital nutrient. They pick up inorganic phosphorus and build it into their cell components. They mineralize organic compounds and release inorganic phosphorus anions. They oxidize and reduce phosphorus compounds. They alter the solubility of inorganic phosphorus found in the several compounds.

These little workers also excrete nitric and sulfuric acids, the compounds needed to break phosphorus from calcium phosphates. This is nature's way of producing mono and di compounds. Some bacteria are like honey bees in that they produce more than they need for themselves, and this excess of solubilized phosphorus is there for plants to do the taking.

3. there is a chemistry

When phosphate loads are rapidly complexed or not made available, fundamental sugar formation continues to function. Symptoms wave a warning flag that can be seen from great distances. Leaves often become reddish and purplish—a lack of chlorophyll—and tips die off. Seeds, tubers, grains, all suffer since all require phosphorus for adequate metabolism. Growth is slowed accordingly. The corn plant has a sign all its own. When there is phosphorus deficiency, the kernels drop off about an inch or two from the cob's end, or they may fail to develop in the first place.

In short, there is a chemistry involved whenever anything is put into the soil, inorganic, organic, salt form, whatever. Rock phosphate is called tricalcium phosphate, and this means it has three calciums, or three negative charges for bonding. This makes it more difficult to disattach from fixation than would be the case with dicalcium phosphate, which has only two charges—thus tri, di! Last, there is the water soluble monocalcium phosphate, which means that as a consequence of acid treatment this form has only one remaining bond.

So if you're over 6.5 pH, and you want to farm organically for good and obvious reasons, you're in trouble. You probably shouldn't be using non-water soluble phosphorus because the soil does not have enough acid to free it up. If a soil system has a pH of 7.5 the farmer probably shouldn't be using the di forms. He should go to strictly mono forms of phosphate. Any farmer who doesn't take into consideration the importance of the active hydrogen ion as being the most important thing to work with thereby authors his own failure.

Acid treatment merely means rock phosphate is being converted from tricalcium phosphate to monocalcium phosphate, and that this highly unstable form is subject to natural reversion back to the stable tricalcium form. The rate of reversion differs. The pH, the free calcium in the soil, the organic matter—all figure in this rate of reversion. But it is safe to say that 75% of the monocalcium phosphate reverts back to stable tricalcium phosphate within 90 days. In some soils the reversion takes place within hours. As soil conditions worsen, release of nutrients from rock phosphate worsens, and the chemical amateur becomes married to buying salt fertilizers, each go-round worsening still further the structure of that soil.

The water soluble phosphates are simply water soluble, not acid. But they are a poor substitute for having the proper pH with calcium, potassium, magnesium and sodium in equilibrium, and from an economic point of view they take on ripoff dimensions.

First, the soluble phosphates come from rock phosphate in any case. By treating, say, 1,400 pounds of rock phosphate with 1,200 pounds of sulfuric acid, the fertilizer industry gets 20% superphosphate—the tricalcium phosphate form being converted into water soluble monocalcium form. This chemical reaction causes 20% superphosphate to be represented by about 45% monocalcium phosphate, and 55% calcium sulfate, or gypsum. This means the bag of 0-20-0 contains about 45 pounds of water soluble monocalcium phosphate, which is presumably desired, and about 55 pounds of calcium sulfate, which may or may not be desired, but which farmers are frequently not aware of.

The fertilizer rating 0-45-0 is quite different material. Farmers who see symptoms of phosphorus deficiency sometimes think a higher rating is the answer, and this one comes styled triple superphosphate. Here the acid used to do the etching is phosphoric. This eliminates the calcium sulfate in the bag, calcium frequently needed to kick up the calcium reserve, sulfate needed to complex an excess of magnesium. By invoking the hotdog concept of plant nutrition a much needed nutrient might be eliminated exactly when it is needed.

Ammonium phosphate such as 8-32-0, 11-48-0, and 80 on, also involve concentrated phosphoric acid in the processing, and this provides a handy outlet for otherwise unsalable fossil fuel company byproducts.

4. *the natural concept*

The lessons seem clear enough. A farmer who is trying to live with the natural concept, and not in the end rely on toxic rescue chemistry, has to keep his soil system down on the acid side as far as pH is concerned. It makes no great difference whether rock phosphate or colloidal phosphates are used.

When the farmer was taught to go high on the pH scale, even if relative balance was maintained, it became mandatory for him to turn to factory acidulated fertilizers. Also, when the farmer quit using more manure, the fermentations made by bacteria no longer helped provide the acids—fulvic acid of compost for instance—that made some of these nutrients come available. This is why biological farming has to be diversified farming, and why there have to be animals on a farm if there is to be anything like a good natural or organic system.

We have mentioned the *di* forms of phosphate, diammonium with a P_2O_5 rating of 48 to 53. What do the microbiologists say about non-straight forms of nitrogen compounded with phosphorus as a mixed fertilizer? Apparently the livestock in the soil work fairly well with this mixed fertilizer form. Di forms are synthetics, of course, and the question surfaces as to whether synthetics have any place in a good biological farming system. To pose that question is to suggest that there is an answer. There is, and it has to be answered when crops are born. If insect, bacterial and fungal attacks threaten a crop, a mistake has been made. Such a mistake is equally damning in its finality whether a soil system is managed according to the precepts of eco-agriculture, or in compliance with the chemical gambit.

Suffice it to say that mistakes in terms of biology are inherent in straight chemical agriculture, hence the highly touted lines of toxic weed control and rescue chemistry.

5. *amateur status*

If a farmer is not going to pay out the price of the effort needed to maintain his soil's organic matter, he cannot compete in the mining game called agriculture without using some of the salt fertilizers, and probably the fossil fuel technology called rescue chemistry. It may be true that imbalanced salt fertilization and toxic rescue chemistry constitute amateur status, but farmers on the firing line have to be concerned about going broke, and not with wishful thinking about an order of economics that does not exist. Eco-agriculture figures some of the synthetic tools can be used because they fit into the ecology, and they do exist in nature.

PHOSPHATES
THE BIO-DYNAMIC CONCEPT

*Phosphates are precipitated in measurable
amounts with snow and winter rain*

*The situation in organic "living" soils:
Phosphate compounds are dispersed in the
soil in living and dead organic matter.
Organic and inorganic acids can render
them available. The more digested organic
matter (humus) that is present, the greater
the availability from particle to particle. To
soil solution, to root area: there exists
"A Dynamic Equilibrium"*

*Because of it,
phosphate parti-
cles do not mi-
grate much. But
under unfavorable
conditions, they
can be precipita-
ted and lay "in
position," but
inert*

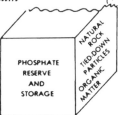

PHOSPHATE
RESERVE
AND
STORAGE

NATURAL ROCK
TIED-DOWN PARTICLES
ORGANIC MATTER

*The natural balance
decides how much is
available or tied down*

*Phosphates are
also contained
in manures and
composts*

*With biological activation and maintenance of soil balance,
natural resources are utilized and only the actual removal
by way of crops, needs to be covered*

E.E. PFEIFFER

PHOSPHATES
THE MINERAL CONCEPT

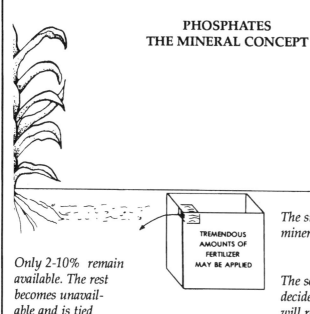

TREMENDOUS AMOUNTS OF FERTILIZER MAY BE APPLIED

Only 2-10% remain available. The rest becomes unavailable and is tied down. With continued unfavorable conditions it will remain tied down and will be eventually lost

The situation in mineralized "dead" soils:

The soil balance decides how much will remain available and not the amount of application. The more that has been applied, the better the chances of tying down result

Acid soils allow better availability
Alkaline soils favor unavailability

In excessively phosphate fertilized soils, large deposits of phosphate have been building up, which can be utilized with the "biological activation" of the soil

If the biological activation is not done and the soil balance is not maintained, then plants can only live from the mineral compounds contained in the soil solution, and the mineral fertilizer theory applies.

E.E. PFEIFFER

Synthesizing something that exists is quite different from synthesizing something that does not exist.

6. *boxscore*

Crops remove hardly half as much phosphate as they remove nitrogen or potassium, 50 pounds for a 100 bushel corn crop being cited as standard. Therefore application of 500 to 1,000 pounds of superphosphate year after year furnishes the proof of its own inefficiency. When the water soluble forms are complexed, they seem to stay complexed. This makes bagged phosphate an uneconomic system when organics, indeed, can slowly release phosphates from the bank account as needed through the years.

7. *imbalance*

Phosphorus imbalance comes in two forms—excess and deficiency. Excess is often associated with poor yield, lack of vigor, missing quality, failure to uptake micronutrients and marginal growth. Luxury consumption is less common than deficiency, however.

Since soil tests cost money, farmers often plant indicator plants—Romaine lettuce and tomatoes, for instance.

LESSON 17

Sulfur

There's a chapter in *Soils and Men*, the 1938 *Yearbook of Agriculture*—now a collector's item—that comes styled *Neglected Soil Constituents that Affect Plant and Animal Development*, and it turns up the proposition that soils are "more deficient in sulfur than in phosphorus." It seems strange, wrote J. E. McMurtrey and W. O. Robinson, "that so little attention has been directed to sulfur as a fertilizer constituent. The fact is that sulfur has been applied more or less unconsciously with phosphorus in the forms of superphosphate and acid-treated bone. In superphosphate, the sulfates commonly exceed the phosphates. Largely because of this, sulfur deficiencies in soil have not appeared. The use of highly concentrated fertilizers in which the phosphates carry no sulfates will create a sulfur deficiency unless sulfates are added."

1. sulfur deficient

Those words were prophetic. Except for the western mountain states, and for an area about the width of Missouri that runs in an upsweep into Maine, all of the U.S. is sulfur deficient, the rainbelt south being the most deficient of all. Even so-called non-deficient areas have spot deficiencies.

During the coal burning era—which may now be returning—farmers got a special benefit without realizing it. Minnesota researchers developed the information that sulfur from the air varied greatly, running at 100 pounds per acre near industrial centers, and as low as 5 pounds per acre in the far northern wilderness counties of that state. It was estimated that rural areas received 15 to 30 pounds per acre annually. Application of manure to farm acres joined the downwash of rain and even residues from some early insecticides in providing farmers with sulfur in those days.

Few farmers who did their homework on such matters figured there would ever be such a thing as a sulfur problem. Then several things happened. Gas and electrical energy replaced coal, and the sulfur rain and snow discontinued. Next, farmers moved from manures and superphosphate to triple superphosphate as the chemical bandwagon raced downhill—brakes out of commission—and this resulted in more sulfur starvation.

2. the early warning system

As usual, the early warning system had to do with animals. A deficiency of sulfur results in shedding of wool in sheep, poor hair coat, poor hoof, hair and horn development—all keratinous tissues. General symptoms include watery eyes, and excess saliva. The *Acres U.S.A.* editor used to work for an old veterinarian, Robert L. Anderes, D.V.M., as a publisher's assistant on *Veterinary Medicine*. Anderes could read these symptoms a mile off and know exactly what was going on in the soil system. He would caution farmers that excess sulfur triggered other problems. It created an acid rumen and increased the need for copper.

Crops take up a lot of sulfur. A 150 bushel crop of corn will tap the sulfur bank account in the soil for 44 pounds per acre. A 60 bushel wheat crop will demand about 16 pounds. A 50 bushel to the acre soybean crop will draw on 10 pounds, and coastal Bermuda grass—at 10 tons to the acre—requires 80 pounds of sulfur.

Livestock farmers who return manure or compost to the soil will not have the same sulfur problem monocrop farmers have. In fact, they will have perhaps three times more of a sulfate content in their soil system than their monoculturing neighbors, same soil type.

Soils vary in their ability to hold sulfur according to how well they maintain their organic matter bank account. Old timers have always preserved their sulfur profile by green manuring, plowing down residues, and adding organic matter whenever possible. Early day agronomists talked as if calcium sulfate (gypsum) were a universal fertilizer. Gypsum added about 32% calcium oxide and 56% sulfate—both nutrients likely to be missing.

But what can a farmer do if he already has too much calcium? Obviously calcium sulfate is not the answer in such a case, even though some purveyors sing the praises of a deposit mined out of the hills and pass it off as good for what ails in every case.

3. the fertilizer industry

The fertilizer industry has provided farmers with 20% superphosphate and triple superphosphate, both treated by acid in the factory rather than in the soil. These materials have obvious disadvantages. Whether they offend the environment in the soil or not, they complex easily and fail in soil systems that worsen by the year with all forms of salt fertilizer and rescue chemistry abuse.

Magnesium sulfate has 16.5% magnesium oxide and 39% sulfate. S-P-M (Sul-Po-Mag) is 18.5% magnesium oxide, 21.5% potassium oxide, and 57% sulfate. It is a natural combination of so-called available magnesium, sulfur, and potassium, with a maximum of 2.5% chlorine. So-called natural fertilizers frequently use this S-P-M material, and it isn't all that bad. Potassium sulfate has 50% potassium oxide and about the same percentage of sulfate. Ammonium sulfates have 20.0% nitrogen and 73% sulfate. Flowers of sulfur rate as 299% sulfate. Sulfates of magnesium, sodium and potassium are of course highly water soluble.

Again, the ability of a soil system to handle acid-treated fertilizers is contingent on the buffering ability of that soil. As with nitrogen, any counterfeit attempt to live without the proper cycle is subject to shortfalls.

Shortfalls generally mean sick crops, insect, bacterial and fungal attack, and toxic rescue chemicals. The upper leaves turn yellow, even the veins take on a sickly hue. Stems stay small and turn hard and woody. Roots become "excessive" and slender. Newest growth presents a tell-tale sign, yellowish, and this extends itself over the whole plant. If this sounds like the symptoms of nitrogen deficiency—well, that is the way it is. As a matter of fact when there is both a nitrogen and a sulfur deficiency, you may not be able to tell the difference.

Some few farmers like to grow indicator plants—orange and lettuce when climate and growing conditions permit. Such plants require lots of sulfur, and if they suffer, it may indicate what's coming down the pike for the rest of the crop. Mustard, cauliflower, kale, turnips and radishes, plants of the lily family such as onions and asparagus, really load in the sulfur—cabbage being the leader, picking up 40 pounds in a 15 tons per acre crop.

4. sulfur is an anion

Remember the discussion on anions, page 123.

Sulfur is an anion. Microbes have to handle the chore of dealing with sulfur, turning it into a sulfate which is just what the plant roots like to accept and transport to a cell building site. There it becomes a component of cystine, a protein constituent, and an aid in the ready synthesis of oils. If sulfur is deficient or unavailable, protein synthesis drags its heels, triggering an even bigger problem, that of aminos and other nitrogen-containing compounds accumulating in the tissues.

When light enters the leaf of a plant, its vibrations kick loose electrons in chlorophyll, the green coloring matter used in photosynthesis. Although the chlorophyll molecule contains no sulfur, sulfur figures in the equation in a manner still to be explained at both the primer and postgraduate level. Thus the yellow instead of the green in sulfur deficient plants.

According to tests across the country, moderate to severe deficiencies of sulfur exist in most of the apple, cherry and pear growing areas. Such deficiencies are frequently associated with manganese shortages. Once this nutrient hunger takes hold, it is hard to correct, even with treatment. And—as eco-growers have discovered—biological control of insects is not entirely satisfactory as a substitute for nutrient balance in the soil. Moreover, balance in the soil, sulfur balance included, permits fruit to withstand cold weather better than imbalanced acres, and keeps insects at bay as well.

Sulfur transformations are largely biological. The decomposers of the biotic pyramid take proteins and organic combinations, of which hydrogen and sulfur are simple forms, and turn them into sulfates and sulfites. The cycle depends on certain types of soil bacteria.

There are imbalance situations in the orchards and fields that admit no quick solution, save the use of flowers of sulfur, or the newer elemental sulfur forms found in some private prescription blends. The wrong sulfur forms have a hard time getting the toehold they need.

A soil audit of scope must point the way. Fortunate, indeed, is the farmer who has composted his manure because such a material does double duty in getting the nitrogen and sulfur cycle working together.

LESSON 18

Potassium

Taking N, P and K folklore for what it is, one should not be surprised to learn that potassium (potash) is seldom absorbed in excess. The pour-it-on syndrome does account for calcium, manganese, zinc and iron deficiency at times, but in the main factory acidulated potassium merely means a farmer is paying for a nutrient that is quickly locked up and fixed in nonexchangeable forms. With the exception of the orange—which sometimes loads in an excess—potassium is better known for being short, rather than in absorbed abundance. Terms such as excess and short, unfortunately fail to properly handle the concepts involved here.

1. the hammer and trowel

Potassium has been characterized as the hammer and trowel in the plant building business. It joins trace nutrients in being a catalyst. It is an elemental greeter of sorts, a Marrying Sam that introduces elements to each other in the plant so that they can form compounds and set up housekeeping or be carried from one part of the plant to the next. Potassium is a prime requirement in the chlorophyll building chore. Plants need it before they can pull free nutrients out of the air—carbon, hydrogen, oxygen. In fact plants can't make starches, sugars, proteins, vitamins, enzymes or cellulose without potassium. They can't even survive summer dry stretches very well without ample potassium.

At the signs and symptoms level, farmers usually talk about shortages, about lodging, about disease conditions. A deficiency causes mottling, brown top, purplish spotting, streaking or curling of leaves—starting at the lower levels. Dead areas frequently fall out, leaving ragged edges. In grains and grasses, firing starts at the tip of the leaves and proceeds down from the edge, usually leaving the midrib green. Premature loss of leaves, reduced resistance to cold and other adverse conditions have somehow conspired to make potassium a legislated entry in the N, P and K fertilizer game, the implication being that potassium is a major nutrient when in fact it hones close to being in the same camp as other catalysts—an engineer in the plant building business.

2. a vital role

Using laboratory techniques, scientists have linked potassium to size, flavor and color of fruits and vegetables. They have assigned this element a vital role in the creation of octacosanols in oil-bearing plants. Using field observations and controlled experiments, agronomists have discerned that potassium governs resistance to certain diseases, that it presides—in a measure—over root growth, and referees the balance between nitrogen and phosphorus.

This is very important because the part of the crop that the farmer sells—the kernels of cereals and grains—are plump or slim according to the balance that management of the soil's life complex can account for.

Unlike soil phosphorus—which is often very limited—most soils have plenty of potassium, and the outside supply in the world at large, oceans included, is considered inexhaustible. Unfortunately potassium is a tough cookie in the availability department. Unless microbes in the soil give a mighty assist, and pH is managed at the proper acid level, potassium availability is a massive problem. It is not helped greatly by the kind of fertilizers available either.

When corn is starved for want of available potassium, it gets plugged up with iron at the nodes. As a consequence, sugars and starches can't travel from leaves to roots. Roots starve. Root and stalk rot becomes the observed result. By the time signs become discernible, yields have been cut irrevocably.

3. a character of its own

The fact is, potassium has a character of its own. It's a lot like sodium except that it is contained in many rock formations, almost always in small amounts. It is a highly soluble cation, a monovalent, and hard to hold exactly where you want it. A 3% saturation of the exchange capacity is the usual eco-agriculture recommendation, and there is a deep seated reason for this. That reason amply illustrates why it seems foolish to even attempt discussion of any nutrient in isolation.

Increasing the saturation of any soil's exchange capacity beyond the optimum carries with it unwanted consequences. For instance, over-sat-uration has a depressive effect on the uptake of calcium and magnesium by the legume crop of, say, soybeans. Yet the presence of boron in the soil reduces the depressive effects of potassium on calcium. Boron also serves to lessen the depressive effects of potassium on the magnesium concentration in soybean leaves.

This inventory of fact caused William A. Albrecht to pose a startling question. "When calcium is a major ash element in the warm blooded animals, and when it is so highly essential to legumes for their greater activities in fixing nitrogen, then boron—as an aid in moving high concentrations of calcium into forage feeds—may be indirectly a very important fertility element."

Thus a discussion of potassium inherently requires discussion of boron, and boron cannot be discussed without a ready index to facts concerning the three major cations—calcium, magnesium, potassium—at hand. Potassium must be considered with this role of the trace element boron in mind. Certainly this extremely small amount of an essential element then takes on a significance within the soil and plant not entirely appreciated.

4. the pH requirement

Further, the availability of potassium invariably returns our thinking to the pH requirement and the friendly environment for soil life that eco-management should account for. Many organic gardeners look to greensand and granite dust for a potassium source, but these depend on pH management for availability. Soils with low organic matter and near neutral pH will not likely tap either locked up stores of potassium, or make use of natural mineral sources that presume to sidestep the harsh realities of factory acidulated fertilizers.

5. the realities are harsh

In the case of muriate of potash, the realities are harsh indeed. At least 75% of all potassium sold through commercial trade channels is potassium chloride, which is sometimes called muriate of potash. This is a salt fertilizer. One atom of the cation potassium is harnessed with one atom of chlorine, an acid element. Placed in the soil, this compound dissolves. The positively charged element potassium is separated from the negatively charged element chlorine.

The now free potassium atom seeks to load over to a negatively charged

POTASSIUM
THE BIO-DYNAMIC CONCEPT

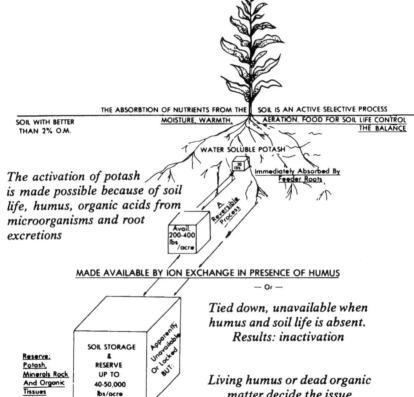

THE ABSORBTION OF NUTRIENTS FROM THE SOIL IS AN ACTIVE SELECTIVE PROCESS

SOIL WITH BETTER
THAN 2% O.M.

MOISTURE, WARMTH,

AERATION, FOOD FOR SOIL LIFE CONTROL
THE BALANCE

WATER SOLUBLE POTASH

*The activation of potash
is made possible because of soil
life, humus, organic acids from
microorganisms and root
excretions*

Immediately Absorbed By
Feeder Roots

A Reversible Process

Avail.
200-400
lbs
/acre

MADE AVAILABLE BY ION EXCHANGE IN PRESENCE OF HUMUS

— Or —

Reserve:
Potash,
Minerals Rock
And Organic
Tissues

SOIL STORAGE
&
RESERVE
UP TO
40-50,000
lbs/acre

Apparently Unavailable Or Locked BUT:

*Tied down, unavailable when
humus and soil life is absent.
Results: inactivation*

*Living humus or dead organic
matter decide the issue*

*A natural dynamic balance is
established—moving upwards
or downwards—according to
conditions. Potash is always
"ready" in a stand-by position*

NO WASTE — NO EXCESS

E.E. PFEIFFER

POTASSIUM
THE MINERAL CONCEPT

The excess consumption of water soluble potash produces toxic effects in animals feeding on such plants. Milk from cows thus fed does not make cheese! Such plants contain 4-6 times the normal amount of potash.

FLESHY, DARK GREEN LEAVES WITH A METALLIC LUSTRE INDICATE THE POTASH EXCESS.

THE EASILY SOLUBLE NUTRIENTS ARE FORCED INTO THE PLANT DISTURBING ITS BALANCE

SOIL WITH LESS THAN 2% O.M.

Part of the excess potash enters the roots and causes physiological disturbance of the plant metabolism

WATER SOLUBLE POTASH IN EXCESS

In a mineralized soil, deficient in humus and soil life, there is only the alternative between water soluble, leaching and/or inactivation. No natural balance exists.

Avail. Potash From Fertilizer

The downward movement is rarely ever reversible, therefore, more and more fertilizer is needed.

UNAVAILABLE POTASH IS LOCKED UP. FERTILIZER IS TIED DOWN. NATURAL ROCK SOURCES REMAIN UNAVAIL.

Excess water soluble potash is washed out, leaches away and is LOST

Even raw manure increases leaching The only protection: to build up soil with composted organic matter.

E.E. PFEIFFER

home base, presumably the negatively charged clay or the negative humus colloid, ready for plant uptake. No problem here.

Dan Skow, one of the chief promoters of the Carey Reams school of farming, best frames into sentences and declares the eco-farming objection to this potassium form. His critique has to do with "what happens to the chlorine." Admittedly, it takes very little chlorine in a stock watering tank to annihilate all bacteria. Can chlorine in the soil be expected to act differently?

When potassium is separated from the chlorine atom, the chlorine element cannot be loaded over to the soil colloid, obviously. It is negative, and like charges repel. Chlorine is therefore left free to leach out of the soil. It goes into the soil solution and finally moves off to the sea, but not without handing off a measure of damage where it was first introduced. Almost all cation elements sold to agriculture as so-called commercial fertilizers operate in this way. They come on in the carbonate, chloride or sulfate salt form. The explanation for muriate of potash is equally the explanation of other salt forms.

Potassium, then, is merely a link in the chain—better yet, a strand in the web—where everything is connected to everything else. It does not function in isolation, and it really can't be considered that way either.

LESSON 19

Compost

So far we've spent a lot of pages taking the interrelated aspects of farming apart. But a farmer has to bring these many items of knowledge back together again. No area of activity serves as well in doing this as compost making.

In her own good time, nature has used the compost process for building soil. Grass and leaves and droppings of animals provided the mix on which biotic life thrived during the eons of time soil construction has been underway.

Just about any farmer knows that disposal of waste organic matter onto land will aid in maintenance of agricultural production. But just as there is this knowledge, there is also general ignorance about health that can be offended by unsanitary utilization and disposal of wastes. There is a reason.

Ehrenfreid Pfeiffer put it into focus in one of those talks that became an article and then a reprint circulated by the thousands.

1. *skatole, indole*

"Some old folks," he wrote, "remember that the odor of potatoes, when boiling in the pot, indicated very specifically the kind of manure—cow, hog, horse or poultry—which had been applied when preparing the soil for the growth of potatoes. Today we still notice that cabbage, cauliflower, broccoli and kale can give off very intensive aromas while cooking. This, too, depends on the fertilization of the soil.

"The reason is that breakdown products contained in the manure, such as skatole, indole and other phenolic compounds are not further decomposed. They enter the soil, are absorbed by plant roots and migrate into the plant tissue. These breakdown products have been excreted by the animal or human organisms, are toxic wastes of their metabolism and imprint their history on the soil and the growing plant.

"Night soil is particularly undesirable in this regard, but even sewage sludge contains many toxic compounds, unless properly decomposed by an aerobic composting process. The difficulty is that these byproducts are rather stable and can survive for a long time. Returned in food to the human upper digestive tract where they do not belong, they can cause many symptoms of indigestion, flatulence and maybe even allergic reactions.

"Biuret is a contaminant or byproduct of urea and urine. It is extremely stable. In greenhouses and field experiments at Purdue University it was demonstrated that it was toxic to growing corn plants. It caused leaf chlorosis, hyponasty and it severely stunted, twisted and deformed plants with their leaf tips rolled together. All fertilizer salts, when placed with the seed, resulted in decreased length of primary roots. The ability to survive germination damage was severely reduced by increasing the biuret content. The rate of growth and the yield were reduced."

2. whole molecules

This long quote was included here to pose a question mark. Whether roots absorb whole molecules or settle for inorganic ions as conceptualized by modern chemistry seems far from settled. Certainly Pfeiffer is right, plants do absorb whole molecules of fece contaminants, insecticides and compounds of organic synthesis. The Finnish scientist, Virtanen, concluded that "in mixed cultures of legumes and grains, the grain can satisfy its nitrogen requirement wholly or partially from organic nitrogen compounds which were excreted by legumes into the soil." Yet it has been the accepted concept ever since von Liebig and Boussingault that only inorganic nitrogen compounds are of importance to higher plants.

That there is contamination in manure goes without saying. That this contamination is eliminated by the composting process has long been a chief tenet of eco-agriculture. Manures and lignated wastes are certainly byproducts of the farming industry. They have been recycled ever since hunting gave way to tilling the soil. It is only recently that science has put a handle on the real reasons for composting vs. semi-fertilization and semi-disposal of raw manure in the field.

Technology has now given farmers inoculants composed of many bacteria types. These microorganisms include actinomycetes, yeasts, molds and the clostridium types. They consume lignated carbons. They digest and excrete a finished product and finally contribute their bodies to the compost heap. The result is a stabilized product that is ready for the field, one without the odors common to phenolic wastes, one without weed seeds—a humus, in short, that is ready to function in the soil as both a buffer and a source of nutrients.

Just what goes on in the compost pile can best be illustrated by this diagram first drawn by Maria Linder when she was professoring at MIT.

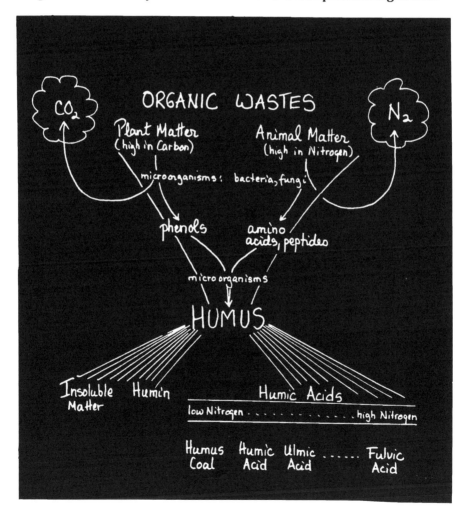

3. *the compost pile*

Organic wastes for the compost pile may be of plant or animal origin. Plant materials will be higher in carbon, whereas animal matter—manures, for instance—provide more nitrogen relative to carbon. Bacteria and fungi break organic waste down from large molecules to smaller units of material—phenols, amino acids, peptides, and other substances such as sugar. Parent materials lose their identity more and more as the process is completed. In a good compost pile, complete transformation takes place. Parent materials become quite different just as proteins in a radish or bean become quite different when cast as part of a muscle or an arm. In composting there is both a breakdown and a buildup process.

Still, humus is not humus and compost is not compost, taking these words to mean machined-alike products. The humus that results from the composting process can be of various compositions. In fact, humus can be quite insoluble. This is called humin. As the diagram on page 219 suggests, there are also the humic acids. These range from low nitrogen containing to high nitrogen containing. And there is quite a range of nitrogen as a percent of humus. In peat it is very low, as an example.

Humus is composed primarily of hydrogen and carbon-hydrogen and oxygen with varying percents of nitrogen attached. The humus molecule is a long chain-like affair. It has the capacity to bind plant nutrients to itself—phosphate, nitrate, for instance. It can coat itself with some amino acids, certain sugars and a host of trace minerals, becoming a veritable sponge for nutrients. In terms of nitrogen, it can't compete with the factory fertilizer in the numbers game, since humic acid does not go beyond 6% nitrogen on a dry matter basis. Of that 6%, between 20 and 50% is in the form of amino acids, and 1 to 10% is in the form of amino sugar nitrogen.

Two sub-classes of microorganisms figure in compost making. The first is the breakdown process. Here parent materials are turned into phenols and other simple organic molecules. The second process is the buildup—taking phenols and other factors into humic acids. To achieve optimal results, a carbon ratio of around 20 is ideal. If the carbon-nitrogen ratio is not right, efficiency of the conversion of carbon and nitrogen to humus is reduced. If the right organisms are not present in the compost pile, nitrogen may well be lost in the form of ammonia or nitrogen gas. Carbon might escape in the form of carbon dioxide.

Ehrenfreid Pfeiffer once published a small manual on composting which lists parent materials for composting, and their possible combinations for good results. It is kept in print and available by Pfeiffer Foundation, Spring Valley, New York.

Modern science has given the farmer compost starter preparations so superb they put compost making on par with beer production and bread making. There is a difference. The brewer is dealing with only one yeast culture, one that is never dry. In compost starters, the bacterial cultures are many, and they are kept growing on agar plates, then combined and propagated in quantity. This involves inoculation, larger and larger production, drying and shipping.

4. starters

Starters are basically difficult to dissolve powders. Placed in water, they hardly break surface tension. It is usually easiest to shake water and powder together in a capped milk bottle before the starter fluid can be diluted further. Introduction to the compost pile requires no more than a garden watering can on the individual farm operation. Spray equipment has been designed for use when long windrows and automated turning machines turn this simple farm procedure into a mass production industrial process.

Barnyard manure can be broken sufficiently on the farm with the use of an in-place manure spreader, taking power off a regular tractor PTO. A typical farm highloader can scoop the material into the spreader bed.

Compost making has only recently emerged from the craft stage employed in backyard gardens. Nevertheless research runs deep. First to surface was the fact that raw manure lost 50 to 75% of its nitrogen in storage and application, according to USDA. Iowa State once turned up the figures of 65% nitrogen, 75% phosphorus and 49% potassium lost when manure is spread direct to farm acres.

And yet the nutrients are there. A 900 pound steer will produce 10 tons of manure in a year, computing these things as averages. Such a manure tonnage contains 140 pounds of nitrogen, 90 pounds of phosphate and 110 pounds of potassium. The figures are just as impressive in a hog operation. A composite 150 pound marketweight hog will produce 2.2 tons (wet weight) of manure a year. This translates to 22 pounds nitrogen, 13.9 pounds of phosphate and 20 pounds potash. The dairy animal serves up similar statistics; 142 pounds nitrogen, 71 pounds phosphate and 142 pounds potash. All this sounds quite impressive. But without a good composting process, approximately half of these nutrients will be lost to crop production.

5. the u.s. manure crop

Acres U.S.A. can estimate the U.S. manure crop as valued somewhere between $8 and $9 billion per annum (figures that may appear foolishly low as inflation continues to soar). Unfortunately much of this value is wasted, just as factory acidulated phosphorus application is largely wasted in terms of the usual 11 to 12% plant uptake.

It stands to reason that if untreated raw manure loses 65 to 75% of its nitrogen in storage and handling, then only about one-seventh of the manure is being utilized fully. Put in another way, it takes only a single ton of well-made compost to be worth as much to the soil system and the crop as 7 tons of raw manure handled in the usual farm way.

Research has not only provided these figures, it has also told farmers what they can do about it. Fletcher Sims of Canyon, Texas, more than any other individual, took the art of composting out of the backyard and put it into a practical feedlot and farm arena. Extension, USDA, land grant universities and farm hostility confronted this transition all the way. As a consequence, most of what composters have had to say has been bypassed.

As a result the farmer has not been thinking of humus as a substitute for factory fertilizers, or about other benefits—aeration of the soil, relief from compaction, moisture retention, or as a sponge for nutrients. Yet it is a fact that humus itself stimulates growth. It changes the carbohydrate metabolism of the plant to increase the amount of sugar production, which in many plants is related to wilting. When there is a large concentration of sugar, there is less likelihood of wilting.

Chlorophyll synthesis is increased. Increased nutrient uptake is dependent on the humus supply in the soil to a marked degree. Some researchers have reported humus as a factor in preventing chlorosis, which stunts plant growth. Chlorosis, as we have seen, is when a plant can't mobilize iron from the root into the leaf.

The benefits of good compost can only be touched upon in this primer. Since compost is ideally humus plus microorganisms plus inorganic elements held in that catchpen, it has the added benefit of having a number of organisms that can fix nitrogen from the air. This means the farmer does not have to apply as much nitrogen in the field per year. As a rule only 30% of the nitrogen applied as fertilizer is actually utilized by plants. In the case of anhydrous, the percentage is probably half, depending on whether the soil is clay or loam or sandy. Not much of the phosphate—probably as little as 10 to 11%, is utilized. Good compost has microorganisms continually fixing more nitrogen as needed and delivering it to plants. Up to 120 pounds of nitrogen can be fixed per acre per year under ideal eco-conditions.

Finished compost acts as a culture media. It can inoculate the soil for efficient decay of crop residue, conversion of nitrogen as well as minerals in the soil, or it can inoculate the next manure pile for compost making. New starter is best, but failing that finished compost isn't the worst starter in the world.

Wet cow platter would probably run 80 to 90% moisture content. The best moisture consistency for feedlot manure composting is 40%. If there is dust, the manure is too dry. As an interview entry—once upon a time—*Acres U.S.A.* and Fletcher Sims covered some of the techniques in windrow compost making.

6. fletcher sims

Acres U.S.A. When do you introduce the starter?

Sims. Well, the starter is more than an inoculum. It consists of different preparations which include chelated trace elements. It would be more accurate to call them a bio-catalyst. There are some things that have to do with the effective functioning of microorganisms. And there are numerous microorganisms involved—some are fungi, and so on.

Acres U.S.A. You get this starter in powdered form?

Sims. Yes, I get this as a powdered unit. It takes an ounce to treat a ton of raw material. I calculate the ratio on the finished product weight. If we have 30% moisture content, we'll lose 20% of the moisture. When we compute it as a 60 ton windrow, we take 60 ounces of this starter. It has, by this time, been activated 12 to 24 hours in advance. I put it in a mix-master and add moisture. The starter does not take moisture readily. I mix it until it is mealy, yet moist. I let it set overnight. Sometimes I get a pasty consistency, but I like it a little drier. There is an extreme amount of investment in one of these buckets of material. If something would happen so that I couldn't apply it the next day, then I could return the bacteria into an inactive state without losing my material and investment. If I allow the material to remain wet, an anaerobic condition will exist, and this will cause me to lose the starter.

Acres U.S.A. How do you introduce bacteria into the windrows?

Sims. First, I extend it in water. For a 60 ton windrow, I extend the bacteria in about 20 gallons of water. I then put it on top of my ricks where they are moist. If it's hot and dry, we play a hose over the ricks. Then we use a pump to apply the starter. We have a turning machine that saddles the windrow and mixes the material as thoroughly as a mixmaster. This machine can turn many tons an hour. Within 15 minutes after the starter is introduced into the pile, we turn the windrow to achieve a uniform mixture. This gives us optimum conditions.

Acres U.S.A. Do you have problems in cold weather?

Sims. Yes, but we have a few tricks. In real cold weather—freezing and below—we never let our ground cool off. We get a slow start sometimes, but eventually low temperature organisms will carry the temperature to where the thermophilic organisms begin to function. We like to achieve a 140 degree temperature. When we're working under a close cycle, we check these piles constantly.

Acres U.S.A. Suppose temperatures climb above 140 degrees?

Sims. We don't have this happen. But we would turn the pile in such a case. If this hot condition persisted, we would probably scatter the material. It takes a certain mass to maintain a temperature, and that is the reason a lot of small, backyard composters never see the heat. They don't have enough mass to allow this to develop. You can get carried away with

the mechanical aspect of composting. Most of the things I have
read about composting have been motivated by the idea that it takes a
gadget to make compost. It doesn't. It takes control of moisture, mass and
bacteria.

Acres U.S.A. How long does it take a typical windrow to become a
finished product?

Sims. Normally, about 30 days. We can shorten that time. We do turn the
piles periodically, or we'd get an anaerobic condition. We try not to turn
anything in less than seven days. There are certain organisms—those that
build colonies—the streptomycetes, the actinomycetes—they have to have
time to grow colonies. Unless they are given this time, they won't function.
If you keep beating them up, they won't function. That's what's wrong
with these gadgets that crank something through in a few hours with
forced air. They mineralize the material in a hurry, and they really don't
allow the humus builders to work.

Acres U.S.A. At the end of 30 days, you simply inspect the finished
product and ship it out?

Sims. Yes, you get a feel and you know the appearance of the finished
product—and, not least, the smell. When you're turning, you can smell
certain yeasts in the air. You can sense the correctness of the situation.

Acres U.S.A. What do you look for?

Sims. A dark brown color. You can still see it was once manure. Some of
the fibers will not have broken down. But this compost is ready for the
soil. I was told by Rudolfs Ozolins that I compost too much, that I ought to
stop sooner. I have been encouraged to introduce the starter, get the compost going, and then turn it over to the soil to finish.

Acres U.S.A. Will plants be able to feed off the protein, hair, and
materials such as that?

Sims. Yes. These proteinaceous materials under controlled fermentation
are turned into polypeptides. The polypeptides are of shorter duration than
the polysaccharides. The polysaccharides are these large, elongated complex sugar molecules that bind the soil. They are much more resistant to
decomposition, but this will go on in the soil anyway. When you introduce
raw manure into the soil—if there is proper moisture, aeration, and organisms present, the elements necessary—calcium, even down to trace
minerals—why some of this can go on. That is why it is possible to get
good results from raw manure if there is time.

Acres U.S.A. What happens when manure is introduced into soil that has
been badly handled—abused with atrazine, 2,4-D, 2,4,5-T, and has little or
no biotic life?

Sims. I have a picture that reveals exactly what will happen in many
cases. In the picture, the man put 15 tons of raw manure to the acres. Application is never uniform. They'll pull onto a field, open the tail gate, turn
on the feeders, and so you have a big burst of manure to start, and then a

slower payoff. Every once in a while a heavy pile falls out. These heavy piles overload the soil. I think that more of this manure mineralizes than is commonly suspected. There are fecal organisms that will do some work. And there are infectious organisms. Even an abused soil has some microorganisms, and under proper conditions they will proliferate. I have seen raw manure dug up in soil with bright organisms I never see in a compost pile. I don't know what they are. It could be determined, of course. The trick is to stabilize plant residue into humus so that polysaccharides are pulled out of the carbonaceous material instead of going back into the atmosphere as carbon dioxide, and polysaccharides aggregate the soil and make it easier to plow and hold water more readily. Proper conditions permit polypeptides to be drawn out of the protein instead of escaping as nitrogen.

Acres U.S.A. What does your compost do for a farmer, specifically? Do you stress N, P and K?

Sims. I'm often asked by farmers, What's the analysis? I say, A billion bacteria per gram, and 10% of those are the actinomycetes and the streptomycetes. These are the humus-building organisms. They are the ones chemical companies make antibiotics with. I'm just putting these farmers on. I know what they want—N, P and K. We run this periodically. We want as much nitrogen and phosphorus as possible. This is phosphatic soil. Much of the phosphorus isn't available and of course with low organic material content in the soil they need as much nitrogen as they can get. Ours runs around 3% nitrogen, 3% phosphorus, and a little higher in potash. The soil here rarely needs potash.

Acres U.S.A. According to the Brookside approach, you don't need a high percentage of P and K?

Sims. No. Calcium is the major nutrient, magnesium second, and if you have a good humus content in the soil, the system can and will fix its own nitrogen supply. We found in one field that Brookside had audited that we made a 6% increase in available calcium on the soil colloid, and we decreased excessive magnesium 7%, and decreased the amount of sodium. We made seven significant improvements in that soil with compost as measured by Brookside soil audits.

7. pre-digested manure

One ton of properly pre-digested manure can replace the function of up to 20 tons of raw manure when applied to imbalanced soils. In addition to the handling and storage problems associated with hauling raw manure direct-to-the-field, disease organisms, fly larvae and pupae are also scattered. These limitations are corrected when manure is inoculated, aerated and composted.

Once in the soil, microorganisms in compost go to work. They eat and digest dead root ends and plant trash. Friendly fungi are stimulated, gaining the edge over unfriendly fungi, nematodes, eelworms, cutworms and aphids.

Calcium, magnesium, sulfur, calcium-phosphate, nitrogen and trace elements can be safely incorporated into manure digestion systems, either to improve the finished product pragmatically, or to answer requirements indicated by soil audit.

Fertilizers made from dried sludge are dangerous. In addition to the fact that they may contain dangerous levels of lead, cadmium, chromium and mercury, they revert to being sludge when moistened and introduced into a soil system.

Listen to Advanced Ag-West, Boise, Idaho. "Soils low in organic matter and humus, or with poor tilth and structure, cannot be rejuvenated with chemical fertilizers. Predigested organic residues can quickly start improvement of such soils since they are immediately ready to function regardless of general soil conditions.

"Humus is the main source of fuel and energy to the soil microbial system. Each ton of manure carries 500 pounds or more of organic matter—and 150 pounds or more of net humus. One ton of compost can support up to 600 pounds of microorganisms per acre. This can be multiplied by the number of separate life cycles in each season.

"Humus carries 50% carbon which is used by microorganisms through fermentation and respiration to convert part of the carbon into higher energy values. Other carbon may also be spent in the oxidation of mineral compounds into simpler and more available forms.

"Humus as a concentrate of carbon and energy compounds aids bacteria to survive cold or dry soil conditions or excess water, enables bacteria to carry out antibiotic effects in the soil, aids bacteria in interrelations with plant roots, which is important to a cold, wet or delayed spring seedbed."

8. pfeiffer's summary

Humus adsorbs the highly active non-nutrient hydrogen to make the humus an acid, basic to the release and conversion of complex soil minerals as well as varied nitrogen sources.

The humus colloid carries three times more adsorptive capacity to hold mineral cations, water, air and carbon compounds.

Density of humus is but 0.8 to 0.9 whereas the inorganic clay colloid has a density of 2.65.

The humus colloid also buffers microbial and plant systems against excesses of sodium, magnesium, potash and other positively charged minerals.

Humus acts to buffer and enhance the four anions: carbon, nitrogen, phosphorus and sulfur in their system of nutritional service to crops.

When associated with organic matter and humus they become dynamic nutrients for both soil microbes and plants.

The humic acids, carbonic acids and carbon energy are activated either through the association of properly digested organic residues or from proper decay of field residues as influenced by soil pH, water and air potential. This requires pH modification of the soil.

The carbon-energy system has far greater productive potential from the molecular energy contained than can be furnished by large amounts of chemical fertilizers.

The use of large amounts of fertilizer and nitrogen stimulates a shock-effect on soil microbes and "fires up" and compels them to consume extra amounts of carbon most rapidly in self-defense. This is explosive consumption of the stored carbon energy—a waste of soil productive capacity.

Another source of biodynamic soil organic matter is made up of cells sloughing off growing roots where soil microflora are most numerous and produce a carbonic respiratory acid which is gently active on the complex inorganic reserve minerals as a new source of essential ionic elements. A healthy growing plant will produce great quantities of this carbon energy during the growing season. This is of prime importance in a well managed soil and crop producing system and as it becomes initiated, lesser amounts of purchased nutrients will be required.

Through pH management, tillage and practices to improve soil tilth and water capacity, and the use of properly pre-digested manures, profitable crop production can be obtained. This is especially important on all field-area limitations low in humus.

Laboratory data seems to indicate that composting is desirable and economical. A number of years experience on farms tend to verify that it is both practical and profitable. Wherever possible, pre-blending the soil pH modifiers with manures can reduce spreading expense and more efficient use of materials can be appreciated.

9. humates

Humates are simply salts of humic acids. This means that ions like potassium (K^+) or magnesium (Mg^{++}) have neutralized the humic acids. In order to have a salt you have to have an acid plus a positively charged nutrient. Remember muriate of potash discussed on page 214. Muriate of potash is a salt. One atom of plus charged potassium is combined with one atom of the acid element chlorine. Dissolved in the soil solution, chlorine is freed to leach from the soil. The potassium atom seeks to combine with the negatively charged clay or humus colloid. Adsorbed to the soil's colloidal system, potassium is now in position to produce a plant's potassium requirements.

The humate salt contains potassium, chiefly. It might contain other elements. It takes the right kind of humic acid in the right soil to do the right job. Not all humates benefit the farmer. This makes incorporation of humate products a problem. After all, it is possible to be misled by a salesman. A farmer simply has to become his own expert.

10. update

The compost story is not likely to be finished in terms of refinement very soon. The lessons Fletcher Sims made a matter of record in his *Fletcher Sims' Compost* stand, a solid fixture in compost science, nevertheless, conditions and available materials always require new sorting out. For instance, Malcolm Beck of San Antonio uses a stockpile method, more time, and the inoculum available in paunch manure for the compost chore. He has experimented with city sludge and pulverized brush (as a carbon source) and has had a hand in the promotion of dozens of compost operations based on the stockpile system. Indeed, compost-making has come a long way since Sir Albert Howard first turned out health-saving compost at Indore, India.

Looping city wastes — I prefer to call them *vivo* (life) back to the countryside remains a development devoutly to be wished. Fortunately the heavy metal bugaboo of yesteryear is vanishing — both because sewer dumping is under better control and because composting *vivo* solids is on the runway. Cities like Milwaukee, Los Angeles and Austin, Texas have led the way.

LESSON 20

Tillage

Just as human nutrition has to start with the soil, so do all the problems that attend farm crop production. In the coffee shops farmers like to talk about drought, about too much water, a pastime known as the blame the weather syndrome. Fertility, tilth, soil physics, all would be better topics on a rain-soaked day, or during a hot-dry spell. Farmers probably would turn to such topics if someone led the way.

1. why, indeed, plow?

Whenever these topics come up at an *Acres U.S.A.* Conference, the crowds come forward. They literally want to know—step by step—what a good eco-farmer should do. Should he use mechanical force to break compacted soil? Should he try no-till farming (with or without chemicals)? What about deep rooting rotations?

Some years ago Edward H. Faulkner wrote a startling treatise, *Plowman's Folly.* Why, indeed, plow? Faulkner could find no real scientific reason for the practice that had become universal. The seedbed—let's say rootbed—didn't require it in nature. And it was sheer nonsense to rely on the moldboard plow to eradicate weeds. Why, indeed?

To all this William A. Albrecht answered, "We need to plow less on some soils. We need to plow more and deeper on others. We need to learn that the differences in degree of soil development according to climatic differences are factors in determining how important the plow is."

Your soil, your equipment, your crop projections all have to figure in what kind of rootbed you hope to prepare. It has become the judgment of eco-agriculture that there is very little room in farming for the moldboard plow. The *Journal of Agricultural Science* once ran a series of articles which traced the development and design of the plow and other tillage tools from the moment of the Norman conquest to date, but no real scientific rationale was presented on why the moldboard was used in the first place.

It came from England and was superimposed on America, a nation with a whole new set of operating principles, weather conditions and soil types. The plow violates a basic requirement for efficient production. It disregards the contributing value of retained organic matter in the soil.

Some seed can go deep in rough soil, such as corn. Alfalfa, clover and timothy seeds are so tiny they need a finely textured bed and hardly a quarter to a half inch of coverage. Time is important. No farmer has time to waste, and therefore the rootbed should be prepared with the least amount of work. The job doesn't start in spring. Likely as not plans will have to have been made months in advance. How each crop fits in a rotation must be worked out on paper. If a general crop plan will not work on paper, it won't work in fact.

2. tillage

But tillage doesn't take place on paper. It takes place in the field, both in fall and in spring, and the two are as the same side of the same coin. The fall tillage package has objectives all its own. It must, first, tear up any compaction barrier that might have developed because of wheel and machine traffic. At the same time, fall tillage must mix trash with air and soil so that pipettes can reach up to trap water and take it below the surface of the soil, thereby preventing erosion. Residues should be mixed in the top five or six inches of soil.

Fall tillage should leave the soil in a ridged condition. These molehill style mountains will prevent wind and water erosion, help heat retention and at the same time set the stage for complete decay of organic matter well before winter closes down the microbial shop. Heavy soils call for a 10 inch high ridge with a wide base. Deep working points on chisel equipment should be positioned up front so as to run both under and between ridges.

The best farmers we've met believe timely fall management that includes cover crops, proper ridging and residue management is the single most important operation. Fall tillage, cover crops and adequate calcium and sulfur join with soil nitrogen to stimulate the residue decay process. The biggest essential here is to hold to a minimum fall trips over the field and at the same time put maximum emphasis on fall tillage management.

3. the soil wakes up in spring

The soil wakes up in spring. If the farmer's work has gone well, trash should be largely gone. And if the soil system has been managed to achieve the optimum, very little nitrogen should be required. Unfortunately, this is not the situation on most of the nation's acres.

The soils are frequently complexed, and instructions that attend fertilizer applications in tandem with tillage are confused. As the several lessons in this primer have already unveiled, nitrogen is a big problem in a system that has no clear cut nitrogen cycle working.

There will be changes for American agriculture. It is not likely that fossil fuel technology can be continued much past 2000, certainly not past the escalated price of natural gas. This means sewage wastes and all manures will have to be looped back to farm acres as a life, nitrogen and mineral nutrient source. In the meantime nitrogen application on organic matter deficient soils remains a must.

4. nitrogen application

The rates of nitrogen application will vary according to soil types, organic matter and humus levels, methods of application, rates and timing of compost applications, cover crops, calcium, sulfur and other nutrient levels uncovered by soil audit, tillage programs and rotation.

In operations that have not abandoned the idea of making a profit, the broadcast system is not a good choice, taking conventional fertilizers for what they are—inefficiency plus. There are always compacted areas accounted for by heavy machinery. Roots will not enter these areas, and therefore factory nitrogen trapped there is wasted. Moreover, nitrogen is often complexed and locked up in the soil organic and decay system. When you're buying nitrogen by the bag, more than enough is amateurism. Roots will use only a minute part of the soil volume.

Roots cannot tolerate direct contact with the harsh factory fertilizers in any case. This is what Albrecht meant when he said that "Fertilizer placement is the art of putting the salts in the ground so the plant roots can dodge it." This is also what he meant when he said "Manure forms an organic shield around the salts. It is a buffer against salt injury. As soils become lower in organic matter we will not be able to use salts so directly."

In order not to use the salts "so directly," agriculture has developed techniques—banding, sidedressing, ridge planting. Thus the technique called surface placed nitrogen near corn at the time of cultivation. Side saddle tanks and rear-mounted cultivators work to accomplish this chore. Hilled corn will grow two or more extra sets of roots. This means that drop nozzles place nitrogen from three to five inches from the row and cultivation covers the nitrogen while hilling the corn. This has proved efficient in terms of fossil fuel efficiency because nitrogen is concentrated and seeps into the buffered area containing the most feeder roots.

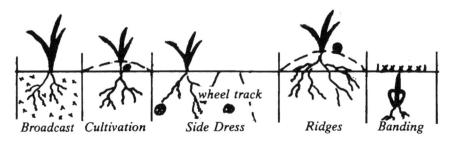

Broadcast   Cultivation   Side Dress   Ridges   Banding

Sidedressing is also efficient—as fossil fuel mining efficiency goes—if knives are kept out of compaction areas, such as wheel tracks.

Ridge planting works best by placing factory nitrogen slightly to one side of the ridge with rolling cultivator covering and adding to the hill.

Banding means applying actual nitrogen directly over a row in a foot-wide band at the time of planting. This requires a saddle tank.

As natural gas prices reach the $4.00 per 1,000 cubic foot level, even the most dedicated "synthetic" farmer will have to do a double-take on this kind of technology. As he does so, his tillage program will have to change.

Even without economics ruling the roost, one of the problems of nitrogen application is the physiology of the plant itself. Using oil company fertilizers, the farmer generally offers a whole meal for the growing season as a start and never returns for a pick-me-up. Nitrogen leaches from the soil. Even urea—which is as good a preplant form as synthetic science has to offer, and releases slowly, and resists wet weather leaching—can't sustain a crop the entire season. Certainly light textured soils profit from split nitrogen application.

So let's backtrack a moment and recap what we're trying to do with a sound tillage program. We've mentioned breaking that plowpan or compaction barrier.

5. trash showing

We've also mentioned incorporating crop residue into the soil, leaving enough trash showing to provide for water insoak, heat retention, and ultimate capillary return, and control of water and wind erosion. Extension says seed germination and seedling growth is slower on mulchtilled fields, but this is true only if herbicides are incorporated on the pass-through, and if decay is not going in the right direction. Good structure and drainage

PRINCIPLES AND GUIDELINES

Since the advent of herbicides, farmers have been conditioned to believe that cultivation is an unnecessary operation. This seemed to be a logical conclusion since cultivation was used primarily as a weed control operation. In many soil types and conditions, proper cultivation can contribute substantially to greater yields. Improper cultivation under similar conditions can also reduce yields. Soil types and conditions vary greatly. Limitation of time and size of operations all enter into the decision making process. Here are some guidelines for your study and consideration.

1. The roots of row crops like natural soil firmness. Roots will not enter a wheel track or any other compaction area. These areas lack sufficient oxygen to permit root extension.

2. Deep cultivation destroys roots. You cannot afford to disturb the roots of any annual plant. Deep cultivation smears the soil colloids, causes excessive drying and creates large air pockets. Roots will not enter these areas until the soil firms up after considerable rainfall. For this reason a rolling cultivator or flat shallow sweeps must be recommended. A roving cultivator puts finer soil around the base of the plant.

3. Soils must breathe. Changes in barometric pressures and the alternating heating and cooling of soil and air causes air to enter or exit the soil. This process is essential to getting carbon dioxide out of the soil and oxygen in for root growth, uptake of nutrients and biological activity. When the soil is short of oxygen, yields suffer.

4. The surface layer, 1 to 2 inches deep, is the area that limits soil breathing. Even though a tight soil has adequate moisture, a rainfall can stimulate plant growth because rain absorbs oxygen and carries it into the soil. This fresh load of oxygen permits roots to immediately increase the uptake of nutrients. Corn standing in flood waters will start to wilt as soon as the oxygen supply in the water runs out. The water or nutrients can't be taken in if the roots can't get oxygen.

5. A tight surface permits the capillary water column to reach the surface and evaporate thus creating excessive loss of soil moisture. A loose mulch decreases evaporation and reduces weed seed germination.

6. Hilling corn or soybeans will stimulate the covered nodes to shoot out two or more sets of roots thus increasing the total root volume. This also permits new roots to bypass any plugged basal nodes.

7. Herbicides have side effects. It has been shown that some herbicides retard the soil bacterial systems. They will also reduce yields and may add toxins to the food chain.

8. Minimize the use of all toxic chemicals.

9. Nitrogen can be applied at time of cultivation whenever necessary. This is the most efficient nitrogen application system when the natural nitrogen cycle is not working at optimum level.

ROW SUPPORT

There is a lot of difference in seed vitality as reflected by the vigor and survival of the seedlings during the early growth period. Two problem areas present themselves.

The nutrient package within the seed is not consistent and will vary greatly depending on the conditions under which it has grown and matured. Seedling vitality and gain or loss of early growing time can be greatly influenced by the quality of the seed.

Soil conditions can be very harsh at or after planting. Cool, damp soils reduce the speed of chemical reactions within the developing embryo and young seedling. Cold soils suppress biological functions and the subsequent release of plant nutrients such as nitrogen, phosphorus, sulfur and key trace elements.

Because of these facts, the row support concept has been developed. Row support uses a combination of essential secondary and micro-nutrients applied directly in the row with the seed. Row support materials are often applied directly on the seed in the drillbox when small grains are seeded. Sometimes— as in planting legumes—beans, peas, peanuts—row support materials are incorporated as a water-based nutrient spray. The same procedure is valid in planting nonlegume row crops— corn, beets, potatoes, milo, etc.

MANURE MANAGEMENT

Manure may be either an asset or a liability, depending on management. The following principles and guidelines must be considered in manure management.

1. Soil microbes use many of the same nutrients that plants use. The decay process must be completed so there is a high level of soil nutrient availability when crop demands are high. Soil microbes and the growing crop must not dine at the same table, or yield will suffer.

2. Composted manures may be used anywhere at anytime. When decay is complete, nutrients are available and competition is eliminated.

3. Light applications of manure over more acres is superior to heavy applications on fewer acres.

4. Spread manure and other soil amendments needed prior to fall tillage. Broadcast rye cover crop ahead of tillage. Use a fall tillage system which thoroughly incorporates residues and positions soil in ridges. Additional manure may be spread down ridges during winter months. If properly managed, this system will greatly increase the rates of manure use. The living rye plant will absorb and hold nutrients, convert and store sun's energy, loosen and aerate the soil and stimulate rapid decay of rye, manure and residues when properly incorporated the following spring.

5. Consider topdressing manure after corn is planted and before it leaves the spike stage. Nutrients from manure will seep into the soil. The manure will decay on the surface and be stirred in during cultivation. Decay on or near the surface won't compete with the crop as much as it would if worked in throughout the root zone.

6. Have a laboratory analysis run on manure. Estimate tonnage per acre and you can then more intelligently plan your soil fertility budget. You can reduce nitrogen and other elements by properly managing manure decay and knowing its analysis.

7. Understand these principles and formulate the ideal plan, then strive to come as close to the ideal as your situation will permit. Remember, you are dealing with either an asset or a liability depending on how you manage it.—*Don Schriefer.*

enhance heat absorption. Heat absorption keeps decay on the right track and moving ahead.

It is not difficult to measure the crop residues being turned back into the soil. Residues collected from a 1 square yard plot selected at random must first be weighed. Taking two or three such samples provides a handy control over the accuracy of the procedure. In fact, three samples can be weighed, then divided by 3. The average weight should then be multiplied by 100. This will give the pounds of residue per acre.

Using a guesstimate system, it is possible to determine crop residues from the production yield. On a pound for pound basis, 1 bushel of corn will leave behind 1 pound of residue; 100 bushels an acre will therefore provide 5,600 pounds of carbon trash. For each bushel of small grains, there are approximately 100 pounds of residue. This means a 30 bushel wheat crop equals 3,000 pounds of residue. Soybean residue has a range of from 1,500 to 2,500 pounds an acre. Hay crops can be computed on a pound for pound basis. In terms of hay 3,000 pounds an acre provides excellent erosion control.

The soil should always be warehoused in a ridged condition over the winter. If the soils are heavy, the ridges should be 10 inches high with as wide a base as possible.

After working with tillage systems half a lifetime, Don Schriefer (Advanced Ag—U.S.A. and Canada) has concluded that penetration, incorporation of trash residue and ridging set the stage for improved insoak of water through both topsoil and subsoil. Residues on top and protruding from the surface will increase water uptake by two to three times over a flat, bare sealed-over surface. By preparing properly for deep water penetration we also guarantee deeper rooting, as well as improved capillary water movement back towards the surface. Both are very essential. We can appreciate the value of a deep root system for expanding both the water and the nutrient reservoir for increased crop yields and drought protection. This procedure will absolutely minimize runoff, erosion and the silting of our waterways and reservoirs. Preparations will have been made for the soil decay systems to begin their important function of decaying crop residues. Soil organisms require warmth, oxygen, a balance of nutrients and capillary moisture to function. They will not function in a cold, dense, waterlogged soil. The decay process is most active at or near the surface where oxygen and the other conditions are more likely to prevail. The ridge and deep tillage guard against waterlogging requiring fewer BTU's to raise the soil temperature. The ridge will also maintain a better air exchange. In summary, the ridge sets the stage for the decay system to run longer in the fall and begin earlier in spring.

6. decay systems

Great emphasis is placed on decay systems for three basic reasons.

1. Decay of residues by soil microorganisms is an important function. Decay activity and body exudates will form soil particles into a crumblike structure. This granular structure assures soil ventilation and builds a natural seedbed quality into the soil. One can recognize the benefits of a biologically built spring seedbed versus one built from cold, dense soils with tillage tools.

2. The decay of residue releases plant nutrients and many unknown factors in organically complexed forms that are absolutely essential to plant life. The decay process releases several organic acids that further etch out plant nutrients from soil minerals. These are nutrients a farmer does not have to buy. The decay process also releases other plant essentials that can't be bagged and sold. So-called pesticide carryover in fact is a cause and a result. It causes decay failure and it is the result of decay failure. A good decay system will go a long way toward cleansing soils of pesticides even if misguided agronomy advice prompts their use.

3. Finally, soil organisms consume and tie up in their body functions many of the same nutrients that plants require. When decay is going at a high rate, nitrogen, phosphorus, sulfur, potassium, calcium, etc., will be at a lower level in the soil. But when the decay process is completed, organisms die back and release these nutrients. Plants and microorganisms—as explained earlier—cannot eat at the same table. If they do, yields suffer. Decay must be managed by the farmer, or it will manage him.

This is why fall tillage sets the stage for next year's yields. Indeed, fall tillage must be considered even before harvest and rotation. It must follow the harvest of each crop immediately.

There is no perfect single tool to accomplish these goals for every farm. Soil conditions vary. But one thing is clear. The moldboard plow must be phased out. It buries residue but does not meet any of the three basic goals contained in the outline submitted above. Several tools are now being used by eco-farmers: the offset disc, certain tool bar arrangements, modified chisel plows, or tools specifically designed to meet the three basic requirements. The Rip 'n Ridger is merely a forerunner of revolutionary equipment that will appear on the market in the next several years.

7. a spring phase

There is also a spring phase to seedbed and rootbed preparation. Once the soil has been stored over the winter in a ridged position, warming rays of early spring heat it rapidly. This creates a biological stimulus and a mellowing out. Several options for leveling ridges now present themselves. The main thing is not to work below the base of the ridge so as to bring up cold, wet soil that might bake out into clods. A disc, field cultivator, or

COVER CROP PRINCIPLES

The total value of a cover crop goes far beyond the prevention of erosion. The root system gathers in available plant nutrients and through the process of photosynthesis converts these nutrients into living organic compounds. By this process and the utilization of carbon, hydrogen and oxygen the plant is able to convert about 5 pounds of soil nutrients into 100 pounds of organic material. Only about 5% of a plant's total weight comes from soil minerals. This tremendous gain of organic matter and its subsequent incorporation back into the soil is important for the soil building processes and good production.

Being high in protein, carbohydrates and other nutrients, the green plant is the ideal food for soil microbes and rapidly stimulates the decay process. Early decomposition and release of nutrients from crop residues must take place before the grain filling process begins. Soil microbes and growing crops require many of the same nutrients. Therefore, a crop will not produce as well if it has to compete with the soil organisms for nutrients during its period of peak demand. The early spring decay stimulus from green manure is important in preventing this competition.

The fiery stimulus of early green decay not only hastens the decay of other accumulated crop residues, but also hurries the breakdown of carryover herbicides that often plague the succeeding crop. Good decay is an excellent soil cleansing process.

Therefore, the green cover crop serves many functions:

1. It serves as a bucket to hold nutrients in a non-leachable form.

2. It stimulates early decay and the subsequent release of nutrients before peak crop needs.

3. Its roots and tops protect the surface from erosion and loosen the soil for better water insoak and aeration.

4. Its rapid decomposition releases organically complexed nutrients, antibiotics and many unknown factors all of which are essential plant needs.

5. It has the potential of turning every acre into a compost heap.

Couple the cover crop with ridging and you will have a good thing going for you. Ridges help in the early management of crops such as rye. Rye is one of the best cover crops; however it must be recognized that it could turn into a liability if it gets away on you in the spring. To prevent this we recommend it be seeded at 1 to 1.5 bushels per acre ahead of ridging and only on well drained soils. A cover crop, as any other good practice, must be a part of a total system and not just an unrelated single-shot input.—*Don Schriefer.*

various tool bar arrangements can be used. The dangers of over water-logging, slabbing and clodding have thus been greatly reduced. Farmers who entertain these ideas will find it more profitable to start spring tillage in a pre-formed seedbed than to begin in flat, dense, cold and often water-logged soil.

MULCHING

The use of mulches to reduce water evaporation is an ancient and respected gardening and farming practice. It may, however, prevent ready entrance of rain water into the soil. Mulches composed of straw, leaves, sawdust and various organic materials carry with them the assumption that soils under them have a finely-tuned fertility level. Otherwise, introduction of these materials might set up microbial competition with crops for nutrients in the soil's supply, according to William A. Albrecht. "It is around this simple principle that the wisdom of the use of the mulch turns for failure or success." Mulching by itself cannot make up for the shortfall of fertility in the soil.

LESSON 21

Rotations

A lot of farming nowadays is little more than mining. Farmers who follow the advice of Extension, USDA and the land grant college ought to get depletion allowances, just as do copper miners and oil companies. Even with fertilization and conservation practices, most farmers take and take, and seldom return on balance with the taking. Crop rotation, at first glance, might seem little more than a scheme designed to buffer the taking and to stretch out the mining process.

1. sound reasons for rotations

This is something less than justice. There are sound reasons for rotations despite the fact that nature doesn't rotate. For a couple of hundred pages we've been taking the pieces apart. Rotation puts them all together again. It takes into consideration the biotic geography of the farm, its cropland potential, its natural pasture lands, its woods—the terrain, whether rolling, steep, shallow, wet or rocky. Possibly half the acres in the United States are suited only for pasture. Many can't even be harvested if animals are not maintained on-scene to handle the chore. Obviously, there is no point at all in going down the tube trying to make corn or soybean land out of acres best suited for pasture.

The name of the game is to use the soil intensively, yet preserve and maintain it, and even improve it year after year while at the same time taking bins and bushels with excellent nutrient loads. Rotation says—other items being duly accounted for—that a balance among row crops, small grains, hay and meadows best accomplishes this purpose. Continuous row crops, corn on corn on corn, or soybeans on soybeans on soybeans, for instance, put the greatest strain on the soil system, its humus supply, its fertility load, balance, tilth and structure.

This strain is lessened when a row crop for two or three years is followed by a small grain crop for even one year, or a green manure crop such as a legume.

2. the pecking order

Next in the pecking order, if ultimate intensity is to be relieved, is the growing of continuous small grains. None of these options are valid for most rainbelt acres, and they present the same general problem in irrigated territories or on dryland acres as well.

A row crop for two or three years, a small grain for a year, and hay for one or two years is considered practical by most midwest farmers, with variations changing from county to county, state to state. This rotation relieves the intensity of strain on the farm acre and sets the stage for management of improved organic matter, tilth, soil structure and conservation for future generations. It harmonizes the reality of soil use with the threat that a farming operation can become a mining operation.

Intensity of stress is lessened further when the rotation becomes row crop one year, small grain one year, hay one year. When the hay crop is stretched to two or more years, man's approximation of nature's cycle closes the circle still further. Semipermanent hay or pasture is easy on the land, but permanent pasture is even easier, and for this reason some soils can be used for pasture when they would probably not prove suitable for other crops.

3. a neat map

In the mind's eye, scratchpad doodling usually serves up a farm as a neat map, each rotation approximately even.

The real world is something else. Most farms have odd shaped fields, acres divided by creeks, hill land, bottom land, areas that naturally divide themselves into odd shapes and sizes. These things make it difficult to maintain a good balance between hay, pasture, row crops, small grain, especially when ground cover and legumes for soil conservation are made part of the plan. Traffic to and from fields must be considered, especially when livestock must rotate with the crops.

35 acres oats seeded with pasture mix	35 acres corn	35 acres corn

		4 acres	7 acres
35 acres pasture alfalfa, sweet clover, alsike		3 acres orchard	6 acres home

40 acres cotton followed by vetch	40 acres corn followed by soybeans or cowpeas	40 acres oats followed by legume hay followed by vetch

A field management and rotation syustem suitable for southern farms needing large amounts of feed and using the maximum amount of legumes.

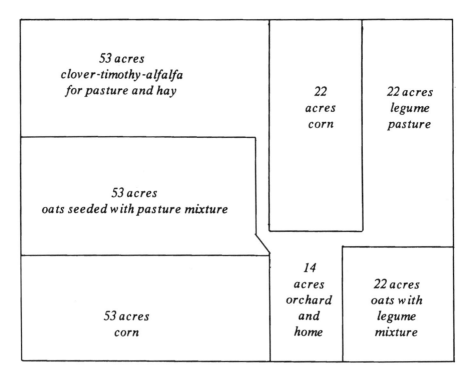

| 80 acres
hay and pasture
alfalfa | 160 acres oats seeded
with sweet clover and alfalfa | |
| 80 acres
pasture
sweet clover | 5 a
home | 140 acres
corn |

A well laid out farm for legumes and livestock in rotation; all fields enclosed with woven wire and convenient to the homestead.

53 acres clover-timothy-alfalfa for pasture and hay	22 acres corn	22 acres legume pasture
53 acres oats seeded with pasture mixture		
53 acres corn	14 acres orchard and home	22 acres oats with legume mixture

An ideal field arrangement for the rotation of crops and livestock. The two rotation systems provide for a greater variety of crops, with smaller fields.

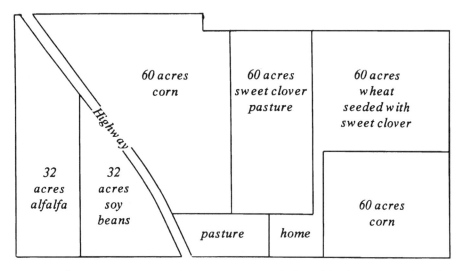

Here is a four year rotation problem well met. Half of the fifth field is in alfalfa for hay, the other half now in soybeans, but rotated year by year.

4. two rotations

As a matter of fact, the entire farm may not gear to rotation as a unit, and the necessity for two rotation systems on the same farm might become mandatory. There is some advantage to two sets of rotation, especially if terrain figures on the one hand, and livestock is to follow a legume pasture on the other hand. Not all farmers like confinement hog feeding, for instance. Hogs, sheep and poultry thrive better when out on clean fresh pasture.

There probably is no general answer to mapping a rotation. Only a few

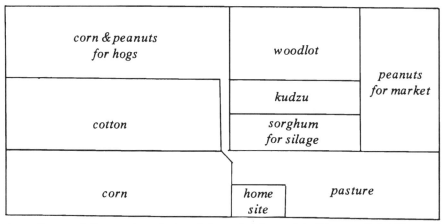

A southern rotation system designed to maintain and build up soil fertility and which also utilizes rough land to good advantage.

A four year field rotation system which answers the objection to fields too large in size and which also permits growing a larger variety of crops.

general principles belong in a primer on eco-agriculture.

Hillsides, even when contoured, probably belong in long-term hay or pasture. Fields with soils that permit a "wet" problem probably belong on a long rotation, just as well-drained fields belong on a short rotation. In addition to improving crop production, maintenance of tilth and fattening up the organic matter bank account are real reasons for being. Eco-farming principles, such as generous use of compost when available, have pushed a lot of rotation lessons out the window, albeit only when those composts are available and plentiful.

5. sod crops

Through the centuries, agronomists have learned that sod crops help restore organic matter and tilth, both of which are intensely stressed by row crops. Thus when organic matter and tilth can be built with legumes, sod crops, grasses, row crops can tap into the nutrient vault and turn in better performances. The idea of vegetables, tubers, corn, tobacco, following good sod crops suggests itself simply because of the high values vegetables and other row crops carry.

It isn't often mentioned in this day and age of herbicides, but crop rotations assist in weed control. The life cycles of weeds are varied and many. Annuals must start from seeds, a reality that makes rootbed cultivation for corn and beans cultivation for weed seeds as well. Annual weed seeds do not stand much of a chance in hay meadows that are well established. With planting they might be a problem, but once turf starts building a weed seed hardly stands a chance.

Pasture scourges are quackgrass, thistles, bindweed, horsenettle—perennials all. They not only grow with a hay crop, but sometimes crowd it out, the anatomy of which is discussed in the lesson on weeds. In any case, perennials come under control when a switch is made to row crop production.

6. ecological fine-tuning

Ecological fine-tuning of fertility loads should cancel out insect and fungal attacks. Unfortunately the theory sometimes falls out of step with hoped-for results. Plants may not achieve fully balanced hormone systems because fertility treatments become complexed because of considerations outside the scope of most soil tests. One great justification for a rotation is the changing of plant requirements. This may cancel out insect attacks by breaking cycles.

Sir E. John Russell, a former director of the Rothamsted Experimental Station, England, has enunciated a proposition that most American colleges reject, that some plants excrete substances harmful to roots of other plants. A good for instance is flax. Some varieties excrete prussic acid, HCN, and this protects roots from pathogenic fungi attack. Certain species of lupins excrete acids that tap so-called insoluble phosphates. Roots in fact put out acids and enzymes. They either excrete or slough off nutrients for microorganisms and account for high microbial populations. Russell cites the wireworm. This economic pest feeds on roots of many farm crops. Beans and perhaps peas appear to excrete something that actually kills the pest, according to Rothamsted observations. The full range of companion plantings that organic gardening folk talk about suggest a validity that experimental science has still to recognize fully.

Rotations can and do affect the pH in the soil. Excretion of carbon dioxide figures, as does differential uptake of cations and anions. Needless to say, it would take an encyclopedia to discuss the vast diversity nature has accounted for.

7. common rotations

Here are some common rotations in the cornbelt.

Corn, oats and clover is common, but corn, corn, oats and clover has the edge when farms are not extremely large. Winter barley or wheat are frequently substituted for oats. Corn, soybeans, wheat or winter barley and clover represent a good rotation containing two legumes. Soybeans has proved to be a sustaining cash crop during recent years. It adds little nitrogen to the soil, but packs a good wallop in the tilth improvement department. Sweet clover for pasture, seed and green manure often fits the bill in place of red clover.

The great plains has its own biotic geography and admits special four year rotations—milo, barley or oats, wheat. Eastern Kansas farmers often settle for corn, corn, oats, sweet clover and wheat. Legumes two out of five years will give an important assist to maintenance of the natural nitrogen cycle.

Many of the northeastern states have excellent pasture areas. Here corn, oats or barley (or a mixture), clover and timothy fit in quite nicely, especially if manure compost is applied routinely. Some of the rotations can be stretched to seven years. Corn is often replaced by edible peas, snap beans, cabbage and the like. A five year rotation of corn, oats, wheat, clover and timothy is more common than other variations.

The great northern wheat belt—the Dakotas, Montana, southern Canada—has several corn rotations, corn, wheat, barley, oats and red clover, or corn, oats, wheat, barley and red clover. Length of growing season and condition of virgin prairie land have a way of evolving rotations simply because rotations have the defined purpose of tilth maintenance, and organic matter restoration for better crop production as a prime objective.

In the cotton belt, rotations are wide, but they almost always include cotton. Cotton, corn, forage covers three years. Here cotton is followed by winter legume and small grain. In the second season winter legume and grain are grazed off or harvested and followed by summer legume for forage, this to be followed by a winter legume. The third year, winter legume is plowed down and followed by corn.

The traditional cotton, corn rotation has cotton the first year, trailed by a winter legume. On the second go-round a winter legume is plowed down and corn is interplanted with a legume. Fall sees the soil tilled and prepared for cotton the next season.

8. economic, mechanical, managerial reasons

There are economic and mechanical and managerial reasons for sound rotations that run beyond the vision—or lack thereof—of monocultures. Family farm operations have to distribute labor. Corn, beans and Florida might sound attractive to the man with several sections of land, a great bank loan, and the hope of trapping a bit of the cash flow for himself, but it fails to fit the backbone of American agriculture. Rotations distribute labor, help stabilize income, and put some chips in different baskets. Weather that may cancel out corn could at the same time favor soybeans or a hay harvest. Late droughts in the midlands can damage corn. Yet this same situation would not reduce a hay or oat yield.

Foliar Nutrition

In the lawmaking bodies of the several states, N, P and K have become holy writ, and the profitable background for the NPK mentality. But that era is ending. Economic management, the international oil situation and the price of natural gas all can be calculated to figure in the final clash of wills.

Foliar nutrition, seed treatments and recaptured values all rate attention as new eco-technology takes hold. The background reason for a change was best stated by Sir Albert Howard in *Soil and Health*. Howard questioned the entire scientific apparatus being used to sell salt fertilizers and toxic rescue chemistry as early as the 1940s.

"The usual sub-division of science into chemical, physical, botanical and other departments, necessary for the sake of clarity and convenience in teaching, soon began to dominate the outlook and work of these institutions. The problems of agriculture—a vast biological complex—began to be subdivided much in the same way as the teaching of science. Here it was not justified, for the subject dealt with could not be divided, it being beyond the capacity of the plant or animal to sustain its life processes in separate phases: it eats, drinks, breathes, sleeps, digests, moves, sickens, suffers or recovers, and reacts to all its surroundings, friends and enemies in the course of twenty-four hours, nor can any of its operations be carried on apart from all the others: in fact, agriculture deals with organized entities, and agricultural research is bound to recognize this truth as the starting point of its investigations.

"The natural universe, which is one, has been halved, quartered, fractionized, and woe betide the investigator who looks at any segment other than his own . . . The final phase has been reached with the letting loose of the fiend of statistics to torment the unhappy investigator. In an evil moment were invented the replicated and randomized plots, by means of which the statistician can be furnished with all the data needed for their esoteric and fastidious ministrations. The very phrase—statistics and statistician—should have been a warning . . . Instead of sending the experimenter into the fields and meadows to question the farmer and the land worker so as to understand how important quality is, and above all to take up a piece of land himself, the new authoritarian doctrine demands that he shut himself up in a study with a treatise on mathematics and correct his final results statistically . . ."

Sir Albert Howard asked a question whether scientists and administrators, in putting themselves "above the public and above the farmer," have not "subconsciously been trying to cover up what must be regarded as a period of ineptitude and of the most colossal failure. Authority has abandoned the task of illuminating the laws of Nature, has forfeited the position of the friendly judge, scarcely now ventures even to adopt the tone of the earnest advocate: it has sunk to the inferior and petty work of photographing the corpse—a truly menial and depressing task."

1. valid technologies

The task may have seemed depressing, but the doers were not depressed. Those who fought back were generally talented amateurs, people who could be put down not because they were wrong, but because they couldn't explain why they were right—or were not extended the right to explain.

Valid technologies have been used to make foliar feeds and liquid fertilizers since 1951. These developments went on virtually without the agricultural community knowing anything about them. Sylvan H. Wittwer, under whose leadership Michigan State University conducted investigations on foliar fertilization of plants, correctly summarized: "There is probably no area in agricultural crop production of more current interest, and more contradictory data, claims and opinions, or where the farmer in practice has moved so far ahead of scientific research." Even today the average land grant college does not have a single scientist who has a working knowledge of using liquid fertilizers in growing crops. There will be those who object to this statement, citing Hanway's Iowa State University research, which came some 12 years after Wittwer's, but this hardly qualified because it failed as a result of using the wrong type of ingredients.

Credentialed and uncredentialed outsiders standing in the wings knew that Hanway's foliar N, P and K efforts would fail because the compound being used failed to answer plant requirements, would burn crops and also because it would cost too much. It involved the application of gross amounts of nitrogen, phosphorus and potassium to soybeans in three applications, using amounts required by holy writ of soil fertilization. As Albert Howard would have pointed out, fertilization of such crops entailed a knowledge of many other things, such as what is a plant's requirement at each point in the growth cycle?

Working with nursery crops, flowers for market, strawberries, apples, pears, cherries, agronomists started answering these growth cycle requirements with seaweed extracts many years ago, all with successes that were spectacular, erratic, and rejected. There were things that could be done with the leaf that proved next to impossible when working with complexed soils.

Some few farmers knew this, but they didn't have a handle on the "why." In the early 1950s Wittwer was working under arrangements with the Division of Biology and Medicine, Atomic Energy Commission. Experiments entailed the use of radioactive isotopes in assessing the efficiency of foliar applied nutrients compared to soil applications of those same nutrients. It was found that the efficiency of the foliar fertilizers was from 100 to 900% greater than the dry applied fertilizer materials. The results became a matter of record under Contract AT (11-1)-888,1969, and soon touched base in select circles as an audio film entitled *The Non-Root Feeding of Plants*. The film has been shown at many *Acres U.S.A.* Conferences, and helped the newer foliar nutrient materials make the transition to eco-agriculture.

Some parts of that summary are of maximum interest. It was noted, for example, that deficiency disorders were becoming more common, that there were soil imposed problems of dilution, penetration and fixation. Foliar fertilization circumvented many of these problems. "A plant's entire requirement for many micronutrients may often be supplied by above ground plant parts because the quantities needed are small and tolerances for the applied materials and rates of uptake are adequate," Wittwer wrote. "But for the macronutrients used in large quantities by plants and for most crops only a part of the nutrient needs are satisfied, but the contribution can still be significant."

Without using atomic tracer studies, T. L. Senn of Clemson University arrived at essentially the same conclusions, working both before and after the Wittwer studies. These studies have not been well received in the main because they involved fertilizers inherently of medium and low analysis according to N, P and K folklore. Since it has been technically undesirable to distort these products so they contain an aggregate of 16, 20 or 24% N, P and K, they have frequently been denied registration. Over 20 states have fertilizer laws and regulations which prohibit registry and sale of such useful materials as commercial fertilizers.

American farmers are going to be compelled to move in this general direction and use some foliar fertilizers in favor of those made out of natural gas. Eco-agriculture has willingly led the way both because the old fossil fuel fertilization system was unecological and uneconomical. The for instances need not detain us very long.

First, it is recognized by all literate farmers that of all the nitrogen fertilizers used, 95% is made with natural gas as an ingredient in that fertilizer. Anhydrous ammonia is composed of nitrogen and hydrogen in the ratio of NH_3. The hydrogen in all nitrogen fertilizers comes out of natural gas. It takes about 40,000 cubic feet of natural gas to make a single ton of anhydrous ammonia, and anhydrous also serves as a building block for making urea, ammonium nitrate, amaphos, and the other forms commonly used in agriculture. That 40,000 cubic feet of natural gas cost about 15 cents or less when *Acres U.S.A.* first started publication. The cost is being projected in excess of $4.00 at the turn of the decade, and this cost will most certainly be built into fertilizer costs. This means conventional fertilizer costs are going to increase another $2 billion a year at the farm while farm parity is being scheduled to go down by the economic managers.

This single factor makes criticism of an emerging technology for fertilizing crops both careless and irresponsible. It illustrates a lack of acceptance of the leadership role land grant colleges should have. News releases and magazine articles down-shouting this new technology may elicit great patience and forgiveness from all those in eco-agriculture, for it is realized that they know not what they do! It is also realized that the university people are going to be compelled to explore new technologies that will unwittingly put them back in touch with the holistic concepts Sir Albert Howard talked about in the first place.

Lee Fryer of Food and Earth Services, Wheaton, Maryland, estimates that by 1985 fully 25% of American acres will be treated with foliar fertilizers. Fryer's *Food Power from the Sea* explained why.

2. a nutrient bank account

Just what is this technology?

The primer lessons so far ought to serve well in helping you comprehend how it works and why it works. First, recall that when you fertilize a crop using conventional N, P and K fertilizers, only 2 to 3% of, say, phosphorus will be present to nourish a crop two months later. Of all the phosphate applied to crops, recovery by plants is less than 12%. The rest reacts with soil chemicals and is leached and washed away or otherwise demobilized. Of all the nutrients applied via the foliar route, fully 80% find their way into crops. Much the same is true of nitrogen. Hardly 30% of all nitrogen applied ever enters a plant. In the case of anhydrous ammonia, the uptake is even less, perhaps only 15%.

All this translates to wasteful fertilization.

Soil fertilization will no doubt have to deliver a fair share of the nutrition to plants—a working formula at this time being 50%. The rest will have to be applied in other ways, the foliar route included.

How can this work? When you plant a corn crop, for instance, a nutrient bank account develops in the plant itself. As the time nears for tasseling and fixing ears, an acre of corn contains some 80 pounds of actual nitrogen in leaves and stalks. Nature has decreed this situation so that when corn moves into its fruiting cycle, it has a reserve to draw on for protein production in the kernels of that corn. Protein is generally computed as one-sixth nitrogen. Thus when a corn crop moves into a critical time—when it is ready to fix ears—the reservoir is in the stalks and ears. This is why agronomists were mistaken when their experiments prompted them to lace the crop with a lot of nitrogen. As corn and other crops move into the fruiting period, plants have already grown the vegetative structure that requires large amounts of calcium, nitrogen, phosphorus, potassium. They have utilized the main amounts of those nutrients required for good crops.

When a crop moves into its fruiting cycle, it needs primarily the whole assortment of nutrients necessary to run the hormone and enzyme systems in that crop. Professors who have spent their lives in conventional fertilizers have a hard time comprehending this inventory of fact. They point to the low N, P and K count in a foliar feed, correctly noting that few of them contain enough "major" nutrients to wad a shotgun. As a matter of fact, one fellow testified before the Minnesota legislature that seaweed and fish fertilizers have no more nutrients than might arrive in a rainstorm.

3. law of the little bit

This is nonsense, of course. Seaweeds contain all the minerals biologically active on our planet—something like 50. Fish contain perfectly efficient forms of nitrogen, phosphorus and potassium—plus magnesium and calcium in the bone. Combined they hold a whole cafeteria of trace mineral keys to enzyme activity. Very few trace minerals are required for plant growth.

There is a reason for this, one dazzling in its purity. The Creator has decreed that rare earth minerals should not be used as cell building materials and remain available for further use and reuse. The scientific basis for all this is contained in one of the greatest books about agriculture even written, *The Diagnosis of Mineral Deficiencies in Plants*, by T. Wallace. It was published in 1951 by Her Majesty's Stationary Office, London, and is no longer available. But at least this quote should always remain available, pasted over the planning desk on every farm. "Since catalysts are not used up in the chemical reactions which they promote, we can understand how it comes about that quite small or even minute quantities of the trace elements, iron, manganese, boron, zinc, and copper, may nevertheless be essential to the plant's health and growth."

Indeed, since catalysts are not used up in the chemical reactions they promote, it is understandable why small amounts stay in harness to do such a fantastic job. The corn crop can take a substantial portion of its nutritional requirement out of the soil, but if the soil is mined out or complexed by the time the critical fruiting cycle arrives, it may need some more. Most of the early foliars accepted this premise.

More recently, formulators have taken to adding a little nitrogen, phosphorus and potassium to take care of shortfalls that might occur. They have added to other chelates and amino groups.

In December 1976 issue, *Acres U.S.A.* published an article by Lee Fryer in which it was suggested that seaweed and fish be combined in a fertilizer for this very reason. Happily several formulators now offer such combinations to the market. This suggestion was not made because *Acres U.S.A.* was romantic, or because the publisher swooned at the sound of ocean water. It was made because it would enable farmers to obey nature's *Law of the Little Bit.*

Formal literature calls this the Liebig-Mitscherlich Law of the Minimum. As stated by von Liebig some 120 years ago, growth and yields of plants are governed by nutrients in least supply, not by those in abundant supply.

As it stands today, it would take a good university research program several hundred years to research and fine tune the trace mineral requirements necessary for hormone and enzyme balance, or to pinpoint the factors behind vegetative cell structure growth that enable a farmer to enjoy a significant yield increase. Yet this technology is available to those who remember the *Law of the Little Bit.*

As a crop goes into its time of stress, the time when it produces the part the farmer sells, it is not the abundant amount of nutrition that decides the yield, but the nutrient in least supply. This would seem to provide an insurmountable dilemma. How is it possible to tell which nutrient is missing?

We can't. The fact is we have to provide them all, albeit in minute amounts. Lee Fryer has related how he sprayed a fine golf green with liquid fertilizer at a point in summer when nutrients had become exhausted. He then sat on an elevated platform and photographed the chlorophyll growing in that grass a half hour later. If the crop is hungry it will produce growth and color in that fast a time.

4. values of foliar sprays

In summary form, here are some of the values of foliar sprays.

1. The nutrients available to the plant are mobilized into the leaves. This is the chief purpose of fertilization in the first place.

2. Incorporated into a regular cultural program, the foliar nutritional spray promotes and maintains a general vigor so that the plant is better able to perform its normal functions and better withstand stress.

3. The term "foliar feeding" implies uptake and utilization of nutrient materials applied to plant leaves. This is misleading to a certain point. All living parts of a plant above ground can absorb nutrients: twigs, branches, buds, fruits, flowers, and stems. Foliar intake of nutrients is similar to absorption by the root system except that foliar applied nutrients are readily available and more easily utilized by the plant than when applied to the soil.

4. Foliar nutrients increase the rate of photosynthesis and by so doing stimulate and increase nutrient absorption by roots.

SHORT COURSE

LESSON 23

Insects

The concept of environmentally sound agriculture took a seven league stride forward during the early 1970s, not so much because agronomists learned more about soil fertility management, but because entomologists learned more about insects. It had been known, according to the give and take world of evolution, that insects made their appearance some 600 million years ago. Some seemingly continued to evolve, some apparently saw no need for further improvement. Almost all had strong exterior skeletons and power out of proportion to size. Some insects discovered flight about 100 million years ago. Some can survive temperatures as low as -30 Fahrenheit and as high as 120 F. These things—including such tidbits as the intelligence that an ant can lift 50 times its own weight in anything—have come to fill volumes on insect lore.

In considering all this knowledge, the farmer's situation is unique. Entomologists frequently study dried, pinned insects. The farmer is in a position to watch them live—a stinkbug draining the juices out of a victim, a scarab rolling manure, a bee bringing nectar and pollen from the field. The cutworm that brings down a corn stalk is a real live menace, not a dead mounted entry under glass. And yet both realms of knowledge have to figure if an eco-farmer is to understand the role of insects in the grand scheme of things.

1. kingdoms

We've deferred the usual classifications of kingdoms of the world until now because we have a hunch insects have the most clout when it comes to putting things in perspective. The kingdoms are known even to children who play guessing games: plant, animal and mineral. (Parenthetically,

Phylum	Common name	Number of species
Chordata	*Mammals, birds, fish, reptiles, amphibia*	60,000
Arthropoda	*Spiders, mites, scorpions, insects, etc.*	713,000
Mollusca	*Clams, oysters, snails, etc.*	80,000
Annelida	*Earthworms*	8,000
Bryozoa	*Moss animals, seamats*	3,100
Echinodermata	*Starfishes, crinoids*	5,500
Trochelmintha	*Rotifers, animiculata*	1,750
Nemathelmintha	*Roundworms*	5,500
Plathyhelmintha	*Flatworms, flukes*	7,000
Ctenophora	*Marine animals related to jellyfish*	100
Coelenterata	*Jellyfish, coralhydra*	10,000
Porifera	*Sponges*	3,250
Protozoa	*One-celled animals*	17,000
Other minor phyla		1,300

there is also a man-made kingdom composed of the synthetics, which we'll discuss later.) Kingdoms always have their subkingdoms. These are something akin to tribes and they are called phyla. In fact that's what the word phyla means. It is Greek for tribe. Phylum is singular, phyla plural. The animal kingdom is usually broken down into 19 phyla.

It will be noted that Arthropoda is the biggest phylum in the animal kingdom. In terms of species, it dominates the next runner-up, Mollusca, by almost nine to one. The fantastic scope of insect life is further illustrated by taking an overview of insect orders, from the lowest to the highest forms.

Order	Common Name	Known World Species	Known U.S. Species
Protura	*Proturans*	90	28
Thysanura	*Bristletails*	700	50
Collembola	*Springtails*	2,000	314
Plecoptera	*Stoneflies*	1,500	340
Ephemeroptera	*Mayflies*	1,500	544
Odonata	*Dragonflies*	5,000	412
Embioptera	*Webspinners*	150	9
Orthoptera	*Grasshoppers, crickets cockroaches*	22,500	1,172
Zoraptera	*Zorapterans*	20	2
Isoptera	*Termites*	1,900	59
Dermaptera	*Earwigs*	1,100	18
Coleoptera	*Beetles*	276,700	26,576
Strepsiptera	*Twisted-winged insects*	300	100
Thysanoptera	*Thrips*	3,170	606
Corrodentia	*Book lice*	1,000	96
Mallophaga	*Biting lice*	2,800	318
Anoplura	*True lice*	300	62
Hemiptera & Homoptera	*True bugs*	55,000	10,200
Neuroptera	*Vein-winged insects*	5,000	338
Trichoptera	*Caddisflies*	4,450	921
Lepidoptera	*Moths & butterflies*	112,000	10,768
Mecoptera	*Scorpionflies*	350	66
Diptera	*Flies*	85,000	16,700
Siphonaptera	*Fleas*	1,100	338
Hymenoptera	*Bees, wasps, ants*	107,000	17,408

These data were extracted from Jaques (1947), Borror and De Long (1971) and Insect Evolution (1971) by Dr. Philip S. Callahan, USDA Agriculture Research Service, Gainesville, Florida.

2. arthropods

In the opening sentences of lesson 4, the point was made that for each 400 foot increase in altitude, flowering of plants is retarded four calendar days. Much the same principle applies to insect life. You can expect the same insects one mile above the sea as you might find 910 miles to the north assuming such a mountain rose straight out of the sea. Moreover, insects travel, sometimes at a fast and furious clip. Grasshoppers have turned up 1,200 miles at sea. The male deer botfly, *Cephanomyia pratti,* has been computed to travel faster than highway traffic.

Some insects hitchhike, not only on animals and birds, but on the wind itself. Young larvae of the gypsy moth have air pockets in their hairs. This makes them float like a winged particle of matter for great distances. Insects breath by means of air tubes and sacs called a tracheal system. Air sacs make flight easier. These items are recited here merely to illustrate the point that insect lore is both rich and everlasting. It has been computed that if all the insect names were printed on pages such as this one in two columns, it would take ten volumes such as this primer to list them. Even so there would be omissions, since new species are being described at the rate of 6,000 a year.

Not all new discoveries have to do with genus and species. Size, shape and favorite eating habits all are entries. Arthropods as a phylum includes crustaceans (lobsters and crabs) and arachnids (scorpions, spiders and ticks). Through eons of evolution the Creator has settled for giving lobsters and crabs five pairs of legs. Millipedes, by way of contrast, ended up with 200 pairs. But all insects, no exception, ended up with three pairs.

Moreover, insects have three physiological segments—head, thorax (or chest), and abdomen. Now we've said everything we can without getting into specifics, and that's where all those new discoveries come in.

3. a molecular bomb

One of the newer items of knowledge evolved during recent years has had to do with the life cycle. Scientific journals had quite a little to say about all this, of course. Popularization of the news did not so much surface in farm magazines, but in *Fortune* as an article styled *A Molecular Bomb for the War Against Insects*. The financial paper told how a new secret had disappeared into the laboratories of Schering, Syntex and Hoffman-La-Roche. All this was of financial interest because pesticide sales had reached a staggering sales total—with insecticide makers such as Dow Chemical, Monsanto and DuPont raking in the profits—but many believed the handwriting was on the wall.

Insecticides such as DDT, wrote *Fortune*, "loosed into the biosphere long-living poisons that have been found to accumulate, sometimes fatally, in all manner of plant, bird and animal life, including man." As a result DDT was turning up in marine plankton, upsetting the basic elements of the sea's food cycle, and of photosynthetic renewal of oxygen. Even more alarming was recognition of the rapidity with which insect generations adapt to the chlorinated hydrocarbons.

Insect fecundity can stagger the imagination. Some species lay hundred of thousands of eggs after mating. Some are born, live out their lives and die in a matter of days or weeks. Generations by the dozen obey the biblical injunction to increase and multiply, harnessed to their chain letter styled activity until food is exhausted or weather takes its turn. In any toxic spray operation, there are survivors. These few insects can make up for lost time and rapidly bring the population back to strength, this time with built-in immunity.

As many as 24,688 aphids have been inventoried on a single tomato plant. It is a matter of record that some grasshopper infestations have come to cover 15,000 acres, consuming everything in sight in less than eight hours. In one instance, some 300,000 pounds of grasshoppers were captured and destroyed.

Of more than 640,000 insect species, only 500 can be considered pests. The rest are beneficial. Of the 26 insect orders, 15 contain predacious or parasitic insects. Not only do insects feed on each other, they provide food supply for birds, reptiles, frogs, fish and other animals, human beings included.

More important, perhaps, is the fact that while many plants harbor a great variety of insects, most insects will dine on only one plant. Almost all species of Nepticula, Phyllocnistis and Lithocolletis have a specific host plant. The boll weevil lives on no more than the cotton boll itself. The nun moth, *Liparis monacha*, lives on oak in one environment, and attacks the pine tree in another. The apple maggot feeds on apples, of course, but it also dines on blueberries. The twig pruner, *Hypermallus villosus*, tunnels its way into wood of the apple tree, the grape plant, and maple wood. On and on.

With their penchant for damage, the insect pests still seem to have the cards stacked against them. When they exhaust the food supply, they die or become dormant. High temperatures kill them by coagulating their proteins. Evaporation and desiccation orbit them as a constant danger. Low temperatures annihilate cell structures, ergo body tissues. Very few insects can survive 31 Fahrenheit. The almost indestructible grasshopper folds up like a wet noodle when the Fahrenheit temperature drops as low as 20 degrees for 48 hours.

The USDA recently has been forced to admit that of the 500 insect crop destroyers, 267 species have built up marked resistance to insecticides. Obviously insecticide makers just don't understand bugs. Thus the importance of certain discoveries in the late 1960s and early 1970s.

4. genetic code

By then scientists had found that insects have a genetic code locked in their hormone systems. This can best be illustrated by examining the gypsy moth caterpillar. This insect has three types of hormones—one governing the larval stages, another for the pupa, one for the adult stage. The minute an egg is hatched, a brain hormone puts in motion another hormone substance scientists now call ecdysone. At this moment in the life cycle of the gypsy moth, a gland near the brain touches off the so-called juvenile hormone, and this hormone represses all but the gene group controlling larval development. During the larva stage, the juvenile hormone acts like an anchor on cellular activity. That is, it prevents the insect from maturing too fast. In the life cycle, the juvenile hormone disappears. This pulls up the anchor and turns on the full force of ecdysone. Come winter, or a change in diet, the brain secretion is shut off and life goes dormant.

The process is complicated in the extreme. For a long time scientists could not find the source of the hormone generation. It was, in fact, a matter of serendipity that accounted for the discovery of a principle folk gardeners had known all along. The paper toweling used to line cages acted on larvae exactly like juvenile hormones. Finally it narrowed down to the fact that fir trees (from which laboratory towels were made) contained a juvenile hormone-like material. Nature had caused these trees to grow their own defense mechanism against insects.

One of the scientists involved in these studies, W. S. Bowers, stated an obvious conclusion. "Insects and plants have undergone co-evolution at least since the lower carboniferous era, and some of the allelochemic relationships between insects and plants are well understood." He pointed out that some plants attract insects, some lure parasites of offensive predators, some repel insects. Dent corn, for instance, produces a compound which inhibits the eating capacity of the European corn borer.

As word got around scientific circles, investigators turned up what they called "analogues of ecdysone." They found the substance in yew trees, in evergreens, in a lot of places. A Harvard man, Lynn M. Riddiford, discovered that mere contact with juvenile hormones fatally deranged embryonic development of life in insect eggs. By juggling hormones, and by making use of analogues of hormones, a super-sterilization effort could be created as a chain reaction. The speed of this process could be compared to the speed at which venereal disease races through a human population.

And with that, research seemed to vanish as each of the companies involved raced to develop salable products ahead of the competition. The big firms were licking their chops. They dreamed of packaging and selling molecular insect warfare to farmers.

5. broadcast short-wave

Astute farmers saw the loophole in all this. They reasoned that while fabricators might make hormones, they could not put them into the sap of plants. The only real way to get insect protection for the crop was to grow it there, a proposition that takes in the full range of technology called eco-agriculture.

It started with genetics—Reginald H. Painter's *Insect Resistance in Crop Plants* remains a classic—and it moved on into nutritional support. To suggest that genetics alone counted would be to imply that plants could be bred to tolerate starvation.

In the meantime, some few other questions about insects have been answered. Farmers have long wondered aloud about how come insects knew just which plant provided the fare they wanted. Watching greenbugs invade one field and ignore another, one could conclude that some sort of communication signal invited the bugs to where the feast best suited their needs. Otherwise, how did a codling moth know where to go? Philip S. Callahan of the USDA and University of Florida, Gainesville, was among the first to recognize the fact that insects and plants in effect broadcast short-wave.

In order to understand all this, it might be well to illustrate here the full electromagnetic spectrum. This diagram illustrates the division in order of decreasing energy and increasing wavelength. The bottom scale has the whole package. Dotted lines take the reader into a breakdown of the visible part of the spectrum—the one we're all more familiar with.

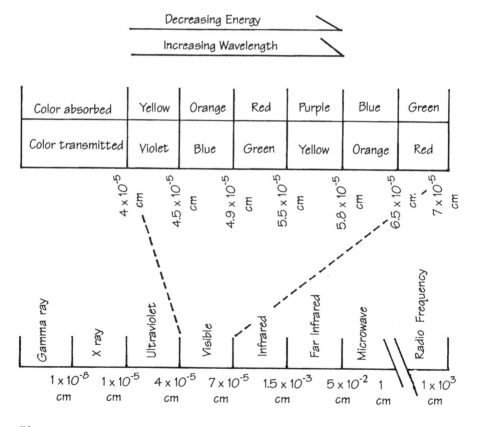

Electromagnetic spectrum in order of decreasing energy and increasing wavelength.

Note that the most energy is concentrated in the gamma ray, x-ray and ultraviolet ray breakdown, just before the visible. The visible part of the electromagnetic spectrum is what it is—the part that rods and cones of our eyes can pick up. As energy decreases and wavelength increases, there are the infrared, far infrared, microwave and radio frequency wavelengths.

6. radiate naturally

Callahan touched on the subject in *Insect Evolution*, in a few dozen scientific papers, and then he issued a popular broadside in *Tuning in to Nature*.

In *Tuning in to Nature*, Callahan spends quite a few pages proving that "all things natural to the earth radiate naturally in the infrared. If we could see in that region of the IR spectrum, the whole corn field would appear by moonlight as a vast array of fluorescent light bulbs sticking up out of the soil."

In Callahan's way of thinking, it is not possible to understand ecology without understanding something about "the part of the world in which life can exist." There is in fact an infrared window—7 to 14 micrometers—that allows IR radiation in all day and all night. The gas makeup of the atmosphere figures in all this, as does humidity and the weather.

The recorded beginning of this line of thinking goes back to 1863, when an Irish genius from Leighlinbridge in County Carlow, wrote a book, *Heat Considered as a Mode of Motion*. His name was John Tyndall. In well cadenced Victorian language he explained how molecules of scent from flowers and plants absorbed radiation. Phil Callahan had been thinking along the same line without ever hearing of Tyndall—to use Callahan's words—"for the simple reason that Tyndall's elegant work was ignored by the modern chemists and olifaction physiologists."

Wrote Tyndall:

> "The sweet south
> That breaths upon a bank of violets
> Stealing and giving odor,
> owes its sweetness to an agent, which though almost infinitely attenuated, may be more potent, as an interceptor of terrestrial radiation, than the entire atmosphere from bank to sky."

This simply means—again to use Callahan's words—that "air space is made up of many, many specific molecules of oxygen, nitrogen, argon, ammonia, etc., but that scattered thinly among those billions upon billions of air molecules are spread very few molecules of scent substances that block infrared light better than air alone." That is why it can correctly be said that John Tyndall invented the instrument now called absorption spectrophotometer.

Callahan's explanation of all this can be read in a hauntingly beautiful book called *Soul of the Ghost Moth.* In it he explains how Tyndall "used his absorption spectrophotometer to demonstrate that when a very few molecules of a scent are released into air space the scent causes the character of the air to change so that it absorbs far more infrared radiation from the sun, or night sky, or stars than it would if the scent were not scattered thinly in the air. The combination of thinly spread scent molecules, plus the gases that compose the air act together, so to speak, to trap the radiation within itself."

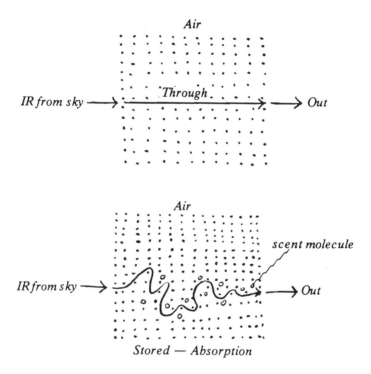

Now the plot thickens. The air and the blackbody conspire to send out coherent waves, much like a radio transmitter.

"Scientists call such heated bodies as the earth, blackbodies. A radiating blackbody is any heated body that absorbs all radiation and radiates infrared wavelengths in a range that depends on its peak temperature. Different blackbodies have different peaks of radiation wavelengths, depending on their temperatures. In general, the hotter the body, the shorter the peak-radiation wavelengths; the cooler the body, the longer the wavelengths. Since the sun is very high, it emits frequencies that peak at very short wavelengths in the visible spectrum.

The earth's temperature seldom reaches an average temperature of more than 70 to 100 F. and therefore peaks at a wavelength of approximately 10 micrometers in the 7 to 14 micrometer window. It is this curve of the peak earth temperature that reradiates back out into space.

"Besides the natural hot and warm bodies of nature, there are, of course, man-made hot objects. Hot automobile engines, factories, tungsten-filament light bulbs, the heated pavement of cities—many more. All are blackbodies emitting an infrared radiation peak at the peak of their own temperatures."

There is a mathematical expression for instant identity of peak radiation for any hot body. It is called Wien's constant—after the German physicist, Wilhelm Wien—and that number is 2897. When this figure is divided by the absolute temperature of any hot body, the dividend is the wavelength. Absolute here has a special meaning. Zero on the absolute scale is -273.16 Centigrade. For ease of calculation, Callahan rounded off that number to -270 before writing this passage.

"A human body with a skin temperature of 96 F. (which equals 31 on the Centigrade scale) will equal 31 C. plus 270, or an absolute temperature of 301 K. The symbol for absolute temperature, now called Kelvin's (after Lord Kelvin), is K. Therefore,

$$(\text{wavelengths} = \frac{2897}{301K} = 9.62 \text{ micrometers})$$

or waves 9.62 micrometers long."

In other words, a human being emits infrared radiation at a peak of 9.62 micrometers. The same computation can be made for "all things natural to the earth," again quoting Callahan, since every "insect, mammal, tree, weed, and living creature" emits its own characteristic blackbody radiation.

Insects do emit a coded infrared signal, and it contains a unique navigational message. When insects fly, they vibrate their antennae at the same frequency that they flap their wings—and those antennae are indeed antenna, just as are the man-made devices that pluck a radio signal out of the air.

Without knowing it, electrical engineers discovered and put to use what nature has been doing all along. The earliest tracking system for bringing in an airplane during inclement weather relied on Morse code, dot-dash (A), dash-dot (N), signals riding each other to create a beam. If the signal got weak, the pilot knew he was moving off beam, and accordingly corrected his course.

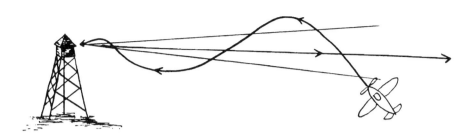

It is equally well known that night flying moths track a pheromone (sex scent) as if they were following a radio beam:

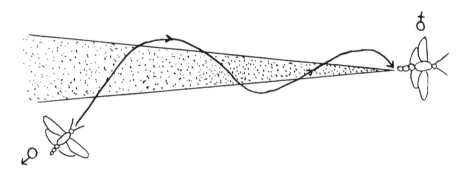

"Every entomologist also knows that the sex odor of the female emits from the scent gland at the top of her abdomen as she vibrates her wings. During emission her temperature increases 6 to 18 F. above that of the air space. This means, of course, that as the scent drifts through the air pushed by the gentle breeze, it cools and decreases in concentration. The scent flows along in the water vapor of the gaseous atmosphere which is the carrier gas." The important thing to remember here is that insect antennas are, indeed, antennas.

As a matter of fact, every antenna ever built by man has its counterpart on insects. Callahan has observed that the pheromone signal (chemical released by a female insect to stimulate sexual attraction) is much better than the old time radio transmitter because it tells the flying male moth how far he is from the pheromone-releasing female as well as the direction.

Close to the female, wavelengths are long. A few hundred feet away they are short. In electronics the system is called "concentration tuning."

Insects use short spines (sensilla) to resonate to short wavelengths, longer sensilla to read the signals as they close in. There are spines for tuning in to sex scents—a part of nature's mating game. There are sensilla for detecting infrared radiations from the food supply. Plants, after all, have odors, and odors are governed by health or the lack thereof, regardless of whether pathology is gross or, say, the corn plant is simply subclinically ill.

If there is a must book among all those mentioned in this primer, it would have to be *Tuning in to Nature*. If there is a must paper, it would have to be *Moth and Candle: The Candle Flame as a Sexual Mimic of the Coded Infrared Wavelengths from a Moth Sex Scent* (both by Philip S. Callahan). Callahan has proved that the moth is programmed like an orbiting satellite to respond to a series of frequencies. There are 17 micrometer pheromone lines present in the candle. "Thus the candle mimics the moth body-wing-modulated signal because the flickering of the flame in the air modulates the candle blackbody radiation between 6 and 25 micrometers—exactly what the moth is programmed to respond to."

Picture the scene. A cabbage looper moth releases a pheromone scent from the tip of her abdomen. It is a complex molecule composed of alcohols and acetates hitched together as a long chain. Free in the atmosphere it is a complex trace element in solvent because of water vapor in the night air. This pheromone is pumped along by other trace elements in the air. Insect sensilla pluck these signals right out of the air. "The drifting plume of pheromone not only give direction to the flying male but also distance from the female by reference to the wavelengths at specific concentrations and temperatures in the air," says Callahan.

7. do plants transmit the signals

Do plants transmit the signals that call in cutworms, cabbage loopers, moths or earworms, or permit them to grow? Even those who agree in principle tend to disagree on specifics, but at least one school believes that daily life rhythms are timed from within the body. If this is so, then the makeup of the plant—the enzymatic and hormonal balance as determined by soil fertility, nutrients and balance thereof—determines whether these microminiature transmitters attract or repel. Indeed, Callahan suggests as much when he writes about the future of agriculture.

"The farmer, instead of going to the store to buy a bag of insecticides, will buy or lease a little microminiature transmitter that either attracts the insect species attacking his corn or cabbage plants, or, conversely, beats against the insect pheromone frequency and jams it so that mating does not take place. No mating, no eggs on his cabbage."

It is well within reason to suggest that a subclinically sick plant will have an altered wavelength. Is this why insects attack the undernourished plant, the one with missing trace nutrients? If so, then the questions have been answered by the eco-farmer who has discovered the bio-chemistry of immunity not only in genetic engineering, but also in fertility management.

Phil Callahan is a bug man and a generalist of sorts. We wish he were also a soils man. We suspect that by marrying soil science, plant nutrition and wavelengths with economics and philosophy, he would come up with even more answers to the baffling riddle the Creator has bestowed on us all.

8. practical researchers

In the meantime, practical researchers in the field have moved well beyond the glass tube and Waring mixer stuff found in the laboratory. Why not grow plants with balanced hormone and enzyme systems so that insects are triggered into destruction or dormancy before they get going? This means moving out of insect lore and back to plant management and soil management.

The most successful practitioners in eco-agriculture have thus come to rely not on poisons at all, but on pH management with calcium, magnesium, sodium and potassium in equilibrium as an opener, using seed treatments and foliar nutrition to achieve healthy plant production as a kicker. The best fine-tuned systems certainly have provided protection from bacterial, fungal and insect attack.

However, if a farm has sick soil, poor management, and wrong management timing, or even a failure to achieve the desired norm, then even rescue nutrition frequently cannot absorb the load imposed upon it.

It has been the general thesis of this entire primer that the best defense against insects is sound nutritional management of plants. This much was hinted at during the late 1960s when William Bowers of Beltsville isolated from Balsam fir an ingredient he named juvabione. Apparently, way back in pre-ancient history, the North American fir developed a mimic hormone for protection from insects, the kind of hormone a healthy plant must have. This hormone inhibited an insect's growth and life much as did the insect's own built-in substance that sees to it that an insect goes through its various life stages in tandem, not atop each other. If healthy plants grow their own protection, then protection from insects—at face value—must be a matter of plant nutrition.

9. pesticide treadmill

But what if the farmer's performance is less than perfect and the insects arrive? Chemical companies insist on answering with unspeakable toxic remedies. Yet toxic genetic chemicals are proving to be a failure because of their "pesticide treadmill" posture. The quoted words have been used by Robert van den Bosch, Ph.D., of the University of California, Berkeley. "You can't beat insects with insecticides, and we are only fooling ourselves if we think we can. They are too adaptable. They have tremendous genetic plasticity. They are prolific as hell and they are mobile. They can move if they have to."

A few other approaches have been used. The classic screwworm program of the early 1960s need not detain us long. In that case USDA released sterile males over most of the southern states. Since the females of the species mate only once, this form of enforced birth control proved effective. California officials recently released some 350 million sterile males in an attempt to control the Mediterranean fruit fly, an insect that operates on much the same sex pattern as the screwworm. Nairobi officials have adopted the same general idea for dealing with the bearer of sleeping sickness, the tsetse fly.

Birth control is simple, where applicable, when compared to synthetic hormone and pheromone approaches. Zoecon Corporation of Palo Alto, California now markets a compound called Altosid SR-10—a compound designed to ape the juvenile hormones secreted by insects. So far that product has been approved for use against floodwater mosquitoes only.

Another wrinkle on the same idea is the antihormone—a substance that causes immature cotton stainers and Mexican bean beetles to become sterile adults, and Colorado potato beetles to enter a hibernation from which they cannot emerge.

Pheromones production is a one-at-a-time business when human entrepreneurs get into the business. One of the USDA scientists at Beltsville, Maryland has identified the pheromones of the American cockroach, Oriental fruit fly, Mediterranean fruit fly and southwestern pine tip moth—which may seem to be a start.

So far 111 insect pheromones have been identified. Over half have involved insect crop destroyers—not a bad start, taking the total of 500 agricultural pests into consideration. In theory synthetic pheromones could be spread over an insect-infested field to so confuse males they might never mate.

That's the theory. Still, it is important that plants produce the hormones required for self-protection, and that they do this internally in the sap system. These hormones might also be produced synthetically and broadcast over the field. And in theory scientists can define which hormone relates to which insect or disease. But in the final analysis they cannot get the synthetic product inside the plant and biologically positioned where it belongs.

10. weeds and insects

Weeds and insects—and bacterial and fungal attacks—all point to the degree of degeneration in the soil. Although everything in eco-agriculture is connected to everything else, the whole fabric of interrelatedness is too complicated to permit straight-line connections (cause and effect) very often. Nevertheless we can offer a few notes.

Fungi involved with plants during the past 15 million years were designed to take down plants that did not deserve to make seed. Such plants do not live up to the Creator's plan, and so they must go. Insects and fungi are the first steps in taking, say, the wheat plant back into the decomposition cycle. If, indeed, insects went after the strongest, most thriving individuals, then there wouldn't be any strong seed—the worst possible natural selection. As it is, insects would rather die of starvation than eat the plants that grow in healthy fields.

Insects relate to the entire combination of factors in agriculture. Any combination has its fallout, and for this reason it is difficult to be exactly specific. There are variables of a weakness from area to area, season to season. Often a bunch of insects will materialize only to be faced by seasonal factors that permit very little damage. Three weeks later the same insect cycle might return with a weather change. With varied stress conditions, all of a sudden the insects have a field day. Prescription foliar sprays deal with this type of stress and buffer a plant's capacity to survive.

It is a fact, nevertheless, that often the same soil conditions that give the farmer foxtail problems also give him corn rootworm and corn root maggot. By way of contrast, it must be noted that cutworm and rootworm do not live together. Soils that support one will not support the other. If they are both found in the same field, it will be because soil conditions vary within the field.

Soils that support the cutworm are low in lime. They are harsh and acidic. Cutworm conditions often follow planting and the first rain. If the soil is not properly buffered against the existence of cutworms, these insects go to work, and they will work to the very edge of the soil type line, always without entering the kind of environment that does not give them permission for life. Farmers who put a little high calcium lime in the seed zone can change the environment so that cutworms stay dormant. Farmers who try to find their answer to the cutworm on labels for poisons merely take out insurance on continuing deterioration of the soil system. That this does not offend many farmers can be explained by the simple fact that many farmers are not farmers at all, but only miners of the soil.

11. japanese beetles

When morning glories, buttonweeds, creeping jenny and weeds of that ilk are a problem, Japanese beetles will likely be a problem, especially in the soybean field. It will be recalled that these weeds are an index of organic matter management and decay, with decay going in the wrong direction. Corn smut is also related to the soil conditions that produce these weeds. In soybean fields, brown stemrot is also a likely problem later in the season, particularly when the late season presides over wet soil conditions. All relate to how organic matter is decayed. To deal with the problem, farmers simply have to look to the calcium level and the cation balan-

LIFE CYCLE OF A CORN BORER

MAY

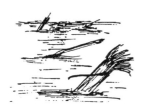

Overwintering borers change to resting-stage (pupae) in old corn and weed stalks and similar shelter.

JUNE

Last wintering borers pupate — pupae change to moths. Moths lay eggs and first-brood borers begin to hatch.

JULY

Some first-brood borers pupate and change to moths. Moths lay eggs. Second-brood borers begin to hatch.

AUGUST

Last of first-brood borers pupate and change to moths. Second-brood borers continue to hatch.

SEPTEMBER

Second-brood borers complete their growth.

WINTER MONTHS

October through April full-grown borers winter in corn stubble and stalks of corn and weeds.

ces that preside over tilth, friendly fungi systems, and trace mineral nutrient release.

12. corn borer

The corn borer has a unique hormone system. When a corn plant produces the hormone dimboa in appropriate concentrations, the corn borer takes about one suck, and that's all. Dimboa, in turn, is a natural phenomena of properly growing corn with. an ample supply of trace nutrients. These allow the formation of hormone systems to work and be present in its normal equilibrium level. Good growing corn has the proper level of dimboa and is never infected by the corn borer. The corn borer will only invade those areas of a field—according to soil type—suitable for its existence, and not cross soil type lines. The same is true of alfalfa weevils.

The corn borer will usually become a problem on lower humus level soils with a lower pH. Or, the pH may be 6.2 or even 6.5, but constructed with too much magnesium—that is, a soil without a true colloidal pH. Under such conditions the grassy type of weed shows up—fall panicum, for instance. It is possible to get grassy weeds in dark soils with more humus, but here again the level of calcium will be too low for the humus system to function correctly.

Sometimes working the soil too wet will create the condition that tells the corn borer to increase and multiply. In the case of lighter soils that are better drained, but still do not have the proper texture and consistency (poor structure, low humus) the corn borer might be a continuing problem, even after high calcium lime treatment. When this happens the soil is saying that some nutrient is missing. Always, the amount of humus is a direct factor.

13. corn earworm

It is quite rare to see the corn earworm present in an area where the corn borer is present. There are, in fact, parts of the country—usually high pH areas—where the corn borer is simply non-existent.

Generally speaking, the corn earworm and the corn earworm beetle come together. During the early silking stage, the corn earworm beetle comes out to attack emerging silks. This little critter starts as a moth. In the beetle stage it'll chew off silk because the silk does not have the right maturing protein capacity to permit pollinization. This is nature's way of saying that the protein structure of the silk vehicle is not mature, usually a consequence of improper nitrogen fertilization.

WIREWORM

Moisture, air and soil texture, particularly at planting time, all figure in wireworm proliferation. Heavy tension of soil particles in the water—if there is more hydroscopic water and less capillary water—figures. The earlier lessons of this primer have detailed the chief culprit—high magnesium! It affects the tension of soil particles and underwrites more hydroscopic water and less capillary water. A good seed base starter, a combination of amino acid complexes and trace minerals to fit the limitations of the soil, enables the farmer to build a defense system around the seedling so that the wireworm won't invade the peripheral area.

A wireworm in a tuber leaves enough emptiness in the tuber to make it unmarketable. When this happens, it usually relates to putting too much water on the soil at the wrong time. This means the tension of the soil is holding the water so that in the early stages of growth hydroscopic water is inviting the emergence of the wireworm. The wireworm will go for sugar in the surface of the tuber or carrot and build fine looking caverns. The wireworm can be stopped cold with amino acid foliar sprays. As these nutrients are absorbed into the plant, they influence the hormones of the tuber. In 24 hours wireworms back off and disappear. Here is a classical case of foliar nutrients influencing the hormonal equilibrium. The best of the foliars can do this very quickly by supplying a complete complex form of every trace mineral, a shotgun mix that can fill voids created by wet conditions.

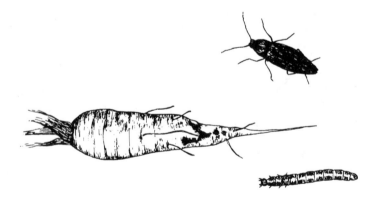

14. bollworm

Bollworm is simply another name for the corn earworm and tomato fruitworm. In fact, cotton isn't even the preferred food plant for the bollworm. The bollworm likes rapidly growing, succulent cotton for egg laying. The small larvae feed on tender buds or leaves before burrowing into the bolls. Here, again, the protein structure of the plant is at fault—better yet, the kind of nitrogen fertilization is at fault, as is the general soil management system.

The pink bollworm—another menace—follows soil with a low level of calcium, and is more severe where the structure is tight or cakey. When soils are loamy and have better organic matter complex, the humus systems are tougher, and therefore the problem is never as severe. Infestations vary according to the individual soil's limitations, either between fields, or within a field. Many times a plant is not only affected by the poor soil conditions in relation to the pink bollworm, but it may also affect certain fungi that prevent the plant from having a normal growth. In the west the pH is usually high, constructed by sodium and magnesium. This is prime pink bollworm territory. In the south pH is usually low. The bollweevil is perhaps more characteristically a southern insect.

pH has a direct influence on the hormone systems that allow the egg to emerge or to remain dormant. And that hormone process in the soil has a very important effect on the cycle of insect or disease that is going to emerge.

It all reflects back to the structure of the soil and its ability to regulate air, water, temperature and decay. These four factors could be different from year to year, and the intensity of the emergence of a specific weed or insect system is going to be influenced by these conditions.

15. the cotton fleahopper

The cotton fleahopper almost always relates to one or some combinations of trace mineral unavailability. Sometimes a certain single molecule of iron or zinc in a certain position in a molecule disestablishes that amino acid character so it is unable to function completely. Trace minerals have to be metabolized. If they do not arrive, the result is a kind of ersatz amino acid complex. It is not able to complete itself and handle the work of forming a finished, mature protein. This is what a higher balance of life has to have. Trace minerals affect the health of animals and human beings based on the complete functioning of the amino acid structure while proteins are being built. Any plant or vegetative system must have the capacity to mature those amino acids into finished, ripe, edible, acceptable exchangeable proteins. And if they're missing a link, nature decrees that they should be taken out of the food chain. That's the function of insects again. The cotton fleahopper has its coded order.

THE CALCIUM-PHOSPHATE SYSTEM

Calcium is an important element not only in the soil, but also in the plant. It is a trucker of proper forms of nutrients.

In the order of degeneration, calcium is prime, potassium is secondary, magnesium is third and sodium is fourth. These four cations govern the pH structure of the soil. They are the pH modifiers that act as the vehicle for carrying other nutrients into the plant's sap system. If calcium is carrying the nutrients, there is a better exchange across the cell wall. Calcium can pass through the membrane and pass back out without building up scar tissue—if one will permit the term. Magnesium and sodium will tend to go into the cell, get trapped and build up an excessive layer in the inner portion of the cell wall, and actually reduce porosity in the cell system of the plant. Calcium is the trucker, the carrier of the ingredients, the prince of nutrients, and that's why we need to have it in good supply in the soil as well. If it is present in the plant's sap, then the hormone processes are now served better. They have more mobility, and as a result they can function to their optimum. Mites and aphids on alfalfa or any crop that needs that calcium-phosphate system working are going to reflect those areas in which the farmer does not have this system working.

When the cotton plant hands off its distress signal, the fleahopper responds. Man can poison the fleahopper, but he hasn't come within shouting distance of the cause when he uses this procedure.

16. codling moth

Anyone interested in scientific insect management simply has got to make some endeavor to understand the relationships between cation groups, such as potassium to magnesium, calcium to potassium, and sodium to potassium. Any of these four cations, which govern the structure of pH in a soil system, have a bearing on the hormonal functions possible in any species of leaf or plant or tree. A lot of the different moths and beetle species particularly related to trees are in turn more related to the calcium to potassium ratio. These insects can appear under cold, wet conditions in spring, and under warm, dry conditions in the late part of the plant's maturing phases. Their function is directly related to moisture or to life in moisture, and their availability.

Often in the spring of the year—under wet conditions—uptake of potassium is minimized in trees. This means an entire syndrome starts to work. In the fall, under dry conditions, potassium again becomes unavailable, or it may be available in complex forms—as is the case with ragweed.

This same syndrome figures whenever there is a pine beetle problem. Moths can lay their eggs. These eggs do not have a threat from the enzyme capacity of the host plant. They become viable. They hatch. And they go to work. But if the right level of calcium to potassium is working—and the other things that go with this requirement are also working—any number of eggs can be laid on the bark of the tree. It hardly matters. The eggs simply won't hatch. All this says that a codling moth problem in the apple orchard means the farmer had better start dealing with the soil according to nature's hard requirements.

Conditions inviting the codling moth surface with the arrival of either irrigation or heavy spring moisture. Soils low in calcium, soils with potassium dominating, add up to having the wrong system in tow. Tilth, fungal systems, even weed patterns will record the day and mark the lesson the codling moth is handing over to the farmer for consideration. Chemical spray programs are no answer for a farmer who thinks beyond this season, this disaster, this crop.

The apple maggot, incidentally, is somewhat related to this same inventory of fact. But, always, it is necessary to go deeper into cause and effect. Under the restrictions mentioned above, root capacity will be poor. Also, unfriendly fungal systems around the roots will become stronger and stronger. They will deteriorate the potential of that tree. Whenever there is a poor root system, there will be all kinds of malfunctions. A root system cannot be corrected overnight. It may take several years to invite and

GRASSHOPPERS

Grasshoppers turn up in every part of the country from time to time. They arrive from drier soil areas where crops are growing under restrictions due to poor water supply. They migrate from their beginning habitat hard on the hunt for more food, but the food they seek has to be produced under the conditions left behind. In irrigated territory, the grasshopper proliferates in unirrigated parts of the field, along road borders and in ditches, but it won't cross the line where water is plentiful and soil and plant nutrition is balanced. On dryland acres, of course, plants are often suffering from lack of moisture and from a whole combination of nutritional problems. And crops suffer according to how well the soil is managed to retain moisture. This makes it imperative that the farmer learn how to improve the capillary moisture system in the soil in spite of the fact that rain is sparse. Nature does not invite grasshoppers into field areas that exhibit growth under well managed conditions. Under dry-cycle conditions, of course, the door is opened to the grasshopper hazard almost everywhere.

MORMON CRICKET

The Mormon cricket is a member of the grasshopper family, but it will not attend the usual family meetings where grasshoppers assemble. There is a reason for this. The Mormon cricket wants a cloudy, dark environment. It wants gas coming from the decay of organic matter on the surface of the soil. When a putrefying organic mass on the soil's surface is delivering gas molecules under moist conditions, that's when a farmer gets Mormon crickets.

The grasshopper is just the opposite. It wants dry conditions, not dampness with methane or carbon dioxide gases coming off the soil. It wants plants stressed for want of water, plants subclinically ill. It does not like plants stressed with too much water, on the other hand.

stimulate the hormone processes necessary to develop new roothair structures. This means it may be necessary to work with the soil for a year or two before limitations can be corrected in the upper parts of older trees.

The same principles hold for citrus groves and other fruit trees. The older the tree, the longer they will have had to build up the monoculture around roots that is now restricting root capacity. It isn't simply a matter of moving in with foliar nutrition to correct the problems.

In fact, a sudden increase in the capacity of leaves and branch systems to demand more nutrients must be backgrounded by a root system capable of supplying those fundamental nutrients, otherwise it is possible to grow a tree to death. After that initial application, the tree may look lush and even start to develop fruit, but then it dies. If a farmer puts extra demand on the photosynthesis capacity, he must always remember that this extra capacity requires balance between roots and leaves. Simple spraying and using hot-shot formulas can be detrimental. Always, it is necessary to work with the entire system. It does not speak well for eco-agriculture when growers take prune, pecan, banana and apple groves and grow them to death.

17. a consequence of toxic control measures

Often insect problems are a consequence of toxic control measures used against some other insect. Witness Imperial Valley's on-again, off-again problem with the tobacco budworm. After attempting to control the bollworm in cotton fields with heavy applications of insecticides, growers kill the parasites that control the tobacco budworm. After that the tobacco budworm prospers and simply takes over the crop.

The stage for this scenario invariably is set by dry winter months. The kind of weather that prevails in the dormant part of the year has a direct bearing on the phylum of life that's going to endure the following year. With dry soil conditions over the winter, soils highly alkaline and with a high pH invite an entirely different system of life to emerge. Southern California ends up with the tobacco budworm. Then when moisture comes to the cotton fields, the tobacco budworm disappears. It moves off to the lettuce fields around Salinas. Here they are overfed with nitrogen from bitter lettuce that is not growing right. The end result—insect populations out of control. Here, again, the insurance policy of imbalanced NPK fertilization delivers a sales ticket for lots of toxic rescue chemistry, the amateur's diploma.

Insect control really has to do with environment management. It is the environment that produces a healthy crop, and in a sense the eco-farmer does not deal with the negative aspects called insect control and weed control at all by thinking about and working with nutritional factors, the eco-farmer cancels out not only insects and fungi and biological predators, but weed problems as well.

WHEAT STEM MAGGOTS

Wheat stem maggots are to wheat what aphids are to corn, and what some of the other fungi are to plants. One and all they seek that excess sugar supply area at the base of the growing plant inside the stem. That's where the sugar is accumulating. Maggots simply drill into the plant, and chew off the beginning of new foliage and new parts of stems growing from the lower level. They'll chew that off inside the plant, and then the plant will never head out.

The cause and effect relationship brings any thinking farmer back to the soil, back to the many relationships that preside over potassium availability and functional enzyme systems.

18. the tiny trichogramma wasp

So far, most non-toxic approaches have had to do with biological control, with using one insect to battle another.

The tiny Trichogramma wasp is the best known of all egg parasites. It attacks over 200 species of pest insects, and has been used successfully against most moth and butterfly eggs in all kinds of habitats throughout the world. Everett Dietrick founded Rincon-Vitova Company, Oak View,

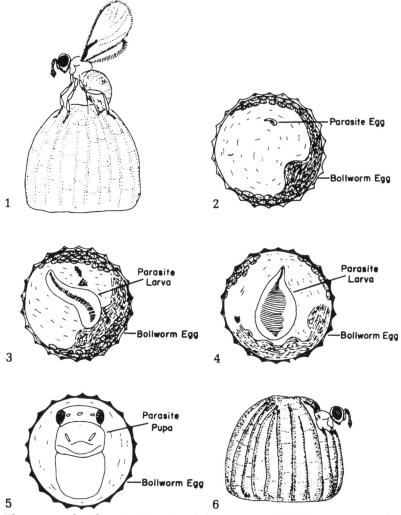

1. *Trichogramma female ovipositing in a bollworm egg. 2. Trichogramma egg inside the bollworm egg, dorsal view. 3. and 4. Of larval development of the Trichogramma, dorsal view. 5. Pupa of Trichogramma, dorsal view. 6. Trichogramma adult emerging from a bollworm egg.*

California, largely on the basis of what the Trichogramma and other beneficial insects could do.

The Trichogramma wasp lays its eggs in the eggs of host pests. In due time the Trichogramma eggs hatch into larva which feed on the immature pest while still inside the host egg. Trichogramma larva completes its development inside the pest egg, and emerges as an adult. Then the cycle begins again.

Depending on the size of the egg to be parasitized, the Trichogramma may deposit one or more eggs in it, and several larvae may develop inside the same egg. A sequence of releases based on weather, planting dates and host egg densities has been worked out by Dietrick and his co-workers. The results depend on timing of the releases and putting the propagated wasps where most moths are laying eggs.

Trichogramma is immediately effective. It kills individual pests before they can damage a plant. A drastic reduction in the numbers of pest eggs leaves fewer pests to survive to adulthood and makes their biological control by other beneficial insects that much more effective.

Black light insect traps are good indicators for the presence of egg laying moths. Trichogramma develop more quickly in summer than in winter. A generation takes ten days at 80 F.

For a moderate sized vegetable and flower garden, 15,000 Trichogramma is usually sufficient for each release. Three releases two weeks apart, starting after the garden is growing well in spring, delivers the best protection. Additional releases should be made if pest problems are severe.

Deciduous fruit and nut orchards can be protected with the release of 15,000 Trichogramma to the acre as soon as leaves are fully emerged. Spring flights of most moths entering an orchard can be covered with repeated colonization. Depending on the balance achieved with soil management and nutritional sprays, additional releases can help fine-tune protection for fruit. Obviously the variations are many. But Trichogramma do work if good judgment is used.

The above recommendations hold for non-deciduous trees and bushes—ornamentals, citrus, avocadoes, almonds—except that the timing of releases should start in early summer when moth activity gathers speed.

Alfalfa, cotton, sorghum, field corn, and grains in general are all infested by various worm pests under imbalanced fertility management, or even when best intentions fail their mark. Weekly releases of 2,000 to 5,000 Trichogramma per acre starting with the first warm weather in early summer have been found to deliver excellent biological control.

19. green lacewings

Green lacewings (*Chrysopa carnea*) are small, pale green insects that serve as all-purpose predators of insects and mites. Lacewings are an effective natural enemy of aphids, mites, whiteflies, mealybugs, leafhoppers, thrips, all sorts of butterfly eggs, and caterpillars. The larva is known as the "aphid lion" because of its voracious appetite for aphids. Just as Trichogramma is the universal egg parasite, lacewings are the broad spectrum predators most useful in biological control of nearly every crop and pest situation.

20. ladybugs

Ladybugs (*Hippodamia convergens*) represent another entry in modern biological pest control. Some insectaries field collect the ladybugs in their hibernation sites and sell them to farmers. This hasn't worked out very well simply because the collected bugs fly away from the area where they are seeded. Dietrick has developed a program of conditioning diapausing ladybugs so as to wake their reproductive activity. Pre-fed and conditioned, these ladybugs then feed and lay eggs in the area they encounter aphids.

Ladybugs are dependent upon aphids to trigger their egg laying response, whereas lacewings are released as hatching eggs that will eat any pest they encounter. The small gardener can get biological control by releasing 3,000 to 5,000 lacewings at pest sites early in spring. Augmentation may be indicated and this can usually be determined with ease by observation. Lacewings also control orchard pests. Eggs and larvae can be placed where tree branches touch the ground or in the crotch of the tree, or by being salted on leaves. Larvae are extremely secretive and hold on to plant surfaces.

21. fly parasites

Fly parasites are tiny parasitic wasps that provide excellent fly control in breeding sites on poultry and egg farms, dairy and beef cattle feedlots, backyard horse stalls, and where there are other domestic animal manure accumulations. These parasites attack flies by laying their eggs in the larvae and pupae. The eggs hatch inside the flies and the parasites eat the immature flies, then emerge as adult parasites. They cannot become pests themselves since they can only survive by finding other immature flies on which to lay their eggs.

The most important pests of tomatoes are tomato fruitworms, and hornworms. These, of course, can be managed with mass releases of *Trichogramma pretiosum*. Green lacewing has also proved effective.

Biological control of late season pests such as armyworms and pinworms can be sustained with releases of *Chelonus texanus* and *Apanteles scutellaris*. The key to success here is regular sampling of insects in the field using a D-Vac vacuum insect net. It gathers in complete collections of all insects, beneficial as well as pests. Entomologists working with farmers can

APHIDS

Aphids attack the sugars of poor-doing plants. They find photosynthesis operative sections of plants that are malnourished, those certainly not supplied with an adequate balance of nutrients for the synthesis process to function properly. Various restrictions in light might figure here. The problem could be related to restrictions in the supply of carbon dioxide. But generally speaking, aphids mean synthesis of sugars due to malnutrition. Here plants do not have the capacity for breaking those sugars as quickly as they are being produced by the heat energy and light energy of the sun. As a result, there is an accumulation of sugars in the upper portion of the plant. Usually this is where malnutrition will first appear. Aphids always find those regions of weakness, which are very succulent. These insects simply suck the sugars. This photosynthesis weakness can be corrected with a supply of nutrients that permits photosynthesis to be more effective.

Aphids appear more commonly during the hot part of the year when the sun is working its best and shade starts to be a factor in the growth of plants. Aphids may be present in fields, and yet turn up only on the lower side of the first two or three leaves of a corn plant—down where the brace roots had developed just below the soil, where the connection of those leaves has been stripped. Such leaves are not working, and they are not feeding back to the plant. They have served their purpose and have no further function in the life of that plant. Their job was to get the plant emerged and out of the ground, and that's what they did. Now the aphids find those parts of the plant and set up shop.

A little later in the season a plant may run out of potassium. This means the enzyme system isn't working properly during the maturing phase. It is then that aphids suck.

thus assess the presence of the natural enemy complex and design effective biological control.

22. biological control of mosquitoes

Mosquitoes are, in part, beneficial to mankind. They are part of the decomposing class of organisms that break down organic matter to simplified compounds so they can be reused in the carbon cycle as plant and animal nutrition. It is only the excess numbers of adult mosquitoes that constitute a pest problem. The eggs, larvae and pupae can exist in large numbers as long as sufficient biological controls are present to prevent them from reaching adulthood. The extremely high reproductive rate of mosquitoes provides for a very diverse and complex set of predators and parasites which derive their nutrition from feeding on immature mosquitoes. Hundreds of organisms feed, directly and indirectly, on the many kinds of mosquitoes in the world. Even the adults serve as food for birds and many predatory insects. When the entire natural control complex is present and at work, a tolerable balance is maintained. There is an equilibrium position around which mosquito populations fluctuate. They rarely increase beyond this equilibrium position to intolerable numbers.

When farmers institute ecologically disruptive control measures that destroy more natural enemies than mosquitoes, they create a mosquito problem, rather than solving one. Once the habitat is disrupted, repeated pesticide treatments become necessary until resistance sets in and the super mosquitoes appear. It is then that the chemical salesmen cry out for stronger poisons and huge eradication efforts, threatening the public with a health disaster unless they are permitted to spray again and again.

Fish are perhaps the most effective enemies of mosquitoes. The following species are listed as most important in the salt marshes of southern Florida: *Cyprinidon variagatus, Mollienesia latipinna, Gucania parva, Gambusia affinia Holbrooki, Fundulus confluentus,* and *F. grandis*, along with transient fish (immigrant young of larger species).

The mosquito fish *Gambusia affinia* is now being widely colonized for biological control of fresh water anophelines. A number of Tilapia species show great promise for use in mosquito population management.

Hydra and Planaria are almost microscopic animals that are useful in controlling mosquito larvae. These may be produced economically and released for biological control. Many small plants have been shown to affect population density of mosquitoes. The main arthropod families known to prey on mosquitoes are dragon flies, damsel flies, aquatic beetles, spiders and predatory mites and bugs. Various parasitic species attack mosquitoes, including water mites and merimethid nematodes.

Sound biological control goes well beyond releases of insects or eggs into

MOSQUITO BLUES

Why do mosquitoes feast on one individual and avoid the next? In Sugar Blues, Bill Dufty claims that mosquitoes like sugar. USDA's Phil Callahan says mosquitoes hone in on the ketones in the human breath. We can, of course, speculate that sugar affects the ketones, and over-sugared people are simply more attractive to mosquitoes. Ditto for animals.

Dufty gave up sugar as part of a health recovery program and now religiously avoids sugar-containing foods. As a consequence, he can lay stripped to his bathing trunks in an area where mosquitoes literally fly off with victims sunbathing a few feet away.

Mosquito control has always proved elusive. Chemicals may annihilate the insects for a time, but the little critters come back stronger than ever. Biological control has proved at least as effective as chemical control. Recently Ohio State University insect pathologists isolated a bacterium from sick Indonesian mosquitoes that has proved fatal to mosquito larvae. The bacterium affects only mosquitoes. It could become available commercially in parts of the United States by 1981.

the pest infected environment. Consultants involved in biological control like to couple releases with sound ecological principles, using pesticides only when indicated to protect the natural enemy complex. For example, waterways should be opened somewhat to keep them free from emergent vegetation. This permits fish and predacious aquatic insects to have access to mosquito larvae and pupae. Controlled burning to clear underbrush in wooded areas can open up these areas to better air circulation so that birds can help the habitat favor biological control. The goal of habitat management should be to create an ecological house for the biological control organisms, and to discourage the survival of mosquitoes to adulthood.

Biological control systems destroy the eggs, larvae and pupae of mosquitoes proportionately in each state until only a few complete the cycle to adulthood. When a complex of predators and parasites is present and at work, very few mosquitoes reach adulthood.

Water pollution is detrimental to biological mosquito control. It causes the oxygen in the water to dissolve. As oxygen decreases, so do certain microorganisms, and in turn the insects and fish depend on these microorganisms for food. Improving the streams to increase oxygen content, and clear the turbidity from the streams, will result in an improved shelter for aquatic organisms and fish which feed on mosquitoes.

23. pests of grapes

The important pests of grapes are Pacific mite, Willamette mite, Omniverous leaf roller, grape leaf folder, and grape leaf hopper. Most of these are pests that have been created by the excessive use of pesticides. Both the omniverous leaf roller and grape leaf folder worms are normally controlled by a natural enemy complex. They only become pests when that complex is destroyed.

The Pacific mite is considered to be the worst pest in western grape arbors, but this too has been greatly worsened by overuse of insecticides. Even leaf hopper damage is over-estimated, since its attack is chiefly cosmetic. It has been demonstrated time and time again that the grape plant can tolerate reasonable populations of leaf hopper with no economic loss in yield or quality.

Good cultural practices are implied, of course, as a part of biological control. Dust control, ample water supply, sprinklers—all are important, the latter because they knock mites off the vine and drown them. One of the most important cultural practices is planting of blackberry patches near the vineyard. There is a parasite called Anagrus which is effective against leaf hoppers. This parasite winters only in blackberries where it parasitizes another species of leaf hopper. If enough blackberries are present to harbor substantial numbers of Anagrus, then leaf hopper damage will be nearly eliminated.

24. fly control

A perfect example of biological fly control is composting manure. All the composting organisms of bacteria, fungi, mites, insects and even small animals, aid in the decomposition of animal dung. Many of these attack the immature flies in manure. Some aid in drying manure.

There is more to biological drying of manure than dehydration due to climatological effects. This can be seen in the shrinking of manure as it dries, rather than a glazing or crusting over as in the drying of mud. Some organisms aid in drying by encapsulating much of the water in manure in their bodies. This activity is important to fly control as well as odor control. Aerobic non-smelling bacteria live in composted, aerated manure as opposed to anaerobic bacteria which live in wet manure.

Manure should never be sprayed with toxic genetic chemicals because natural enemies will die in greater numbers than flies. In order to derive maximum benefits from biological fly control, existing predatory and parasitic fly enemies must be preserved in the manure. Once the biological control complex is established in the piled manure, it is important to culturally manage this dung in order to avoid driving away the beneficial insects during cleanout.

Certain management techniques suggest themselves. One is alternate row cleanout, leaving a dry pad. The best time for cleanout is early summer and after the spring increase in flies. When all manure is cleaned out at one time, almost all the beneficial insects are either removed with the manure or driven away. Cleaning out only alternate rows of manure preserves half the habitat of natural enemies. They are concentrated and ready to move into the new piles of manure that are beginning to form in the cleaned out rows. After a month or so, fresh droppings will have built back to a control complex so that the remaining rows may be cleaned out with minimum disruption. Leaving a 6 to 8 inch pad of droppings under cages at cleanout time further aids in providing a base to start drying fresh droppings and preserves the habitat of the biological control organisms.

There was a time when biological control appeared set for the great leap forward. USDA *Technical Bulletin No. 1139* was published June 19, 1956. *Use of Diseases to Kill Plant Insect Pests* was printed by USDA Agricultural Research Service in October of 1961. This fundamental work was picked up, expanded and communicated by entomologists such as Dr. Robert van den Bosch and his associates—individuals interested in putting biological control on the front burner. Paul DeBach's *An Analysis of Successes in Biological Control of Insects in the Pacific Area* became a part of Proceedings of the Hawaiian Entomological Society the year Rachel Carson published *Silent Spring*, 1962. It told of 125 successful control projects in the Pacific area. Involved were 87 pests in 23 regions, or more than 50% of the 221 world cases of successful biological control in some 65 countries.

BACILLUS THURINGIENSIS

One of the new biological insecticides now available to eco–agriculture contains viable spores of Bacillus thuringiensis. There is general agreement that the product is harmless to man and animal. Food and Drug Administration has exempted it from the usual tolerance claptrap used to rate poisons. The glossary of this primer amply explains the LD50 concept. On the basis of tests with mice, dogs, rats, rabbits and guinea pigs, Bacillus thuringiensis came up with flying colors. Extremely large oral doses produced no toxicity. Inhalation studies revealed no ill effects. Only mild, transient irritation followed derma and ocular application. No sensitization or anaphylactic shock was produced in tests with animals. This one is usually available under the Dipel label.

DIMILIN

Dimilin, often touted as a bio-control for the forest tree destroyers, is really a carcinogen. It continues to cause cancer of the spleen in test animals. In fact Thompson-Hayward's registrant information revealed tumor production in test animals. As a cancer causer, Dimilin is a hazard to wild life. It also causes methemoglobinemia, a condition in which the blood cannot transport oxygen. This is instantly fatal to infants. Dimilin has been used to control the gypsy moth. It was perceived to be a bio-control because it inhibits production of chitin in animals. Unchallenged, this conceit could be used to characterize almost any of the man made poisons. Almost all inhibit something, damage something, destroy something. Dimilin is in fact made from benzene. This substance has been banned in Canada. In the U.S. officials are using Sevin to combat the gypsy moth. Repeated use of Sevin has turned Maine and New Jersey into a cancer corner.

SHORT COURSE

LESSON 24

Weeds

It is a common experience these days to take a field trip across the corn and wheat belt and bring home field notes about mustard taking over whole fields of grain crops, or weeds of several kinds crowding out pastures, row crops and orchards. As with insects, the great overload of chemical poisons and herbicides seems to produce more weeds than it eliminates. In the early days of publishing *Acres U.S.A.*, there used to be long conversations about weeds with William A. Albrecht at the University of Missouri. And each time Albrecht would simply reiterate his observations—that failure of the chemical bandwagon was coming on strong because we refuse to look into the history of plant domestication. He said that domestic plants in their virgin condition grew to a climax stage of a clean stand with exclusion of competing weeds.

"Growth is possible only through the balance of the several fertility essentials in the soil because better nutrition protects the crop to nearly complete exclusion of all other crops," Albrecht said.

Albrecht and his associates came by this knowledge on the university's facility, Sanborn Field. There they learned a very simple lesson about nature's answer to the problem of noxious pasture weeds. Two plots had been maintained in continuous grass, namely timothy. One plot had been dressed with barnyard manure at the rate of 6 tons an acre. The second plot received no treatment. Crops had been removed regularly from both plots.

The untreated plot often became so infested with weeds that plowing and reseeding was required. Accordingly, both plots were similarly treated to plowing and reseeding, regardless of the weed free grass stand on the manured companion plot. The manured plot always delivered a fine sward with early spring and late fall growth of quality hay—2 to 2 1/2 tons every summer. The untreated plot exhibited snowy white, wooly tops of broom sedge during seed production time. This attracted a great deal of interest.

Soil samples were collected and sent to B. M. Dugger, one of the research doctors at Lederle Laboratories in New York. These samples made possible the isolation of *Actinomyces aueofaciens*, the source of the antibiotic aureomycin. Isolation was effected from the weed infested plot. The same fungus apparently could not be isolated from the manured companion plot, the one that grew excellent timothy hay routinely.

"The important thing to note," said Albrecht, "is that broom sedge failed to invade the nearby plot, yet windborne seed was scattered extensively. Seeds were particularly numerous on the manured plot because winds blew from the west, or from the infested plot."

Examination of the 40 other plots on Sanborn Field revealed that they, like the manured plot, were immune to infestation by the weed. Broom sedge was a threatening and dangerous weed only to untreated plots and to grass-sod roadways between plots.

It may be that Professor J. W. Sanborn, who laid out those Missouri University plots in 1888, knew what he was talking about when he said, "Fertilize the soil so it will grow grasses which are healthy plants and thereby nutritious feeds, so consequently the troublesome weeds will stay away."

Albrecht scoffed at laws that hoped to prevent machines from carrying noxious weeds from farm to farm. He viewed soils of weakened fertility as the culprits because such soils invite the invader. As for costly weed poisons, they were that: costly. They could not control weeds in the fullness of time. They could only deliver side effects it would take generations to understand fully.

1. an index

Weeds are an index of what is wrong—and sometimes right—with the soil.

There are over 400 plants in the United States that have been characterized as weeds, it being understood that weeds are weeds only in man's eye. Nature has assigned special roles to weeds. Seed producing weeds—especially the annuals—produce a far greater abundance of seeds than so-called cultivated crops. Herein is both a clue and a logic entirely consistent with nature.

Few agronomists have picked up either that clue or that logic. Exceptions are people such as Newman Turner (writing in *Fertility Pastures*) and Ehrenfried E. Pfeiffer (writing in *Weeds and What They Tell*).

2. pfeiffer saw weeds

"Weeds are specialists," wrote Pfeiffer. "Having learned something in the battle for survival they will survive under circumstances where our cultivated plants, softened through centuries of protection and breeding, cannot stand up against nature's caprices." Pfeiffer also noted that weeds "resist conditions which cultivated plants cannot resist, such as drought, acidity of soil, lack of humus, mineral deficiencies, as well as one-sidedness of minerals. They are witness of man's failure to master the soil, and they grow abundantly whenever man has missed the train—they only indicate our errors and nature's corrections."

Pfeiffer saw weeds as a great teacher, "indicating through their mere presence and multiplication what is wrong."

There were the weeds that lived on acid soil—the sorrels, docks, fingerleaf weeds and lady's thumb—and in some cases horsetail, hawkweed and knapweed.

One group suggested crust formation or hardpan, or both—field mustard, horse nettle, penny cress, morning glory, quack grass, the camomiles and pine apple weeds.

A third group included the weeds of cultivation, or farmers' weed—lambsquarters, plantain, chickweed, buttercup, dandelion, nettle, prostrate knotweed, prickly lettuce, field speedwell, pigweed, common horehound, celandine, mallows, carpetweed.

Pfeiffer's several other classifications were also fairly perceptive. He defined the salty soil weeds as shepherd's purse, Russian thistle, sea plantain, sea aster, artemisia maritima. He called attention to weeds brought on by inept fertilization, such as excessive use of salt forms of potash—the wild mustard plants and weeds of the Crucifaerae family, a prolific and troublesome tribe—charlock (*Brassica arvensis*), Indian mustard (*B. juncea*), ball mustard (*Neslia Paniculata*), wild radish (*Raphanus raphanistrum*), the peppergrasses both common (*Lepidium virginicum*), and green flowered (*L. apetalum*), and field peppergrass (*L. campestre*) as well. It is a matter of record that seeds from plants such as charlock and wild radish can lie inert in the soil for up to 60 years. Now and then a farmer will plow up a pasture, forgetting that it was once a grain field. A swift reminder can come on in the form of the wild radish, its seed coming to life after several decades in an underground bunker. Also a beneficiary of heavy potassium overload in the soil are marsh mallow, wormwood, knapweed, fumatory, and the opium poppy. Pfeiffer once pointed out that red clover simply disappears when there is a shortage of potassium, hence its value as an indicator plant.

Here are a few more Pfeiffer observations.

Sandy soils often deliver lush growths of goldenrods, flowered asters, arrow leaved wild lettuce, yellow toad flax, onions, partridge pea, and broom brush. Steppe type pastures are often infested with Russian thistle, sage, and loco weed. Alkaline soils of the west deliver a fair complement of iron weed, sagebrush and woody aster under most circumstances.

Limestone soils, such as those along the Missouri River bluffs and the race horse country of Kentucky, identify readily with pennycress, field peppergrass, hare's ear mustard, wormseed, Canada blue grass, cornelian cherry, Barnaby's thistle, field madder, mountain bluet, and yellow camomile (*Anthemis tinctoria*). When lime is absent or perhaps locked up and complexed, yellow or hop clover, rabbit's foot clover, fox glove, wild pansy, garden sorrel, sundews, white mullen, scotch broom and black vetchling surface and sour the situation.

Badly drained soils develop quite an inventory of weeds. Other factors figure in just which ones survive and prosper. Generally speaking the inventory will include smartweed, mild water pepper, hedge bindweed, silverweed, white avens, swampy horsetail, meadow pink, hedge nettle, stinking willie, Canada and narrow leaved goldenrod, tradescant and purple stem aster, Joe-Pye-weed, March foxtail and rice cut grass.

Grain fields are not limited to these, but include them all: wild buckwheat, all the mustards, wild radishes, pennycress, grass leaved stickwort, mouse ear chickweed, morning glory, purple cockle, bachelor's button, tansy, poppies, chess.

Buttercups, dock knotweed, all fingerweeds, white avens, grassleaved stickwort, St. John's-wort, pokeweed, milkweed, briar, wild garlic, and thistles, all are pasture weeds, albeit not limited to that location.

3. first principles

To put all this in another way, there are reasons why these several weeds tend to grow where observations have placed them. Some weeds announce completely worn out soils, acres so fully deteriorated that they no longer have a living function. Eroded hillsides with a lack of organic matter, and a pattern of nutrients being leached away due to excess rain, both set this stage. Or it may be a case of excess calcium or sodium flowing out of one area and settling into a lower basin area, filling up the catch area with a high sodium content.

Plowing up and down a hill during the fall when there is no body in the soil can trigger erosion. Under such a circumstance the farmer should not be fall plowing at all. He should be mulching instead. Moreover, machinery systems can be either a plus or a negative, and this can lead to degeneration. The weeds say it all when they tell about completely deteriorated soil.

Imbalanced pH soils have no body structure. As a consequence soil particles are subject to wind erosion or leaching or water erosion. These are causes of breakdown in the soil system. All, of course, are related to how a farmer manages his soil.

There are weeds that talk about pH character. Soils with a pH in disequilibrium do not have the right structure or body or porosity or capacity to hold or regulate the release of nutrients. They might have an excess of one cation over another and exhibit a domination effect. Weeds have a lot to say about imbalanced pH—much more, in fact, than the entire lesson on pH in an earlier part of this primer.

One weed might grow three feet tall under one condition, and grow hardly a foot under another. The same weed may be thick stemmed and have fine leaves in one environment, and it may have a thin stem and a lot of spindly branches in another. Stunted weeds speak of soil limitations and genetic arrangements of the plant itself.

A lot of weeds point to drainage and how drainage is directly influenced by pH structure. Indeed, pH character, makeup—the inner reactions of elements within pH—all have a bearing on function and what it does to change soil structure. pH determines whether the soil drains, whether it holds water tightly, how it aerates, and how it compacts. Soils that clod together invite certain weeds. As respiration continues within a clod as organic material ferments, carbon dioxide may be accumulated and trapped within the clod. The presence and concentrate of carbon dioxide tends to stimulate a specific group of hormones, the kind that wake up foxtail, nut sedge and watergrass, and tell them, this is your year to increase and multiply. In fact, germination of each weed species is dependent on specific hormone-enzyme systems, presence or absence in the soil seed zone.

Weeds speak in terms of pH, and pH speaks in terms of weeds—if a farmer has the wit to read the message. Poor pH structure has a negative effect on water management. As a result poorly drained soils are easily mismanaged. They are often plowed too wet and driven over too early and compacted. Almost all of the weeds on which some farmers use chemicals such as atrazine are of this character. Atrazine is designed to deal with weeds that grow on poorly structured soil. It does absolutely nothing to correct the pH shortfalls in the soil that brought on the weed problem in the first place.

There are weeds that are related to the direction of decay of organic matter—just as Pfeiffer noted. How residues, roots and crop materials are processed in the soil—and the direction of decay they take—determines whole classes of weeds. Soils that are not balanced properly in terms of cation exchange capacity do not permit the proper decay of organic matter. Thus they either produce hydroxide, alkaline solutions or different gases— methane, ethane, butane, all byproducts of decaying organic matter in an undesirable environment. The wrong direction of decay produces alcohols which depress biological processes in the soil. In the absence of biotic activity, hordes of antagonistic hormone processes go into motion, giving farmers buttonweeds, jimson weeds, in fact all the weeds in that general family. Without exception they are indicators of improper decay of organic matter or, if you will, "soil indigestion." These weeds say the soil is not able to manage decay of organic material into a good, virgin, wholesome, brand new supply of fresh humus, and without this good governor of a living soil system, problem weed types announce the degeneration of soil potential.

Morning glories and bindweed have rhizomes. Under conditions of improper decay there are no fungi antagonistic to those rhizomes. It takes correction of pH, air capacity and moisture control to set in motion a soil environment that assures the proper decay of organic matter. Under such soil conditions a whole new array of fungal systems can start to break down the rhizome seed stock and eventually correct the soil system for desired crop growth—yet one antagonistic to the bindweed and morning glory. These basic fungal systems can make short work of rhizomes and roots, digest them and remove them.

Weeds have a lot to say about minerals. Whether there is a mineral imbalance, or whether there is a potential in the soil system to release, complex or simplify minerals held in inventory, all are more important than the actual mineral content of the soil itself. Ragweeds grow behind the grain harvest in very dry soils, or they grow when stretches of dry weather makes the soil's surface over-dry. Under dry conditions biological processes are slowed to a snail's pace. Potassium depends on bacterial processes to make it available for utilization. Sluggish bacterial activity in these dry soils causes potassium to be unavailable or available in its properly processed form, and this is what ragweed is telling the farmer. Ragweed grows late in the season under complexed forms of potassium. It also grows in over-dried soils—or in soils flushed with water, then over-dried again. The sequence of rain—the wilting and drying of soils—determines how nutrients are processed, and this then sets in motion the hormonal environment that stimulates the awakening of various species of weeds. This is the reason weeds come and go. One year may be wet-dry-wet, the next simply wet, the next dry all the way.

Variables of weather figure because weeds are often the consequence of stress. Fossil fuel based fertilizers are made water soluble. When the soil dries up, these forms of fertilizers are unavailable and often are totally complexed into the soil base. They change their form and do not remain available for root uptake. This sets the stage for a different plant to function because each weed species often requires a certain form, character and quality of nutrients.

Weeds almost write a biography about organic matter management on a farm. They tell quite a bit about how residues of previous crops are managed, whether they're plowed under, disced or mulched into the soil surface or reduced in size, or chopped up to expose the stalk to freezing and thawing on the theory that hazards of weather will kill pupae and eggs—the latter an asinine proposition. Often the consequences of organic matter mismanagement arrive six months later. Residue left on top of the ground frequently begets a slime mold. This seals off the soil and makes it anaerobic. Over the winter months rain and snow do well to tuck things in and pack down a good residue seal over the soil. And if cattle are allowed to glean crop residue all winter, they are almost certain to be forgotten during a wet stretch. It is then that they do the damage, packing the ground the way a sheepsfoot packer works over a roadbed.

When water can't pass through the variable density layers of the soil, spring tillage is usually delayed. Most farmers, however, are envious of a neighbor and eager to impress those who have a windshield view of the farming operation. This means they often rush the spring tillage anyway. Tracks beget compaction and compaction begets weeds. Then comes the spring ritual in which farmer bragging reaches an art form—how timely he was in applying a herbicide, modern agriculture's idea of holy water to forgive the sins committed in the name of early soil tillage.

Elsewhere in this primer are given temperatures at which certain seeds come to life, details carefully worked out by researchers over the years. No such list for weed seeds exists. In terms of ballpark figures, it can be noted that many weed seeds germinate at 40 to 50 F. and start to stake out territory immediately. A great deal of commercial crop production runs best when soil temperatures match the optimum temperatures for bacteria, the general range for bacterial activity being between 65 and 70 F. It should be noted, however, that even under lower temperatures, other bacterial and fungal systems continue to function, and each of these systems leaves its mark and eventual effect on the next generation of plant life.

4. clean tillage

Clean tillage has a nice ring to it, especially in a day and age of toxic chemical worship. The idea hasn't much of a press largely because TV and magazine ads tout herbicides that last (without carryover, of course) and keep that combine from clogging. The clean tillage side of the fence seemingly has nothing to offer but blood, sweat and tears. Indeed, many weeds are like a Hydra. Cut off one head, and another one appears. Mow down some weeds at the wrong stage of growth, and rather than controlling the pest, it responds with renewed hormonal vigor to quickly replace the flowers and "sneak through" a new crop of seeds. After all, reproduction of the species is the prime function of a plant. Wild carrots, as an example, require fine-tuned mechanical combat. If they are mowed when they begin to blossom there will be two to five new blossoms where mowing removed only one. If they are mowed after seeds are matured, the farmer's activity will serve as a veritable drill, scattering seeds to the soil, and sweeping them under with the tractor and mowing machine pass-over.

Some farmers use fire to burn away unwanted weeds. As a consequence, weeds rise like a Phoenix from the ashes more suitably fed than ever before. Fire visits destruction on organic balance in the top layer of the soil, coagulates humus colloids, and worsens the situation, but seldom kills the actual seed.

Eradication before and after weeds emerge isn't all that simple either. In theory frequent mowing and tillage starves roots, taproots and rhizomes of biennials and perennials, the kind of weeds that propagate from root splinters and plant parts. Judgment and insight have to govern the tillage procedure. Harrowing usually does little more than damage rhizomes and taproots, and if the season is wet the end result will be even more prolific growth. If the ground remains dry, such a frontal attack on biennials and perennials can enjoy a measure of success. Usually deeper working implements are necessary—a spring tooth harrow or duckfoot cultivator. Unless the root is dug up and removed—often an impossibility—there is always a danger of tillage serving to enhance, not inhibit biennial and perennial weed growth.

Annual weeds complete the cycle from seed to seed in a single season. Much like garden crops with small seeds, these weeds germinate very near the surface of the soil. Such weeds are easily disturbed by cultivation. When moisture conditions intervene to prevent reestablishment, annual weed crops falter and die while field crops flourish.

The single point in the question is always the same: what kind of weed are you dealing with, how does it propagate, and what are the conditions for its growth? In the final analysis, the biochemistry of weed control lies in fertility management, and not in coaxing a more deadly goodie out of the devil's pantry.

5. broom sedge

As noted above, broom sedge crowds out crops grown on poor or sterile soil. It is almost always a serious pasture weed on marginal land. The map on the following page of broom sedge distribution was constructed by USDA as the cutting edge of the chemical era got underway. It reveals the fact that there were more poverty soils in the chernozem soils of the rain-belt south at that time, and that California also held in escrow its fair share of broom sedge acres.

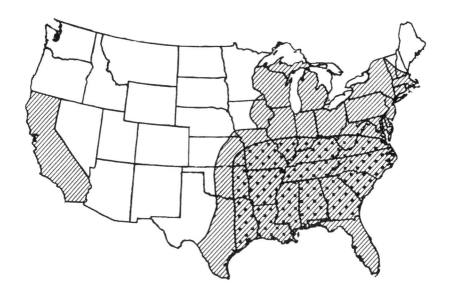

Broom sedge (*Andropogon virginicus* L.) is a perennial. It grows erect and in small clumps, is reddish-brown when dry, and roots very deeply. It has densely fibrous roots. Generally speaking, broom sedge means worn out, burned out, oxidized soil, soil without a governor, namely calcium. In many places it also means a lack of both calcium and magnesium. As soils get depleted and eroded, as topsoil vanishes and the humus supply disappears, real soil poverty sets in. And that is what broom sedge means—a poverty grass for poverty soil. Usually such a soil will have degenerated over many years as a consequence of breathtaking mismanagement. Such a farmer will likely have used great overloads of salt fertilizers, destroying soil structure each season of his tenure. Erosion is usually heavy on poverty soils. Since most of the topsoil is gone, the soil has to be rebuilt before broom sedge can be controlled. This is not easy.

A good starting point is calcium. The addition of calcium can quickly allow higher phyla of life to endure. As soon as the calcium in colloidal position moves up, life in the soil begins to recover. New hormone processes then start to stimulate higher forms of plant life, leaving the broom sedge behind as a dormant generation, the result of a drastic change in its biological environment. Calcium is the thing, but the right form of calcium is a must. Coarse gravel may take five to seven years to break down for release of enough calcium to influence the balance. It is a slow process in any case. It is slower in degenerative soil than in soil with a good organic matter inventory. For this reason fine calciumlime—modest amounts over a period of years—will best monitor the recovery process.

Magnesium is usually ample in poverty grass areas, although there are variables—Kentucky or the southern areas of California and Virginia, for instance. An application of dolomite or Sul-Po-Mag might be required in some cases.

Broom sedge.

Here, as with most weed situations, the calcium, magnesium, sodium and potassium equation must reign supreme to effect recovery. How these basic cations construct pH remains both the flywheel and the balancewheel of weed control.

Unfortunately soil test readouts generally available through Extension and most laboratories do not provide the figures needed to draw the appropriate conclusion.

How can a farmer get rid of this weed? First, it will be necessary to bring the pH back into better equilibrium. The calcium level must be brought up. This will change the hormone and biological processes in the soil. At first these adjustments will do little more than keep the broom sedge weed seed dormant and allow a better quality weed to grow so that it can return more humus to the soil. The comeback trail will permit other crops to grow and sustain the repair job at the same time. If, however, chemistry is used to get rid of the weed, attempted control will simply perpetuate mediocrity and take out insurance on an even lower phylum of life moving in and setting up shop.

Withal, broom sedge is what it is—a poverty crop for depleted, degenerative soil. It reigns supreme when a soil system has reached a depression of depletion, and stands as a flag waving its signal of having achieved total failure.

Broom sedge should not be confused with broom brush (*Sarothanmnus vulgaris*) because of the similarity in name. Broom brush is a soil builder in rough, sandy and worn out soil. It concentrates calcium carbonate, bringing that essential nutrient from areas not generally available to farm crops, then depositing calcium in the top few inches of soil via the mechanism of decay of leaves and stems. It can shade other plants, but it can also choke them out if allowed to proliferate.

6. giant foxtail

Giant foxtail (*Setaria faberi*) is an annual and is therefore reproduced by seeds. Weed experts make much of the fact that this weed was originally introduced into the United States from China, possibly in millet seed, although this is only of academic interest to eco-agriculture. There was a time when giant foxtail was a serious problem only in Missouri, Illinois, parts of Tennessee, Kentucky, and Indiana. Since hard nitrogen has invaded the scene, giant foxtail has expanded its geographical domination. Mechanization and an improved capacity to destroy the basic character of the soil have speeded up the processes of deterioration, hence foxtail over most of the country. Foxtail is a relative of pigeon grass. And like pigeon grass it grows more in an organic matter soil where there is a surplus of humic acid (because the content of humus is so much greater).

Although pH adjustment has been front burner stuff in another lesson of this primer, the topic has to surface in any discussion of the foxtail weed problem. Following faulty advice, farmers have been attempting pH adjustment with no more than ag lime for years. Most of the materials used have been taken out of quarries in the gravel business. Ag lime has almost always been a byproduct, a very coarse byproduct at that. These lime particles endure for many years, some of them for 25, 30, even 40 and 50 years. When farmers have paid little attention to the character of the lime, they have often ended up with dolomite, and this has increased the concentration of magnesium to a point where it tightened up the clay particles. As a result soils hold water longer. They do not warm up. And they are subject to more complexed conditions and thorough compaction.

High capacity machinery is almost always a contributing factor. When a farmer works the soil a little on the wet side, heavy machinery tends to compact the soil underneath the surface. Even working with a light disc, tractor wheels will compact the soil six inches to several feet deep. It is this compaction that sets in motion the environment which induces the foxtail seed to germinate. The foxtail seed germinates in the absence of air and in the presence of an accumulation of carbon dioxide which cannot escape its airtight chamber.

Magnesium is a contributing factor. It tends to keep the soil wet and it gets the clay particles to electrically bond themselves together tighter. The chemical-electrical bond on clay particles is very high when magnesium is in excess, and calcium is in deficiency. As a result of this, soil tends to hold itself together. Tractor traffic compacts it and presses out the air and as the soil begins to dry it cements together into a hard and germinous clod. This is the environment for the foxtail. This will partially explain why foxtail usually emerges a little later in the season, often after the final cultivation.

The solution is obvious. Fertility management has to be used to regulate a compaction propensity out of the soil system.

Lime in the right form is the key. But there is more. Weed seeds go into dormancy in the fall of the year. The length of day and night has a bearing on their life cycle. That is how nature controls all the living processes simultaneously—with kind of light, wave lengths of energy, angle or incidence to the sun, and how light factors are absorbed by hormones. Hormones must have enzyme partners, and enzyme systems must have trace minerals in the right form. Trace minerals are fundamental in catalyzing these systems to work. Those same trace minerals govern bacteria and fungus and hormones. Hormones are the time clocks that regulate the sequence of life. They do this by absorbing light wave energy. They tell the weed seed to slow down or change after the middle of August. If they get into a good, intense, efficient dormancy, they are easily susceptible to fermentation and digestion, especially if one or two tons of good compost is mulched into the yawning soil of the fall. Bacterial activity stimulated by the compost will digest the dormant weed seed, turn it into humus, and start the cleaning process.

This only happens when seeds go into dormancy, and only if compost is on the acre 30 to 40 days before a heavy freeze and winter. Foxtail seeds

Foxtail.

can be digested in this way. There are other benefits. The soil will drain faster. There will be more air. It will be possible to plant on time. With a commercial crop off and running ahead of weed seed competition, the kingdom for the year is established.

A final caution. If the soil is disced in spring, allowed to set for a week, then disced again, new bunches of weed seeds are given a wake-up shock from light. It takes only a millionth of a second of light to trigger the process. Spring tillage often wakes up weed seeds. The more times it is done, the more difficult it is to manage weeds. One extra pass might make a fine seedbed, but this may not be the ultimate compatible objective of wholesome crop nutrition and production.

Reason with us for a moment. Who doesn't remember planting, then praying there won't be a gulley washer that night? Who doesn't hope it stays warm? Yet many times after planting there is a cold or wet weather break. The soil gets waterlogged and stays cold for three weeks. This sets in motion a whole series of hazards. That last pass over the field may have fined the soil too much. Now it crusts over quickly. Pigeon grass, foxtail, watergrass, nut sedge—all grassy problem weeds now receive permission to take over. As a matter of fact it might pay to reread the lesson on tillage when considering weed control. Proper methods of fall tillage and a reduced number of trips in spring will permit the crop to be planted before many weed seeds wake up.

7. fall panicum

A tough grass that jams up machinery and cuts cornfield production is fall panicum (*Panicum dichotomiflorum*), an annual that branches from the base and nodes. It likes moist ground, and can therefore be found along streams and waste areas where compacted, wet anaerobic soil conditions prevail. If a field provides the tight compacted soil that puts foxtail on the march in early summer, it will likely support fall panicum later on in the year if herbicides manage to erase foxtail weed.

Fall panicum indicates ongoing soil degeneration. It represents a lower phylum of life than foxtail. As a matter of fact, if foxtail are allowed to grow, there will be no fall panicum. But if foxtail are attacked by herbicides, this takes away the auxins from the foxtail that usually cancel out fall panicum. Foxtail can dominate the fall panicum through its auxin root residues. This makes fall panicum a herbicide-caused weed—a consequence of a living system building resistance to herbicides.

Fall panicum is worse than foxtail. Since it depends on higher temperatures, it usually arrives after corn is cultivated in mid June or July. It dominates the environment in which it grows. Since it is a knarled and knotted and disjointed plant, it has the angles to go anywhere and choke anything. It pulls down corn plants, messes up snappers and cutting bars, and often claims life and limbs when farmers try to cuss it out of harvest machinery. We curse fall panicum, often forgetting that this is nature's way of preventing sick nutrient material from getting into the life stream. Corn that grows up with fall panicum is bound to be sickly or at least subclinically deficient. Usually it will die early and fail to mature its grain.

Every phylum of life has a means through growth substances—or auxins—emitted by roots, sometimes in the first 24 hours, to permeate the soil as much as a half inch from the seed. Auxins require every other weed

Fall panicum.

seed species in that area to stay dormant. Auxins define which type of growth can happen at the time of germination. The kind of such auxin hormone reactions that will occur are usually directly related to absence or complexity of basic cations and trace minerals.

This is the arena in which weed control has to be fought. Herbicides can never win. They can merely devoid the soil of other bacterial processes and worsen the battleground for crop production next year. The remedy for fall panicum is exactly the same as recommended for dealing with foxtail.

8. common burdock

Burdock (*Arctium minus*) is a biennial herb that reproduces by seed only. Its existence on a farm is generally a signal that the soil system is dominated by iron and is in need of calcium, and has a low pH. It is found all over the United States. Burdock has fantastically enormous roots, big leaves and burrs that stick to animals as well as clothes of human beings. Common burdock (*A. Minus*) is smaller than great burdock (*A. Iappa*). Gypsum soils frequently exhibit burdock and it can be suggested that burdock is often a signal that soil has been incorrectly limed with magnesium carbonate or subdued with great applications of sulfate of ammonia and lime, giving the effect of calcium sulfate in the soil.

Burdock is a good phosphate feeder. It has to have a good available source of phosphate. In the midwest there are a lot of soils low in calcium and high in magnesium, pH reading being on the low side. Many times farmers apply gypsum or ammonium sulfate on soils reading below pH 6.2. Unfortunately this has the effect of agitating the construction of the pH, driving it down to pH 5.8 or 5.7 in terms of a colloidal position. As explained in the lesson on pH, a colloidal position pH can be lower than the pH in terms of the entire environment, taking both colloidal and soluble form pH ingredients into consideration. Remember that magnesium will raise pH 1.4 times as high as the same amount of calcium.

When gypsum (calcium sulfate) is put on soils at pH 6.2 or 6.1—even though such soils are low in calcium—the resultant pH structure will still be too low to govern the proper release of trace minerals regulated by calcium. At such pH levels there are mineral releases on the surplus side. Often there is a surplus of aluminum.

Moreover, there is a difference in gypsums, depending on whether and where they are mined, or whether they are byproducts. As a rule byproduct gypsum is more easily broken down to influence the reconstruction of the pH equilibrium. It has a more fluffy particle surface (like a snowflake), which is quite different from the usual product which is textured much like a pearly smooth grain of sand. Many mined gypsums take a decade or more to break down.

Heavy acid flushes both trace minerals and heavy metals out of the soil. When calcium levels are low, aluminum is flushed out very easily, as is iron. Burdock can tolerate more iron and aluminum than other weeds and crops, but it is still a good phosphate feeder. It has a capacity for tolerating this inventory of conditions without becoming toxic. By way of contrast, corn and grain crops cannot tolerate the releases of aluminum and iron made mandatory by very low pH systems.

Burdock will proliferate in an environment of iron and aluminum that is

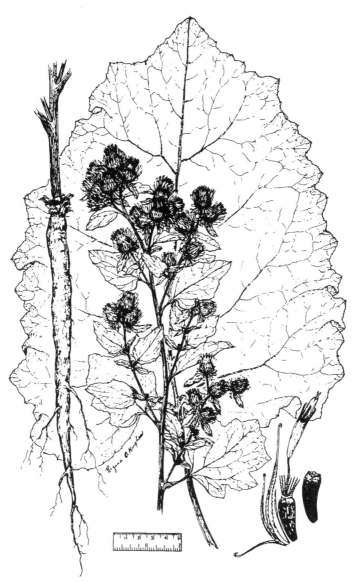

Common burdock.

killing off higher plants. In soils that are low in iron, or do not have a daily release of iron, burdock will not grow. Soils that have a proper decay system going—and this will not happen at low pH—will not give burdock permission for life. If a low pH soil is releasing a full complement of iron, burdock will function.

Iron in relationship to manganese is also an important factor. High iron in the presence of low manganese regulates the kind of fungi that can operate in that soil. This iron manganese posture also affects the bacterial processes that can occur, again determining which hormone systems are going to function and allow a weed seed to wake up and grow.

9. russian thistle

Song and legend tell about tumbling tumbleweeds, which is what Russian thistle (*Salsola kali L. var tenuifolia Tausch*) become when old plants break loose and roll before the high plains wind. Nevertheless Russian thistle is spreading to other parts of the country. It is an annual with a spreading taproot.

Experienced field consultants can generally tell the soil problem before the tests arrive back from the laboratory just by looking at the weed situation. Russian thistle likes an environment in which iron is depressing manganese availability.

It might be noted that an environment that will support burdock is exactly opposite of the environment that will support thistle. Thistles have prickly leaves. Biennials of the species send down great taproots and concentrate potassium. They grow almost everywhere. Canadian thistle (*C. arvense*) have long, creeping rootstocks. The field sowthistle (*Sonchus arvensis*) has very deep creeping roots that put this weed in a class by itself. Common sowthistle (*S. oleraceus*) is edible and is cultivated in some parts of Europe. Blessed thistle (*Cnicus benedictus*), as the name implies, is used as a bitter extract for Benedictine Liqueur.

Thistle likes low iron, low calcium, and a poor organic matter structure in the soil. In other words, chemicals that poison merely destroy the living weed plant, and do little to stop more from coming. To deal with the problem, it becomes mandatory to deal with the first cause.

If the calcium is inadequate to govern release of trace minerals, and the colloidal pH is improperly positioned, metals are released in excess amounts. High amounts of nitrogen applied to soils low in calcium leave the soil without a governor, and the inevitable result is an excess flushing out of minerals. This sets in motion a lot of hormonal processes that affect the character of weeds that can grow and regulates their ability to germinate and achieve their genetic potential. If a weed is missing certain nutritional requirements, this affects its genetic mutation.

Organic material decaying improperly with the wrong fungus systems in tow and a lack of actinomycetes molds all figure in setting the stage for Russian thistle. In the west, where pH is often built with high levels of sodium and potassium, but with low levels of calcium and magnesium, the pH may read near neutral, but still provide the environment for Russian thistle.

Russian thistles have a tremendous root system. These rhizomes build

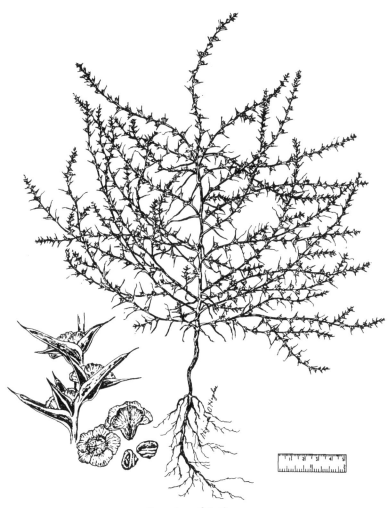

Russian thistle.

themselves in a mat, much like a strawberry plant, and dominate the environment. They have a fantastic auxin product production which subdues anything else, including grain crops. They grow rapidly with big leaves and shade out any other crop established at the same time.

Here, again, construction of pH figures first. Generally speaking it is necessary to change the construction of pH by increasing calcium and magnesium, and complexing sodium and potassium, all to put a governor in charge again. This adjustment will regulate the iron to manganese relationship, and the release of other heavy metals. It is not necessary to measure cadmium, lead and chlorine. It is enough to control the practices that tend to flush them out.

In the presence of high iron, low manganese, a whole strata of fungi that like mineral imbalance increase and multiply, and these are favorable to Russian thistle root development. Thistle roots are also rhizomes that function well below the plow level. This means it is not physically possible to tear them out or penetrate them with herbicides. They can be subdued only by increasing the supply of manganese and regulating the manganese to iron profile. This much accomplished, a whole different strata of bacterial mold systems is invited, and these are antagonistic to the Russian thistle rhizome. They rot it out, usually in a period of three to six months.

Control of the Russian thistle is not simple, and farmers are an impatient lot. They want action now. Thus the love tryst with toxic sprays.

Deionized sulfur is one of the things that will permit regulation of the manganese to iron ratio. Ordinary sulfur or sulfur contained in salt fertilizers is quite soluble and can be obliterated quickly. Ammonium sulfate will provide sulfur that has only a fleeting acid action. It is quickly absorbed and neutralized. The moment it is made soluble it becomes quite active and agitates the imbalance even further. If there is an excess of iron, it will increase the excess many times over.

Sulfate salt or ammonia can flush out a lot of iron, and it is this flushing out that creates toxicity to many plants and sets in motion hormone processes that invite the weeds that thrive in the type of environment in which they find themselves.

Poison sprays just kill the tops and never get to the roots. As a consequence, auxins remain to do their handy work. Clean tillage is sometimes effective if the thistle can be cut at a certain stage of its metabolic life system. Folklore observers have noted that the position of the moon and its magnetic effect have a bearing on how plants react to being cut. Although this inventory of observations has never achieved high science, many successful farmers have kept their own records and picked up patterns they believe to be successful. Richard Thompson of Boone, Iowa, an environmental farmer with many successful years of operation under the belt, cultivates according to the "right phases of the moon."

10. wild oats

Wild oats (*Avena fatua*) are hardy annuals, and a most troublesome weed crop reproduced by seeds. Seeds usually ripen before the wheat and drop to the ground before harvest.

Since they are not taken off the field, they perpetuate themselves. Wild oats are a problem over most of the west, especially California. This weed relates to soil areas that are overly wet and poorly drained. Cold seedbed temperatures usually figure, and there is usually sedimentation problem

Wild oats.

due to poor surface porosity or surface tension of soil particles. These particles tend to filter down causing sedimentary barriers in the lower levels of soil surfaces. Likely as not they've been agitated by working the soil on the wet side, working the ground early while trying to get the wheat planted early. It is almost always a cost of rushing field work before many areas in the field are ready for machine tillage.

Draws, swells, low pockets, poorly drained areas, sometimes a slope with a seep coming from above—all need air to warm the soil. It is usually a case of a farmer working a ridge, discing mealy soil—then the disc hits that wet spot. Almost every field has susceptible areas that are wild oats soil conditions.

Here again pH figures in preventing the target area from getting proper drainage. Soils planted on the wet side seal over easily following a shower. The problem can be corrected by adjusting the calcium load in seal areas. Anyone who goes to the trouble of getting a soil audit will usually discover a deficiency in calcium, low magnesium, high sodium and high potassium. There are plenty of farms with 1,000 to 2,000 pounds of sodium per acre. These acres exhibit a high pH—sometimes pH 8 or more—but this reading represents a false character pH. Sodium is an active element with a high position on the periodic chart. It can be make soluble with good ionized active sulfur and another cation injected to take the place of sodium—calcium, magnesium, potassium, or humic acid, for instance. The sodium or excess cation will be flushed off the colloid and mobilized to a water soluble form. It will end up as dry powder on top of the soil. Or it will leach into the subsoil and reduce the concentration and saturation of the colloidal remains.

Plants live off the colloidal position, and elements that are water soluble won't interfere. Plants depend on energy exchangeable nutrient forms. Water does not wash nutrients into the plant. The water soluble fertilizer requirement of state agencies is—as Bill Albrecht said—"a damn fool idea."

Another way of dealing with oats is to use good, ready-to-go-to-work compost materials, or special soil wetting agents such as Amway, WEX or Basic H. These things will reduce surface tension and permit the soil to get rid of excess water.

It might be redundant to note that if a farmer would wait until the soil warmed up in the super-wet areas being discussed here, he would not irritate the system, and there would be no wild oats. The seeds would simply stay dormant. Thus the general remedy: treat those areas in the fall of the year with surfactants, pH modifiers and compost, and correct the structure of the soil so that it drains better. In the spring, simply delay tillage of those areas until they are dry and ready. If the field is to be plowed or deep tilled, it would be well to disc and scratch those areas before plowing to take off the surface tension. Sealed areas will always suck water to the top. As soon as surface tension of the sealed over area is reduced, water tends to sink and go down faster.

11. stinging nettle

The stinging nettle (*Urtica dioica*) is a perennial reproduced by both seeds and creeping rootstocks.

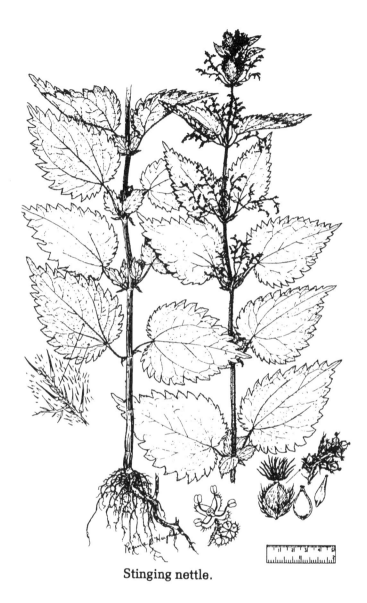

Stinging nettle.

There are a lot of nettles: the burning nettle (*Urtica urens*), the slender nettle (*U. gracilis*) and the non-stinging or false nettle (*Boehmeris cylindrica*). In the case of the stinging nettle, the sting comes from fine hairs on stems and leaves and formic acid, with perhaps an unidentified poison thrown in for good measure. A plowpan barrier contributes to their existence in cultivated fields.

The stinging nettle is a direct indicator of yet another variable in decay of organic matter under soil stress conditions. Involved here are almost the same principles that govern buttonweeds—yet not quite the same principles as the ones in charge of morning glories and bindweeds. It would be fair to say that when stinging nettles grow, the soil has an indigestion problem. The breakdown of organic materials in a poorly structured soil is likely to produce complex gases. Heavy applications of chicken manure, concentrated droppings with a lot of urine, too much raw manure, and the wrong temperatures—all figure in setting the stage for stinging nettles.

Decay of organic matter in the absence of oxygen and certain actinomycetes molds allows methane to ferment into the hydroxide direction. Or the decay direction may be putting too much hooch into the soil, alcohols breaking down into ethane, methane, butane, all complex carbon type gases. These gases permeate the decay area and affect the type of bacteria that can live there. Thus organic matter needing a decay system is deprived of its balance of microbial workers.

Eventually such a system ends up taking a formaldehyde form. This sterilizes the soil and preserves the very materials that ought to be decaying, pickling them, so to speak. The counter effect on hormone systems can be conceptualized. On the surface they activate stinging nettle weed seeds.

All this adds up to the stinging nettle being an index of an accumulation of excesses. The problem can be solved only by dealing with pH, soil structure, air, compaction, that overdose of organic matter, and the quality of decay. Small variables trigger a weed system this way or that. But the basics are almost always the same.

12. cheat

Cheat (*Bromus secalinus*) is also called chess. It is a winter annual reproduced by seeds, a superb pest in grain fields and meadows. It has its hotspot in the border states, but can be found throughout most of the country.

Farmers say seeds for this infestation arrive in impure rye or wheat seeds, and if the year is wet cheat takes hold.

This is an alibi. In loamy, energized soil, these seeds never work, whatever the source. Cheat is a shallow rooted plant that grows on the surface of the soil under dry conditions where there is compaction due to traffic or heavy pasturing. Light tillage in spring, when the ground is a little too wet, often figures. Many farmers like to fine-up a seedbed by going over the land with points removed. Often a heavy shower follows. Over-worked, fine-textured soils tend to puddle and cement together. Moreover, without porosity or structure, the soil cannot handle the stresses of compaction. Once it dries, it tends to sediment on a shallow basis. There is little biological activity in such a soil. Cheat moves in when this degeneration occurs.

Cheat.

The hazards of weather frequently figure in the establishment of cheat. This explains why a farmer will have the weed one year, not the next. After spring tillage, there is a very strategic time when rain on the soil is bad news. For instance, the first 12 to 23 hours after tillage are very critical. Often a farmer discs in the morning and finishes late at night. Then comes a 4:00 a.m. shower. The part of the field disced at 8:00 the morning before is probably okay. It may have melded together and positioned itself for rain. Not the afternoon acres. A rain within 12 hours will destroy the biological process in the stirred soil. That is why the time in which water falls can set up the conditions that permits cheat.

13. johnson grass

Johnson grass (*Sorghum halepense*) is a perennial with both seed and rhizome reproduction. Its most vicious reproduction cycle is in southern states, although it seems no less troublesome elsewhere.

Johnson grass grows everywhere, everywhere being defined as where soil conditions permit. It has no geographical limitations, although the weed appears less in northern climates. It has the greatest concentration in the rainbelt south, where low organic matter, worn out soils prevail. Johnson grass tends to grow in soils with a low calcium level regardless of what the pH is. If the biological environment is favorable for crop production and has the proper decay organisms present, Johnson grass will not last. It can be given competition by planting small grain at the time an adjustment in pH is made.

To control Johnson grass, it is necessary to set in motion biological processes that will attack the rhizome and make the root weaker. If a different strata of bacteria is put in motion with timely application of well digested compost, they will etch out the rhizome system and ultimately clean out the weed. This may take a period of a year.

Johnson grass grows where there is a reasonable level of available iron. The pH character of the soil governs iron release. Thus when small grain seeds are introduced to give Johnson grass competition, and pH adjustment is started at the same time, trace mineral release will invite a strata of bacteria inimical to Johnson grass.

Johnson grass is of the same family as quackgrass. Thistles have the same type of rootbase. Each is related to a different supply of trace minerals, the complex of trace minerals being what determine which weed family is to endure. Trace mineral release may shift an acre to quackgrass, and of course quackgrass and Johnson grass do not grow together. All these weeds are related to the iron to manganese ratio as released by other soil factors. If iron to manganese are kept in equilibrium, then the environment for almost all rhizome type weeds is removed.

Basically, rhizomes have to be broken down by soil fungus diseases. When weeds such as Johnson grass grow, it simply means the soil does not have compatible fungus control for commercial crops. This means the crop is on the weaker side, and Johnson grass takes over. It sits at the table first and becomes a tremendous competitor to any grain crop or pasture grass. It also grows in sugar cane fields. Most sugar cane acres have imbalanced pH systems, excess iron, low manganese, or iron to manganese out of ratio to each other.

Johnson grass.

The best way to control the situation is to introduce the type of byproducts that will energize bacteria—excellent composts, for instance—thus subduing the soil life forms that are permitting that type of weed to dominate.

14. lambsquarters and rough pigweed

This annual herb (*Chenopodium album*) generally known as common lambsquarters reproduces from seeds amid cultivated crops, in grain fields, and also in waste areas. It thrives in a surprising range of climates and geographical locations. Potato acres in particular become infested with lambsquarters once the potato plant's specific requirements become unavailable. Otherwise the massive root structure of potatoes prevent lambsquarters growth.

As with most weeds, lambsquarters has a lot of colorful names: fat hen, goosefoot, muckweed, dungweed. It grows in high organic matter soil

Lambsquarters.

Redroot pigweed.

where there is good decay. It indicates rich, fertile soil.

Rough pigweed (*Amaranthus retroflexus*) is an annual from seeds. In fact the seed can lie dormant in the soil for years. Potato crops can be annihilated by the pigweed if it is allowed to proliferate.

Pithy names tell something about how plants grow according to available light, temperature and geography. Lambsquarters sometimes grow five feet tall. It harbors a lot of nutrition for certain insects and stores a high quality of phosphate. It reflects the availability of good nutrition in the soil. Used in silage, it supplies needed nutrients. As a matter of fact, both pigweed and lambsquarters can become excellent silage. Both have a high protein content.

Lambsquarters grows in a soil that has an appropriate decay system going, which of course has a direct bearing on release and catalyzing release of a good array of quality mineral nutrients. This weed does not depend on phosphate or potassium from fertilizers.

It might be helpful to remember that something is almost always growing on soil. Soil in fact is intended to grow vegetative matter. Lambsquarters and redroot pigweed stand at the top of the ladder of the phylum of weeds that can grow under different soil conditions. They reflect almost ideal producing functions in a active biological soil system. There is no water insoak problem where lambsquarters and pigweed grow.

Both of these weeds are easy to control. Neither have serious negative effects through auxin emission to growing crops. Indeed, there is a symbiotic relationship between redroot pigweed and lambsquarters and most plants. These weeds are the best possible laboratory analysis for phosphate being available on a daily meal basis.

It might be that competition for moisture or available light could be a factor where lambsquarters and redroot pigweed grow, the matter of early dominance being first and foremost in any farmer's consideration. In a cornfield, pigweeds grow spindly and thin. It is a case of the corn crop dominating light.

One thing is certain. Where these weeds are growing year after year, there is no need to buy phosphate fertilizers. There is no concern about buying potassium or about raising the levels of potassium, even though a soil audit might suggest that the till is not quite full enough.

15. common milkweed

This perennial (*Asclepias syriaca*) propagates from both seeds and spreading rhizomes. It turns up in soils with a near perfect cation exchange balance.

It is seldom a problem in growing fields. Borders, fencerow waste areas, areas where there is not a lot of tillage—these are the sites where milkweed usually appears. It is here that the soil is constantly being challenged in its biological process, where it is not disturbed by compaction, tillage and traffic. Milkweed sometimes appears in ideal soil that is permitted to lay fallow for a period of time. It is easily managed by tillage.

16. jimsonweed

Datura stramonium, or jimsonweed, is famous in song and legend ("the cattle feed," Gene Autry sang, "on the lonely jimsonweed"), which is more

Common milkweed.

poetic than accurate for this seed produced weed is neither lonely nor good cow feed. The plant is narcotic and poisonous. According to variety, stems of this plant are green, corolla white (*var. stramonium*) or stems purple, corolla lavender or pale violet (*var. tatula*).

Jimsonweed is the nemesis of every soybean grower in the nation. It may be that northern areas—because of sun angle—are less likely to have jimsonweed.

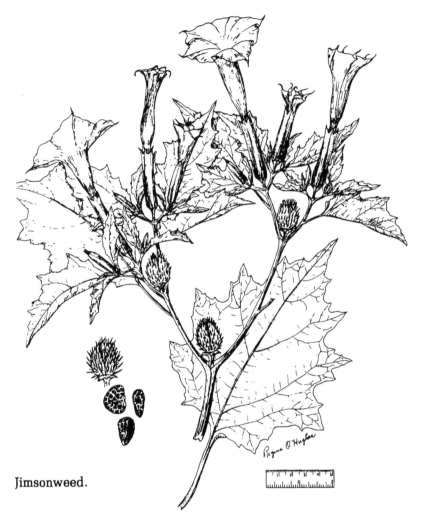

Jimsonweed.

It will grow in a very actively decaying organic matter complex, the kind of system one might relate to buttonweed. Buttonweed grows where there is methane gas production. Jimsonweed wakes up and grows where gas production is ethane. Decaying organic matter can sequester and accumulate and govern the availability of cobalt. Improper decay will allow these things more freedom. Again, when conditions are adverse to the life style of actinomycetes molds, decay takes the hydroxide direction.

Typical is the case of soybeans following corn, lots of undecayed stalks being left over. It is frequently a case of working the soil a bit too late in the fall, or late tillage in the spring. Under warm conditions with excess organic matter—and in the absence of bacterial activity taking decay in the proper direction—jimsonweed seeds are fired up and allowed to dominate.

Of course if the calcium level were adequate, this alone would guarantee a different decay direction for trash. Thus the remedy—an improved calcium level in colloidal position, a regulated pH, and proper decay of organic matter. An injection of compost sees to it that decay proceeds in the right direction.

17. wild mustard

Wild mustard (*Brassica kaber*) is an annual or winter annual capable of prolific seed reproduction. It grows well in grainfields, particularly amid oat stands. There are in fact several mustards; Charlock (*B. Arvensis*), Indian mustard (*B. Juncea*), ball mustard (*Neslia paniculata*), and all inhabit grain fields, as indeed do wild radishes (*Raphanus raphanistrum*), the peppergrasses (both *Lepidium virginicum* and *L. apetalum* and *L. campestre*).

It is, however, in oat fields that seeds of these plants wake up and grow, first exhausting the food supply in the seed, next invading the nutrient bank account of the commercial crop.

A lot depends on how trash behind the harvest is handled, since this will have a bearing on which mold and fungus will be invited to that table. Also, a lot relates to the supply of juices, vitamins, minerals, sugars present in the organic material returned to the soil. Prompt tillage of harvested residues will capture all these goodies and stimulate quick fall fermentation, and a cleaner rootbed the following spring.

Wild mustard is usually related to a field planted to small grain. Most grain crops follow crops that leave a lot of stubble from the year before, and frequently the field is not worked in the fall. It is usually a case of mulching in the stubble during a spring when soils are cold. That's a common habit in putting out small grains.

An accumulation of trash that winters over on top of the soil is conducive to slime mold production. These slimes make the soil sour and waterlogged. Even if that soil is given a shot of air, slime mold works its mischief for a week or ten days. Warmer temperatures might come along and invite a new mold to take over. Nitrogen use figures in all this.

Nitrogen goes to calcium first, then to phosphorus, taking the nutritional supply away, and this situation has a bearing on which of the Fusarium molds will function. If the undesirable Fusarium molds take over, they put into motion hormone processes that fire up mustard, wild radish, and weeds of that ilk. This is the reason that stover must be turned in as soon after harvest as possible in the fall.

Slime mold is not likely to take hold in trash on the surface if the decay system is functioning under proper aerated conditions.

It is almost impossible to find wild mustard in the better soil areas of a field unless spring tillage jumps the "start." More commonly the weed grows in an area stressed due to poor drainage and poor structure. It is

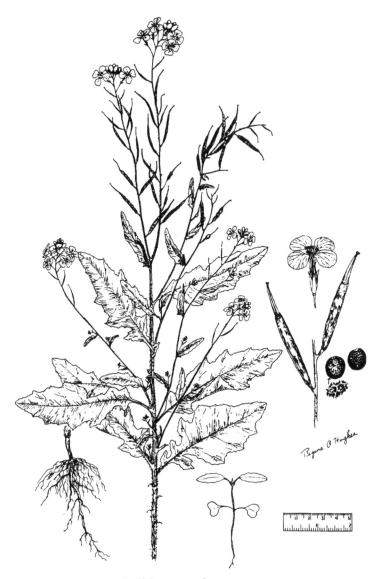

Wild mustard.

usually a consequence of working the ground early under poor drainage conditions. Everything has been done pretty wrong when mustard covers an entire field.

It might be pointed out that small grains have a different auxin effect than coarse grains. As soon as they germinate they put out their auxins, and these are sometimes complimentary to stimulation of Fusarium molds. Stem rust, root mold and the like are always associated with the same areas as wild mustard because both are manufactured by a sequence of weather factors.

The advice on wild mustard comes down to this. If good compost is not available, it is best to use some nitrogen to get the decay process moving. This should be applied with a fall mulch of crop residues. Nitrogen is not necessary if compost is available because good compost can hand over something like 100 pounds of nitrogen per ton—all of it in colloidal form, not leachable, never wasted.

18. sowthistle and bull thistle

As the illustration readily depicts, this annual (rarely biennial) called Sonchus oleraceus, has a deep taproot. It is reproduced from seeds, and the inventory of its habitats includes grainfields, vineyards and orchards as well as lawns and roadside areas.

Bull thistle (*Cirsium vulgare*) is a biennial reproduced by seeds. The bull thistle is an aggressive weed, but it will not easily survive cultivated fields.

Sowthistle and bull thistle often appear on the farm in a few scattered spots, usually in protected areas—along fence rows, near fence posts, along the sides of pathways, rarely in cultivated areas. Seeds can fall on the soil, but if something else is growing there the sowthistle and the bull thistle seed simply won't get started.

Sowthistle and bull thistle can grow under variable soil conditions. It is difficult to isolate any one set of circumstances that contributes to these thistle proliferations. Traffic and competition are their enemies, and a lot depends on the kind of light availability these plants enjoy. Both of these thistles are high protein building plants and depend on a good nitrogen and phosphate source, and the availability of nitrogen and phosphate balance has to be accompanied by reasonable levels of zinc. These three elements have to work together.

If there is a restriction or phosphate limitation, the size of the plant will vary accordingly. It is possible to have weeds of the same species that are small and depressed and some that are giants. Sowthistle and bull thistle are a type of weed that are giants. Sowthistle and bull thistle are a type of weed that can indicate fairly good nutritional systems in protected areas— and in no-till areas, along fence rows where there is a lot of trash not too high in carbon so it takes less nitrogen to break it down. They grow where more cellulose type residues are going back to the soil, rather than grain stubble or corn stubble. Stress and tillage can wipe them out.

19. chickweed

Chickweed (*Stellaria media*) is an annual or winter annual. It is weakly tufted. It reproduces by seeds and all sorts of creeping stems, rooting at the nodes. As a universal weed, chickweed grows from coast to coast, border

Bull thistle.

to border, seemingly impervious to heat and snow. Once established, it can cover the soil like an army blanket, keeping out air and sunlight and water.

Chickweed is the antithesis of crabgrass. Chickweed grows where there is a good amount of working organic matter on the surface of the soil—more so than at great depths. It is frequently a case of too much grass clippings not being consumed, a lot of peat or manure on a garden, with acids coming off organic matter flushing out minerals that become a little too hot for grasses or vegetables. This situation issues an invitation for

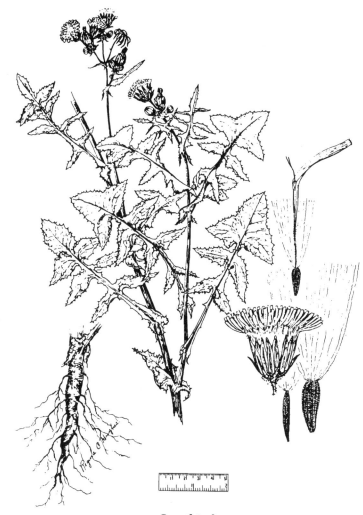

Sowthistle.

chickweed to set up shop.

Sometimes organic matter on a lawn decays for a week, dries up, gets wet again, and once more dries. With organic matter decaying in a de-toured media, chickweed takes off—if the pH is right, that is, not too low or too high. By way of contrast, crabgrass reflects a higher pH, the kind that does not permit actinomycetes molds, and it reflects soil that is tight, crusty and sedimentary. Acids from partially decayed organic material in-

Chickweed.

fluence release of excess amounts of trace minerals, often a combination of many of them. When organic matter decays properly and completely, it has the power to release trace minerals that are needed, and it does this in a good colloidal way. It is possible to release the minerals but not have enough colloidal material in the soil to hold those minerals. They become water soluble and disperse, and the next stage of degeneration begins. Chickweed, crabgrass, quackgrass, nettles, plaintain, buckhorn, dandelions and others now grow the "green flags" of nature's system of limitations.

20. dandelion

Of all the weeds, this one is most recognized by laymen who know no other weeds. It is a perennial. It is usually a symptom of overgrazing. Lawn *aficionada* go after the dandelion (*Taraxacum officinale Weber, T. vulgare Lam*) with a vengeance, as though it offended both the eye and the psyche of mankind. Yet on balance, the dandelion is a friendly weed. Its roots penetrate some 3 feet deep, transporting calcium and other minerals to the surface. Earthworms like the vicinity of the dandelion. Each time such a plant dies, remaining root channels become a conduit for worm travel and also a colloidal source of worm nutrition.

The dandelion is a monoculture weed, an index of sedimentation, a biography of rain, root webbing, organic material forever in place and unstirred. In every case calcium is colloidally weak or absent. The dandelion simply says organic matter residues are musty and are barricading the warehoused supplies of food.

Dandelion.

As the season goes along, dry weather usually weakens and subdues the dandelion. But if a quicker remedy is sought, the soil should be six to seven inches deep with a supply of calcium lime. Anyone really desperate will have to replant, but chances are the dandelion won't be back for a long time.

It might be added parenthetically that astute lawn managers often regulate the availability of phosphorus to reduce the growth of grass and yet keep it strong in the absence of dandelions. The object is to restrict photosynthesis capacity of leaves by managing the conversion and release of phosphorus on a diet of "one meal" per day. As one agronomist put it, "I don't want a lot of soluble phosphorus on a lawn. It needs a good supply of zinc regulated by a calcium base, not too much nitrogen—just enough to etch the phosphate needed to grow a day at a time. I want the phosphorus to be lazy and slow, and at a balanced pH of 6.5. I would never apply nitrogen or phosphorus fertilizer but would use gypsum and sulfur as needed to have a slow growing, healthy and vigorous lawn and watch my neighbors wear out their lawn movers."

A pH range of 6.2 to 6.6 happens to preside over actinomycete mold balanced activity, mineral release—and, as with all other weeds, there has to be a reference back to calcium and its colloidal function.

21. hedge bindweed

This perennial (*Convolvulus sepium*) reproduces from both seeds and shallow creeping roots. It is a blood brother to *field bindweed* or *morning glory* (*Convolvulus arvensis*).

It grows almost everywhere, even on eroded hillsides that are well drained. Bindweed is a typical reflector of an improper decay of organic matter and excess accumulation of heavy soil metals. It may be that a bale of hay broke open in an area, and was left there. It may be that cattle were fed, and much organic matter was stomped into the ground. Morning glories function best in the presence of ample humus materials and an antagonistic decay system. Bindweed tend to flourish more in eroded low humus soils, which cannot suppose corrective decay systems for soil restoration.

Most creeping vine type weeds have extensive and fast growing rhizomes that develop to completely entrap the soil nutrient system in and around all the clusters of organic residues. The biological energies contained in these foul rotting residues support numerous dominating hormone enzyme systems that are "just right" for the vine weed families, and "not just right" for other species of soil and plant life. Such conditions can occur within soils of high exchange capacity (clay) or low exchange capacity (sandy)—with low or high organic material content—always in soils that impose limitations on ferment and breakdown of organic residue

Hedge bindweed.

in the desired direction. Such soils are unable to govern the humus system. They also lack the capacity to support the right kind of nutrition needed for better plant and animal food, chiefly because of the imbalanced hormone-enzyme system that is sustained by improper decay.

Field bindweed, morning glory, creeping bindweed all dominate the plant kingdom decause of a short circuit in the energy release of fouled decay systems. These limits are generated by an accumulation of dry-dead organic substances either under dry fall condtions, or in wet spring soil—with compaction, sedimentation, and improper tillage timing figuring in the equation. Cultural practices that relate to stress systems are greatly influenced by the pH character of the colloidal system involved and by the effect of drainage and air capacity of the decay medium. Correct these soil limitations through pH management and the bindweed-morning glory syndrome becomes completely dispersed. No herbicide chemical or fertilizer material can replace good soil management. Such treatments only camouflage the signs of illness and prevent a farmer from coming face to face with the truth. Herbicides may control the weed for one crop, but return to haunt the next one. If necessary to get a farmer's attention, the bindweed-morning glory complex can mat over the soil like a carpet. Nature in her wisdom and patience uses weeds to challenge man's intelligence and his ability to recognize the truths of biology.

22. western ragweed

This one is a perennial or annual with reproduction geared to creeping roots and seeds. Hayfever sufferers swear at western ragweed (Ambrosia psilostachya), and in defiance ragweed proliferates on prairie land, sometimes in uncultivated places, sometimes in crop areas.

Ragweed makes its appearance later in the season; very little germination takes place during small grain growth. It usually happens this way. June and July bring hot weather that is superbly dry. Under very dry conditions biological processes in the soil fail to release the right form of potassium.

Farm crops, by way of contrast, do best when they can get their potassium in a simple colloidally energized form. But when conditions of soil dryness prevail, when the soil's refinery process isn't hitting on all cylinders, potassium comes off in a hard to take colloidally complexed form.- This form is just right for ragweed.

It might be noted that smartweed has an opposite index card. In the early spring—and sometimes shortly after harvest, cool, cloudy weather sets in. Fields become waterlogged. More complexed forms of potassium take over. Colloidally energized potassium is not available for good crop growth. Under wet conditions, smartweed gets its signal to germinate and grow, and under dry conditions ragweed declares its domain.

It is the Creator's rule that potassium must be activated and released and catalyzed by bacterial processes. Under dry conditions this is not possible. So when the potassium supply is minimal, enzyme processes trigger hormones that tell the ragweed seed to move along. One can conclude from all this the significance of having a good capillary water system. When a living soil gets too wet or too dry, too cold or too warm, it means

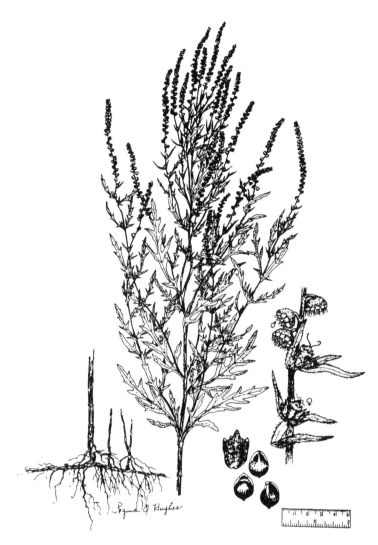

Western ragweed.

problems, and specific weed families tell all about the symptoms. We have only to make the proper diagnosis to determine an enduring corrective treatment. However complex nature's message seems to be, one lesson at a time should accumulate into wisdom and good judgment. A study of nature can teach us to cope with any cause of limitation that permits one weed to dominate or limits good crop growth.

23. *red sorrel*

This is another perennial that reproduces by shallow creeping rhizomes and seeds. Sorrel (*Rumex acetosella*) is a fertility level weed. As pastures become acid because of faltering fertility, sorrels and docks find an environment in which their seeds can take off. The distance from seed to rhizome is very short, measured on any scale. Sorrel is rich in oxalates, or salts that are not readily digested. Because of this sorrel has a reputation for being poisonous to horses.

Redwood forests always have a companion weed called oxalate. The term oxalate also designates a mild organic acid that keeps etching mineral nutrients out of native rock and lignated carbon residues to sustain red-

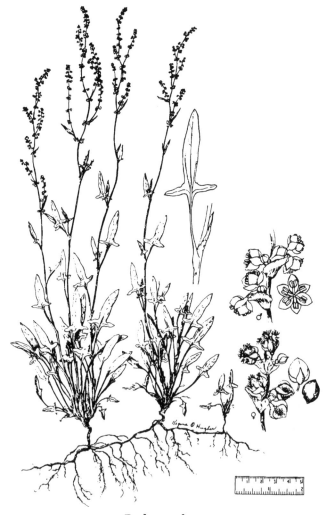

Red sorrel.

wood nutrition. Only lignated leaves and bark contribute the organic matter. The function of the oxalate root—and its auxins—is to produce oxalic acid needed in that decay process.

High carbon lignins in a redwood forest need to be recycled. Oxalic acid is required to keep the mineral supplies coming, the ferns and fungi working. Some of those redwood trees are 3,500 years old, having been born well before the time of Christ. So it can be seen that the whole system works, year after year, without the aid of synthetic nitrogen, bagged fertilizer or toxic rescue chemistry.

Red sorrel grows in acid soils, and it contains oxalic acid. The weed suggests soils that have not been cultivated or aerated or maintained properly, soils with a subdued biological process, certainly not with enough biotic activity to stimulate a good rotation of life. Often as not wood and high fiber organic matter figures in this weed pattern. Under acid conditions with low calcium, there is little actinomycetes activity.

Buckhorn is a prolific weed, albeit in aerated soils that would otherwise support sorrel. It all relates back to wood, wood decay—usually wood shavings in the soil, old bedding, accumulations of heavy amounts of horse manure, which is high in fiber. When the soil is aerated, buckhorn takes off, and sorrel remains subdued.

One of the problems with red sorrel and buckhorn is that they can't be killed with either a hammer or hoe. Tear off a stem, then shower it with a little water, and it will stand up like a member of the guard. Like a cancer it is almost always impossible to get it all, and the part that survives grows in wild proliferation.

The answer is to get the decay process going again with managed levels of calcium, and a balance of the other basic cations—magnesium, sodium and potassium.

24. foxtail barley

Foxtail barley (*Horedeum jubatum*) is a perennial with roots that are densely fibrous. It is troublesome to cattle since bristles often injure soft mouth tissue. Conditions of wet soil early in the growing season, lingering layers of water and waterlogging—all figure in the proliferation of foxtail barley. This weed says mineral conversion is negative, that potassium is complexed, that trace minerals are either not released or governed in their release. Foxtail barley grows in low pH soils or in low calcium soils with higher pH. Soils in which this weed grows could have a good humus body, yet be low in calcium. Since waterlogging prevents drainage, bacteria necessary for mineral uptake conversion fade away. When the soil finally dries out—usually in a week or so—it takes a while for bacteria to get reestablished. If there is another rain and the loggy condition is extended, the environment for foxtail barley is perpetuated.

Foxtail barley.

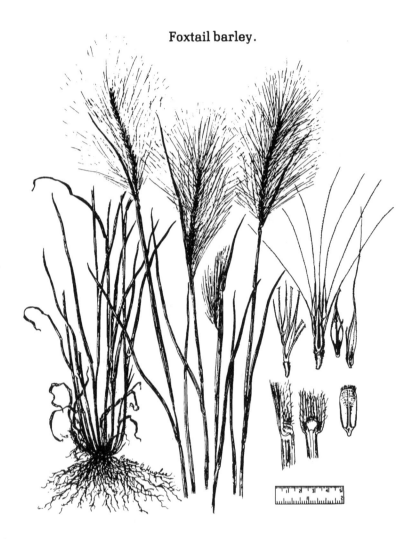

This weed is commonly seen where cattle and hogs have been fed, where hoofs have packed the soil below, where a sequence of warm soil conditions and heavy rains have kept the soil on the wet side.

Even well drained soils can have their balance upset. Too much irrigation is a common culprit, and an open sesame for foxtail barley.

25. tumbleweed

When tumbleweed (*Amaranthus albus*) carries the Latin designated here, it is really tumble pigweed, and not Russian thistle, or some other kind of tumbling plant carrying a colorful local name. It is an annual seed produced plant with pale green stems that are diffusely branched. It is not to be confused with redroot pigweed or rough pigweed (*Amaranthus retroflexus*). Variety albus possibly got its start in the great plains, but is now distributed over the entire nation. It is more commonly a problem along fence rows and in waste ground.

Tumbleweeds grow at the edges of pivot circles on irrigated acres. Usually they break off and blow into the field. They tolerate low moisture.

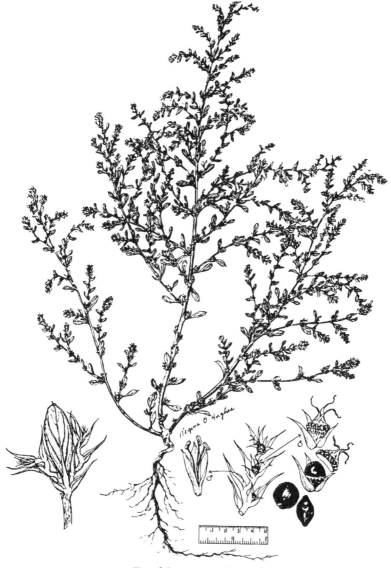

Tumble pigweed.

Tumbleweeds can be regulated very easily by supplying humic acids to the soil. The same general rule can apply to sandburs. Humic acids from humate materials or decayed organic materials—compost and so on—can suppress and dormantize the seeds of this weed. Indeed, this weed gets its order to grow when humic acids are not being released from organic matter and bacterial systems are slow and non-productive. The best control approach is to put on humates at a reasonable time in the spring after considered pH adjustments are made, or to serve up a seedzone meal of humic acid, a little inoculant of bacteria and an amino acid flavoring of trace elements.

A warning is in order. In areas of excess rainfall, where there is plenty of leaching and soils are low in humus, an impoverished governor presides over release of trace minerals. In high acid soils there might be an excess of such minerals. Humates can release the goodies on a slow, gradual basis—never in excess. Young plants take that extra load of excess minerals and promptly plug up the stem nodal system. Here, as elsewhere, a good level of calcium is mandatory.

26. horsetail

Horsetail (*Equisetum arvense*) is a perennial fern ally. It reproduces by spores and rhizomes attached to small tubers. Rhizomes are usually deep seated and long running. This one grows best in sandy or gravelly soils with a high ground water level, but it can also stand a dry summer. In general it suggests improper drainage.

Horsetail gained much attention with the publication of Louis Kervran's *Biological Transmutations* since this weed has silica skeleton, as high as 80% of the ash. According to Kervran, silica transmutes into calcium, making the organic silica in horsetail very valuable indeed.

It will probably be years before science formally identifies with what Kervran is saying. For now it is enough to reiterate his assertion that "Fresh green vegetables (young plants), radishes, etc., contain a large amount of silica. We now know it is due to the ingestion of fresh grass that milk cows can excrete more calcium than they ingest, without decalcifying. The mother who breast feeds her baby may correct her diet by adding a small amount of horsetail to her food in order to avoid decalcification."

Silica contained in horsetail has long been used as a tea remedy for various ailments, and biodynamic agriculture uses a 0.5 to 2.0% solution made by boiling such a tea for 15 to 10 minutes as a spray against mildew and fungi on grapes, vegetables, rose and fruit trees. Although not as strong as Bordeaux mixture, the biodynamic people report it swift and effective and kind to soil life. They point out, however, "that it should not be applied until the first course of fungus infection, too moist a stand of the infested plant, is taken care of, too."

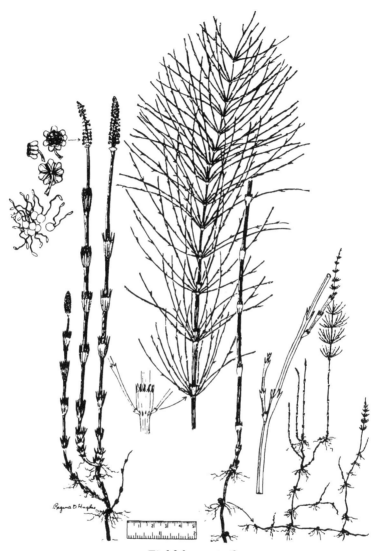

Field horsetail.

Many books list horsetail as poisonous to livestock, including USDA's *Common Weeds of the United States*, causing equisetosis in horses, yet Ehrenfried Pfeiffer used it for kidney disturbances in cattle, horses and dogs. The "arvense" variety is quite different from other varieties. The "useless" varieties have a collar formed by black leaves. Field horsetail is green or perhaps slightly discolored.

Horsetail is seldom much of a weed problem. Silicon could be available in good balanced soils. There are areas of the country, such as southwest Minnesota, where nitrogen will complex the silicon in the soil and leave weaknesses in the plant and there are areas where lack of organic matter and a poor calcium governor has a bearing on excess boron uptake.

27. quackgrass

Quackgrass (*Agropyron repens*) is also known as couchgrass. It is a perennial that reproduces by seeds and extensively creeping underground rhizomes, according to botanical manuals. In fact seeds are relatively unimportant for reproduction much of the time. Examination of ears often reveals seeds to be nonexistent or sterile. Although considered a noxious weed in the United States, rhizomes from this plant have herbal qualities and are used as extracts in treating urinary disorders. Rhizome propagation makes clean tillage hazardous. Each little broken piece of root or stem promptly becomes another thriving plant when moisture conditions so decree. It has been suggested that when Sisyphus was sentenced to repeatedly roll a stone uphill in Greek mythology, he missed out on an even greater punishment—weeding quackgrass. Dealing with this weed means structuring a proper decay system in the soil, one that invites actinomycetes and several other species of molds, acidiomycetes included. In order to endure, these beneficial molds must have a well aerated soil and adequate calcium.

Quackgrass soils hold water. They never have enough air. Wetting agents can make a great difference. But most important is calcium. It takes calcium to manage aluminum. Too much aluminum can dominate the availability of all other trace minerals. Aluminum cannot bother quackgrass, and for this reason quackgrass is an indicator that the colloidal pH is too low, and that release of aluminum is too great.

Excess aluminum release is a problem in high pH soils of the west because pH readings are usually high due to the high content of sodium and magnesium, when in fact the soils are low in calcium.

Quackgrass can be dealt with to some extent by invoking biological controls: planting soybeans, cowpeas, millet, so as to choke out the grass and subject it to auxin attack. Two successive crops of rye have been used to cancel out quackgrass, according to Ehrenfried Pfeiffer. The use of well finished compost, particularly in the fall, along with calcium or a calcium-sulfur material can cleanse the soil system and invite active decay of the quackgrass root.

28. wild onions

Wild onions grow in cold, wet soil, usually in spring. A slime mold develops due to a lack of drainage. Residue on the surface sets up an an-

Quackgrass.

aerobic condition, and this is the environment wild onions like. Also, an improper pH is a standard finding when wild onions appear. The obvious remedy is to deal with the causes, both in fertility management and in physical conditioning of the soil system. Here again the lessons expressed in other similar weed patterns come into play. The common denominator is pH management, pH modifiers, all to manage tilth, moisture and fertility balance.

Crabgrass.

H. Allen Smith could well write, tongue in cheek, "let the crabgrass grow," and the advice might be good for a lawn owner with a sense of humor. Yet the fact remains that crabgrass (*Digitaria sanguinalis*) is a serious weed of cultivated ground. It is an annual that is reproduced by seed, branching and spreading, and literally blankets the United States, sparing no geographical area, yet limited to environments that are right.

Crabgrass simply says that the soil is low in calcium and that it cannot support decay starting with actinomycetes molds. It is usually possible to adjust the system in a year or two, putting a little calcium lime and sulfur or gypsum on the soil in order to restructure the pH. As this is accomplished, most of the infestation dissipates, virtually rotting away. Usually applications can be computed on the basis of 1,500 pounds of high calcium lime per acre, 30 to 40 pounds of processed and active sulfur per acre. If ammonium sulfate is used, a little more calcium is required. On certain soils a mixture of calcium lime and gypsum is also good. As calcium exerts its adjusting capacity to correct the soil, a whole new array of rotting and decay organisms start eating at the roots of the crabgrass. Desirable humic acids are made available to release soil nutrients for lawn grasses. These can now begin to fill in eroded patches formerly occupied by crabgrass.

30. cocklebur

Cocklebur (*Xanthium pensylvanicum*) is an annual that can grow in an environment with a high level of phosphorus available, and a reasonably good pH system. There is great variation in shape, hairiness and spininess of mature burs. Accordingly the names X. strumarium L., X. orientale L. appear in the literature, as do X. canalense Mill and X. commune Britton. Phosphorus can only be available at good quality when the pH is somewhat in line. At the same time, this high level of phosphate tends to complex the zinc. And it is in this kind of environment that the cocklebur finds its hormone system activated, and nature telling it that it is its turn to grow.

Taking one good look at this weed, a farmer can know enough not to fertilize with phosphate.

31. an experienced teacher

Every weed is an experienced teacher of very specialized subject matter. And the college of nature has a complete staff of such specialists. It is a pity that the classrooms are not well attended, and that farmer students are allowing their attention to be diverted from the lessons at hand.

Surely we now know some of those lessons.

Weeds evidence the total character of soil nutrition and the management thereof. They measure man's effect in terms of what he does to the soil and they record a host of processes occurring therein. They depend on relationships of nutrient supply, the presence of biological processes, the amount of energy and food materials that may be present, and the eventual hormone-enzyme system that emerges from the soil cauldron at any given time of the growing season.

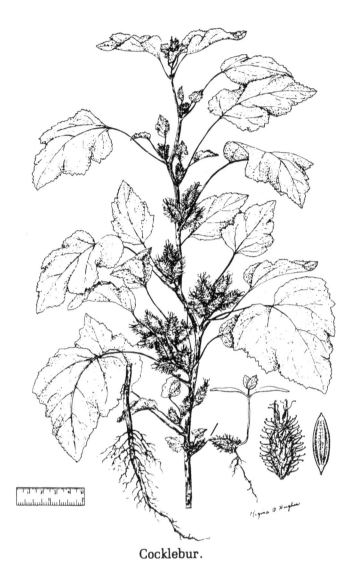

Cocklebur.

Weeds are a product of the nature of things present, but more often they are the consequence of factors not present. Nature is still building, and many soil areas do not have a balanced pantry. Such soils cannot supply the essential ingredients that may be needed. Weeds signal these deficiencies if we have the wit to read what they are saying.

Each weed species is an historical monument to either the recent or the long term past. Each marks the soil for what it is in the immediate area around each seed.

Weed crops directly measure the factors of time, light, temperature, moisture, biology and the genetic factors that have been or are about to be. And science has found it difficult to even begin to comprehend what weeds are all about. This is understandable, and we can forgive it because science cannot lift these facts out of their related context and come up with a composite conclusion. On the other hand, all soil managers must continue to seek a common understanding of nature's laws and exercise good judgment based on our track record of farming practices. Frequency of specific experiences can lead to a better understanding of nature's teacher—the growing plant. Weeds can never be noxious to nature. This is man's pitiful thought, and it needs modification.

SUMMER SESSION

LESSON 25

Animal Health

How necessary animals are to an ecologically sound farm depends on what other sources a farmer might have for animal manures, or their equivalent values. It also depends on how a farmer can best utilize the quality feeds his brand of soil management accounts for. It does not fall within the purview of this outline to debate the matter very much. It is probably enough to point out that fully half the acres in the United States couldn't even be harvested without animals.

For the purpose of this primer, it might prove helpful to set down a few facts about animals and the ailments that often hamper economic animal production. It should prove equally helpful to draw attention to the role of nutritional support in animal management and its necessary relationships to proper soil management. Basic to our considerations here are at least four animals.

	Cattle	Swine	Sheep	Horses
Adult male	bull	boar	ram	stallion
Adult female	cow	sow	ewe	mare
Young male	bull calf	shoat	ram lamb	colt
Young female	heifer	gilt	ewe lamb	filly
Newborn	calf	piglet, or pig	lamb	foal
Castrated male	steer	barrow	wether	gelding
Puberty (female)	4 to 8 months	3 to 5 months	first fall	one year
Pregnancy Average (days)	283	113	150	336
Cycle (heat period) (days)	21	21	17	22

The veterinary journals and texts can extend this body of information ad infinitum. A *shoat*, for instance, is a pig of either sex if it is under a year of age. The term *barrow* means a castrated pig if the operation was performed before sex characteristics appeared. If castration was delayed until the animal was mature, the proper term is *stag*.

1. the awesome roster

The awesome roster of ills that animals' flesh is heir to must rate consideration, else it becomes too easy to think of a remedy in terms of crisis medicine, and not in terms of good soil management. William A. Albrecht used to say "the quality of the bones determines the quality of the horse because bones depend on breeding and the quality of the feed. The quality of the feed, in turn, depends on the efficiency with which the plant factory uses air, rainfall and sunshine to bring bone making minerals—along with others—from the soil into the vegetative bulk. We feed our horses and other animals accordingly as we provide fertile soils."

If Albrecht was right—and all the evidence suggests that he was—then the imposing list of animal ailments presented here all call for one remedy. "We may well give emphasis to feed quality for disease prevention, rather than go to the veterinarian for a cure," Albrecht said.

When the *Acres U.S.A.* editor served as assistant publisher of *Veterinary Medicine* magazine, the following notes were kept within reach. *Veterinary Medicine*, after all, dealt with crisis problems, and this meant recognition of signs and symptoms more than knowledge of the first cause. Here are a few problems and what the veterinarian has to say about them.

Acetonemia. This condition is akin to low blood sugar in human beings. Dairy men often call it ketosis. Heavy milker dairy cows are often affected by acetonemia during lactation periods. It can be spotted by watching for an off-flavored milk, a sweetish, offensive breath, bellowing, insane actions, weakness, trembling, sometimes collapse. loss of appetite, listlessness, indigestion and constipation also figure in the blanket roster of signs and symptoms associated with acetonemia. The best milkers require extra nutrition and any deficiency is greatly emphasized.

Acidosis. This is a pregnancy disease of sheep. It is also called acute hepatitis and ketonuria. Ewes about to twin are often victims. It is important in sheep growing areas because the condition is associated with a 90% fatality factor. Affected sheep can be spotted because they stand behind the flock, grind their teeth, urinate often and tremble when forced to exercise. In the latter stages they refuse to eat and drink and lie down on their breastbones, head to one side. Upon post-mortem examination the liver turns up yellow, crumbles easily and has a lot of fat. Kidneys are pale and softened.

Anaplasmosis. It affects cattle, albeit seldom very young ones. Signs and symptoms include anemia, jaundice, frequent urination, rough coat. Death usually takes from 25 to 60% of the older animals in a few days. Those that recover continue to carry the disease. At necropsy, skin, teats and mouth parts exhibit a yellowish color. The liver is enlarged and yellow. The spleen will be soft and not unlike blackberry jam. Gall bladder contents will have a dark green hue and exhibit the consistency of cool gelatin.

Anthrax. This hunts out almost all warm blooded animals, all ages, man included. Cattle, horses, sheep and goats suddenly die in acute cases. Others run high temperatures, suffer failing appetites, have bloody discharges from natural body openings and exhibit soft swellings that pit upon pressure. Hemorrhages under the skin are common, as are hemorrhages in the throats of swine. Fully 90% of all infected animals die. It may take soil processes more than 30 years to rid itself of this disease.

Atrophic rhinitis. Often a farmer will notice young pigs sneezing and discharge coming from the eyes. There are nasal hemorrhages and irritation of the nose. At necropsy snout and face bones will be observed to be wasted away. No fever here, just a death rate of 20 to 30%, enough to rip the profit right out of a hog operation. This one is a real profit stealer because symptoms are seldom observed before pigs are three weeks of age.

Blackleg. This is a 100% killer of cattle, sheep and goats. The symptoms are high fever, loss of appetite, swelling under the skin of shoulders, hip, under breasts, on flanks and thighs. Post-mortem findings are more definitive. There will be bloody froth out of the mouth, nostrils and rectum that smells like rancid butter. Bloat comes on shortly after death.

Bluetongue. This one affects sheep of any age, and delivers death between 10 and 40% of the time. As the name implies, it starts with inflamed mouth and nose, frothing and labored breathing. The inflamed parts become blue, and there may be ulcers and bloody spots as well as a nasal discharge changing to catarrh with crusts on the upper lip. Necropsy reveals blood and fluids in the lungs as well as muscular degeneration.

Brucellosis. Better known as Bang's disease, brucellosis has a second tag, contagious abortion, although this disease is about as contagious as the stomach ache. It affects cattle, hogs, goats and human beings (undulant fever). Signs are abortion, retention of placenta and reduced milk production in cattle; in hogs, abortion, arthritis and inflammation of the testicles.

Coccidiosis. This one is sometimes called red dysentery. It affects cattle, hogs, sheep, goats, poultry—any age. Very young calves and pigs in particular are affected. Outside of general unthriftiness, there is always a bloody diarrhea, anemia and general emaciation. The death rate is not high, but the economic impact is. Coccidiosis exhibits unmistakable gross pathology at post-mortem examination: a rectum wall two or three times too thick; contents of large intestines and rectum, chiefly blood.

Cholera. It affects hogs of all ages bringing death nearly 100% of the time. In a highly acute form animals just die suddenly, no symptoms. In the acute form, there is a loss of appetite, general depression, fever, purple patches on the abdomen and ears. A discharge often makes eyelids stick together. Because of a weakness in the hind quarters, affected animals walk with a wobbly gait. Examination of organs after death serves up signs, but these can be mistaken for other diseases. Included are a spleen full of blood that is darker than usual, kidneys with pinhead hemorrhages, and raised ulcers in the large intestines.

Diphtheria. There are several handles for this disease condition, namely necrotic stomatitis, gangrenous stomatitis, necrotic laryngitis, malignant stomatitis and sore mouth, all affecting young suckling calves. Several symptoms follow general depression, notably drooling, swelling at the side of the throat, wheezing and coughing. There is almost always a yellowish or greenish-yellow sticky discharge from the nostrils. Odor from the mouth will be quite offensive. Tongues stick out. Death rate is high. At autopsy a cheesy, grayish-yellow mass will be present in the upper windpipe. The same mass is often present in the stomach, lungs, intestines and liver.

Encephalitis. This disease condition is also known as listerellosis and circling disease. It is found in animals with inadequate nutritional support—sheep, goats, cattle, hogs at any age. As the popular name suggests, symptoms include staggering, pushing into fences, circling, and general paralysis. Hogs in particular exhibit hind end dragging, trembling and paralysis. Practically 100% of affected animals die.

Equine encephalomyelitis. This is also called sleeping sickness, and it affects horses, mules and human beings of any age. The death rate is very high. In all cases fever is followed by sleepiness, grinding of teeth, a wobbly gait and a general difficulty in swallowing and chewing. Sometimes animals become wild and unmanageable. Lips, tongue and cheeks are often paralyzed. Sleeping sickness is often accompanied by pneumonia, gangrene or edema.

Johne's disease. Also known as paratuberculosis and chronic specific enteritis of cattle, horses, sheep, goats and deer. Johne's disease starts with a general loss of condition—thirst, diarrhea, rough coat, dry skin, but no fever. In the final stages animals simply refuse to eat. Signs and symptoms are difficult to discern clinically. Much the same is true of necropsy findings in younger calves, which exhibit nothing more than swelling of the small intestines, often in a small area. In older animals the ileum, cecum, large intestines and rectum are thick and have red patches. The thickened areas can be up to five times the normal thickness.

Leptospirosis. It affects cattle, hogs and horses, killing at least 25% of affected animals when they are bovine. In dairy animals there is a drop in milk production, milk becoming thick, yellow and blood tinged. During pregnancy the disease causes abortion. In hogs the usual symptoms include circling and meningitis. Calves exhibit fever, prostration, labored breathing, anemia, jaundice, red urine. As one might expect, necropsy findings always include urine in the bladder approximately the color of port wine. Anemia and jaundice are also typical. Kidneys have white spots or reddish-brown ones.

Malignant edema. It affects horses, hogs, sheep, cattle, any age. It is also known as gas phlegmon and as braxy in sheep. It is 100% fatal. Veterinarians usually associate it with trauma—nails, castration, docking, shearing. High fever, loss of appetite and swelling touch off the symptom parade. Pressure causes a thin, reddish fluid to flow from swellings that— in any case—make a crackling sound when touched. Swellings are usually found in the lungs and elsewhere, as in blackleg.

Milk fever (parturient paresis). This disease condition is usually associated with dairy animals, although female sheep, goats and hogs can be affected. It occurs after newborn are delivered. Symptoms include refusal to eat, trembling, staggering. Without treatment such animals die within hours.

Navel ill. This is a problem with newborn horses, sheep, cattle and pigs. It is sometimes called joint ill because hot swollen joints are a common symptom. The death loss is usually near 100% when the condition comes on within a few hours after delivery. Otherwise the death rate drops to between 30 and 75%. Necropsy findings include abscesses in the spleen, liver or lungs. The navel is generally inflamed. Joints often contain pus, particularly hock and stifle joints.

Necrotic hepatitis. This is called black disease of sheep. It has no symptoms except that death comes swiftly. Sometimes a bloody foam will run from the nose. Bones of the snout and face are often wasted away.

Typhus. Infectious necrotic enteritis or "necro" is a disease condition that affects two to four month old hogs and delivers a high death rate. At the onset there is fever, loss of appetite, diarrhea and general unthriftiness. It is easy to mistake pig typhus for hog cholera. This is important because vaccination with the wrong serum will not halt the death march.

Pink eye. Sometimes known as infectious keratitis and infectious conjunctivitis, this one affects cattle and sheep of any age. There may be occasional blindness and some few deaths, but the main result is loss of milk production or unthriftiness. In all cases eyelids become red and swollen. A yellow deposit forms over the eyes.

Pneumonia. Inflammation of the lungs affects all warm blooded animals of any age, human beings included. It is often associated with other disease conditions. In animals as in human beings, symptoms include dullness, high temperature, rapid breathing, hard pulse. There is often a discharge from the nostrils. Wheezing or gurgling sounds can often be detected in the breast. Cattle breathe through their mouth and extend the tongue. When examined after death, lungs exhibit reddish or grayish-red patches. Air tubes of the lungs are filled with yellow or gray pus. Often serum is present in chest cavities and lungs are full of pus.

Red water disease. It affects cattle and sheep of any age. Bowel movements are scanty, then bloody diarrhea follows. The most common symptom is foamy urine the color of dark red wine. Death comes within 24 to 36 hours fully 95% of the time. Severe anemia and jaundice are part of the clinical picture at autopsy.

Scrapie. Scrapie of sheep is fatal nearly 100% of the time. Post-mortem examinations yield nothing unless fine pathology instruments are available. In the field a farmer might note fine tremors that produce a nodding movement. Intense itching starts on the rump, then travels over the entire body. Emaciation follows. Sheep tend to step high in trotting. This condition runs its course in six weeks to six months.

Shipping fever (hemorrhagic septicemia). A lot of farmers call shipping fever stockyard pneumonia or slobber disease. The death rate can reach 10%, sometimes more. Cattle, sheep, and hogs are affected. In almost all cases there is a hacking cough, watery eyes, a discharge from the nose—and that swollen tongue that causes animals to slobber and drool. On post-mortem examination lungs turn up phlegm coated. Lungs have a reddish serum. Lymph and throat are swollen.

Sore mouth. This disease condition of sheep and goats affects chiefly young lambs and kids. As a precursor of other disease conditions, it can account for very high economic losses, often up to 50%. Symptoms are reddening and swelling lips and gums, and blistering of these parts and the swollen tongue. Pus breaks out in a few days. Scabs form on the resultant raw spots. Scabs usually fall off in a few weeks.

Strangles. This is really distemper, and affects horses and mules. The death rate is usually low, about 5%, and is restricted to colts over six months of age, and horses between two and five years of age. A reduced appetite, a discharge from the nostrils, snorting and coughing, plus swelling under the jaw and in the throat—all mark the etiology of this condition. The swollen area becomes filled with thick yellow pus, and this interferes with breathing—hence strangles.

Swine erysipelas. It affects hogs and human beings, primarily, but is often a problem in sheep and turkeys as well. The death losses are very high in acute cases. In a way, swine erysipelas resembles hog cholera. Animals in seemingly good health die quickly. Some animals have reduced appetite, a stilted gait, red patches on the belly, and suffer vomiting, constipation, then diarrhea. Post-mortem examinations generally reveal swollen lymph glands and spleen, hemorrhages in the stomach, intestines and kidneys. Joints often contain fluid. Joint bones are frequently wasted away.

Tetanus. This is simply old fashioned lockjaw. All warm blooded animals, homo sapiens included, can be victims. The problem can almost always be traced to cuts or puncture wounds. Chewing and swallowing become difficult. Muscles attain an alarming rigidity. Animals take a stiff-leg stance, tail raised. The mortality rate is extremely high, sometimes 90% in the case of eastern type infections.

Swine influenza. This one often infects an entire herd. It starts with loss of appetite, coughing, discharge from the nostrils, and progresses to red eyes, labored breathing and high fever. Hogs squeal mightily when handled. Post-mortem examinations always reveal hemorrhages in several organs of the body. Lymph glands contain blood. Body cavities and lungs contain fluids. Bleeding around and in fatty tissue is symptomatic.

Trichomoniasis. This condition of mature cattle results in abortion. It is different from Bang's disease in that the dead fetus remains in the uterus for some time. Sexually mature animals generally survive, but the condition takes a toll in fetuses.

Vesicular exanthema. This is an economic disease, one that costs in heavy losses of young pigs. It can be recognized by blisters on the snout, in the mouth, between toes, the soles and dewclaws, which are filled with a clear fluid. Unless ruptured blisters become infected, they usually heal in a few weeks.

White scours. Sometimes known as infectious diarrhea or acute dysentery, white scours affects calves up to five days old. Fully 90 to 100% of badly infected animals die. The chief symptom is a yellowish to white foul smelling diarrhea. Often temperature drops below normal. Examinations after death reveal a reddish serum in the body cavities. The liver, spleen and kidneys are almost always at least partly wasted away. The digestive tract is always inflamed.

X disease. This disease of cattle is also known as bovine hyperkeratosis, and it affects all ages, delivering a death rate of near 80% in calves that are very young. The death rate usually goes down for older animals. In X disease, raised, rounded bumps appear in and around the mouth. There is a watery discharge from the nose and eyes. There is loss of appetite and diarrhea. The gall bladder, liver, kidneys and pancreas are affected. Abortion and mastitis in cows are commonly a part of the X disease condition.

2. health out of control

Even the cursory examination of these disease anomalies suggests that one would be thrice a fool to attempt raising animals. The veterinarians have vaccines and serums and needles to cope with many of these problems, all of them a part of the practice known as crisis medicine. Crisis, however, means what it says—health out of control! Each of the conditions previously noted in outline form represents a health profile out of control, or physiological bankruptcy. It may be that some of the problems can't be prevented even with good nutritional support. It is safe to say that 95% of the crisis medicine can be sidestepped by dealing with the soils first, feedstuffs second, finally with the animals themselves.

Albrecht said it all when he wrote in *Good Horses Require Good Soils,*

"When a plant can make much forage yet deliver no seed, the wide fluctuation in chemical composition of the vegetation should become evident. Grass crops that are measured in terms of tons of forage in place of seed yield per acre may be growing on soils too poor to make seed, yet we accept their forage without suspecting defective composition and poor feeding quality. Such soils have mainly a site value and serve largely as plant anchorage. It is this soil property that makes forages deceptive as feeds."

Albrecht went on to note that "Wayne Dinsmore (an associate) reported *that periodic ophthalmia, commonly called moon blindness, frequently is seen in states east of the Missouri River and occasionally west thereof.* In that statement he revealed the possibility of the dominance of deficiency diseases within regions of heavy rainfall, or on the humid soils and their decrease with less rain, or regions of arid and less leached soils." Or, as an octogenarian Hereford breeder in Missouri once said, "They aren't doing on this land what they did here 50 years ago." In short, as nutrients run out or become complexed the general imbalance conspires to debilitate the animal. Pathogenic organisms are opportunists. Much like weeds or insects, they select the weakened body for their handiwork.

Viewing this awesome list of disease problems—all of which can cut the profit picture to pieces—one must ask the obvious question: *just what does it take to keep an animal in good health?* We have already suggested that crisis medicine is a poor answer. Then what is a good answer?

Warren Spring of Milledgeville, Illinois once provided *Acres U.S.A.* with an uncanny summary of research work that consumes countless pages in the scientific literature. "What you should have for an animal is a balanced ration," he said. "This is usually taken to mean so much protein. Yet the first thing you have to get in your mind when you balance a ration is that the most damage you can do will be with an excess of anything."

This is true. When an animal has an excess, it has to borrow from the energy source to get rid of it. Ruminants, for instance, live on the bacterial crop in their unique chambers. The more efficient the production and the better the crop, the more near optimum is the nutrition of the ruminant. What is optimum? Optimum is not a minimum daily requirement, or being certain that there is plenty of a certain element, but the best possible level of each nutrient in respect to other nutrients required by rumen microflora.

A well grown plant, even though stressed by the hazards of nature, will supply the best possible level of nutrients—all biologically processed into the colloidal content of stems, leaves and seeds. It is this form of nutrient material that will supply the entire animal system with the digestible ingredients that assure ingestion into the colloidal fluids and blood supply. At least 70% of the mineral intake must be in colloidal form for optimum health and body function. Too many of our synthetic nutrients never reach the blood system, and more often than not they merely pass through the animal.

If we are to avoid health problems, we must grow or select food materials that are physiologically mature and ripe, and these materials must be obtained from plants that are grown on soils in basic equilibrium. It is such feedstuffs that stimulate optimum hormone processes essential to support of the biotic system involved. This balance, we have pointed out, is best achieved with an array of essential amino acid nutrients and protein of the right character.

3. a mathematical formula

Some parts of nature's equation have been reduced to a mathematical formula by Philip C. Anderson and Janet L. C. Rapp. Their breakdown was first proved out at Panhandle A. & M. College, Goodwell, Oklahoma, and Wisconsin Alumni Research Foundation and the University of Nebraska.

Any system for feeding farm animals properly has to deal with quality. It has to scale the measurable intervals between severe deficiency, hidden deficiency, optimum excess and severe excess. In terms of a pie chart, here are the basics—just five simple isolated groups of elements extracted from each other. And unless they are brought together by the systems of life, energized by sunlight, processed by photosynthesis in the leaf of a green plant, assembled into molecules of nutrients by "cell workers," and packed into a carbon-oxygen bundle by the forces of hormone and enzymes in a living plant, they cannot sustain higher forms of life. And if any of these systems are short changed either in form or amounts of each ingredient, consequences of malnutrition will be the result. Can we simply say that animal health depends upon soil that is healthy, and a healthy soil depends on a healthy microbial environment? There is a biotic relationship that is essential to life, and man—and production animal—cannot endure a folly of substitutes.

A pet antipathy of many eco-farmers is the feed trade's preoccupation with protein. In 1928 *Morrison's Feeds and Feeding* started reporting protein in feedstuffs. They took samples from all over the United States, with eastern states weighted heavily. Each year more samples were added, and yet the old ones were not dropped. Thus there exists in Morrison's a cumulative figure with marginal relationship to the fact of crop nutrient content.

For instance, the fiction of total digestive nutrients has survived. Yet years ago Purdue found TDN passe. They proved that three corns could test the same TDN-wise and yet littermates would do differently on it. Second, TDN takes no account of oxygen at all. Excess oxygen in the feedstuff on a dry matter basis lowers energy. Energy is the flywheel and balancewheel—but not the spark plug—of animal health.

Here are a few notes on feedstuffs. Observe how feed imbalances quickly relate to many of the disease conditions discussed earlier. In a few cases we

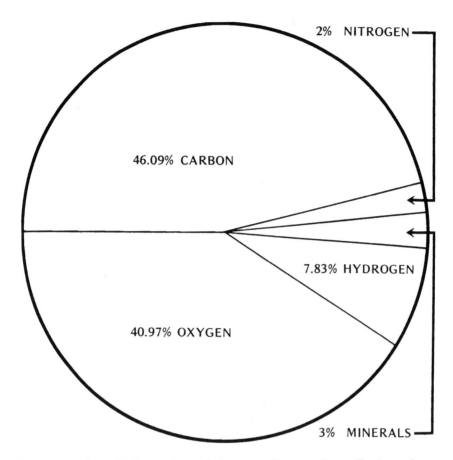

2% NITROGEN

46.09% CARBON

7.83% HYDROGEN

40.97% OXYGEN

3% MINERALS

know exactly which nutrients bring on disease—brucellosis and grass tetany for instance. In others we can only hint at the mix of imbalances that cause pus to flow, urine to turn wine red, scours to debilitate, internal organs to waste away. Moreover, arrival of the synthetic kingdom—the herbicides, the pesticides, all the toxic genetic chemicals—ape and mask disease conditions much as syphilis imitates many of the common disease conditions in man. Unfortunately modern science often pauses when it discovers a proximate cause, and forgets to pursue the foundation cause—an oversight that has kept the white light of analysis away from nutrition all too long.

4. content of a ration.

Dry matter and the ash content of a ration govern the solubility value of a ration and the rate at which it moves through the digestive tract. Dry matter should not be less than 20% for, say, the ruminant. The ash content—which measures the oxides of minerals in the ration and provides an index of fiber content—should be between 4 and 6%.

As illustrated on the previous page, the carbon content should be 46.09%, according to Anderson and Rapp. Carbon governs the need for hydrogen, oxygen and nitrogen, an excess increasing the requirement for these elements, a deficiency decreasing that same requirement.

Note the posted figure for hydrogen—7.83%. An excess over this figure means the energy value of the ration is in excess. Bloat is a likely result. Deficiency means muscosal disease, watery eyes, a hacking cough, undigested feed in droppings, unthriftiness. It takes little imagination to start matching a nutritional lack to the specifics veterinarians try to remedy with crisis medicine.

5. the oxygen content

If the oxygen content of a ration is in excess of 40.97% to 41.5%, animals founder and exhibit a stiffness of gait. The depth of the body becomes shallow or "tucked up." Roughage and grain in droppings increase. Milk and butterfat production drops, and the rate of gain in beef animals drops. If there is an oxygen deficiency, butterfat production may stay up, but milk production drops. The rate of gain is slowed. A warning sign can be seen in droppings that stack up like canned biscuits because the ration is not soluble.

6. the nitrogen level

The nitrogen level has to hone closely to that 2% figure—never to exceed 2.88%. An excess means ketosis, scouring, mastitis, a general decline in milk production or rate of gain. At the other end of the scale, deficiency accounts for retarded growth, irregular heat periods, poor rate of gain in beef animals and poor milk conversion in dairy cow.

The lexicon of animal husbandry holds to the equation 6.25 x total nitrogen = protein equivalent. The shortfalls of this formula have been discussed on page 172. Proteins are a major component of every living cell. Protein as a 12.5% component of the ration is optimum for all ruminants. But these proteins must be ripe and mature proteins—not just a laboratory measurement of nitrogen content. Too much protein is a contributing factor in ketosis, scouring, poor feed conversion. A deficiency soon translates to retarded growth, poor wool or milk production, and irregular heat periods.

Needless to say, problems do not generally arrive as a consequence of single factor imbalance. Often several parts of the equation are out of whack when animals become the focal point of exotic veterinary talk. Sulfur, potassium, calcium, magnesium, phosphorus, chlorine, silicon and the trace minerals, all figure when they account for an imbalance—either too much or too little.

7. sulfur

The optimum amount of sulfur should be .14% of the total dry matter. An excess accounts for an acid rumen and increases the need for copper. Too little sulfur causes sheep to shed their wool and results in poor development of keratinous tissue—hoofs, horns, hair. Excess saliva and watery eyes become a legacy. It may not be possible to tell just which other imbalance joins a sulfur imbalance to trigger anything from acetonemia to bovine hyperkeratosis, but the foundation cause can easily be seen.

8. potassium

East Texas farmers still talk about the potassium shortage in native grasses that once had cattle either dying or becoming shy breeders. In fact potassium should account for .93% of the ration on a dry matter basis. Too much slows down bacterial growth in the rumen. Too little results in retarded sexual maturity and difficult breeding. Because carbohydrate utilization is slowed by too little potassium, growth is also slowed.

9. sodium

Sodium also has its index. The optimum amount for a ration should be .27% of the dry matter. Possibly 90% of all rations—even many "scientifically balanced" rations—are low on sodium. Too much, of course, figures in swelling due to water retention, and inhibited bacterial growth in the rumen. Too little short circuits utilization of protein and energy. Rough hair coat, retarded growth, poor appetite and poor reproduction all point to sodium deficiency.

10. calcium

Calcium is the prince of nutrients in the soil. It should occupy .48% of the total dry matter. It happens also to be basic in the animal ration. Too much increases the need for phosphorus, vitamin D-2 and zinc. It decreases the availability of protein, phosphorus, iodine, iron, manganese and zinc. It figures in birth paralysis and depresses the rate and economy of gain. A deficiency, on the other hand, impairs bone growth. It increases the need for vitamin D-2. In the feedlot or on the pasture the calcium deficient animal will appear listless, often exhibiting an arched back and a depraved appetite.

11. magnesium

Magnesium should be .29% on a dry matter basis. In excess it increases the need for phosphorus. As a deficiency, it accounts for grass tetany, a condition that sees animals blinded, turning in circles until balance is lost, and frothing at the mouth.

12. phosphorus

Phosphorus walks hand in hand with calcium. When calcium is excessive, cattle will eat phosphorus in excess and then excrete both calcium and phosphorus down to the optimum level. Phosphorus, of course, is acid in nature. An excess of this nutrient increases the need for iron, aluminum, calcium, and magnesium. The observed result is poor skeletal formation. Deficiency means poor fat assimilation, either a delayed or missing heat period, prolonged intervals between calving. On the economic front, there is poor rate of gain, faltering milk production, general unthriftiness. Just as an excess of phosphorus increases the need for iron, aluminum, calcium and magnesium, a deficiency of phosphorus can be created by excess iron, aluminum, calcium and magnesium.

13. chlorine

Chlorine is another element that must figure in the ration—this one to the tune of .42% of the dry matter. As with sodium, excess brings on swelling due to water retention. It also increases the iodine requirement. A deficiency results in loss of appetite and poor weight gain, poor hair coat, hyperalkalinity. Tetany and death are often a result.

14. silicon

Silicon has its optimum figure in the ration, .33%. Any excess slows passage of food through the rumen. It also decreases digestibility and palatability. A deficiency slows growth and multiplication of rumen bacteria, causes poor fill, and a depraved appetite.

15. trace minerals

In addition to the above, there are the trace minerals, most of which have not even been investigated in terms of animal health. Individually and as combinations, they probably contain more answers to health problems than all the medicines ever synthesized by man.

Trace minerals are reported as parts per million (ppm) on a dry matter basis. Optimum amounts of trace elements are expressed as follows:

Iron	100 ppm
Aluminum	60 ppm
Manganese	60 ppm
Zinc	60 ppm
Boron	10 ppm
Copper	10 ppm
Molybdenum	1.0 ppm
Iodine	0.5 ppm
Cobalt	0.5 ppm
Nickel	0.5 ppm

16. iron

An excess of iron interferes with phosphorus absorption. Sodium or potassium bicarbonate are required to precipitate iron excess.

A deficiency of iron results in anemia. This is most likely to occur in calves because milk is low and little iron passes across fetal membranes. Cow and calf operations can exhibit anemia and are more susceptible to disease conditions.

17. aluminum

An excess of aluminum increases the need for phosphorus.

18. manganese

An excess of manganese increases the need for iron.

A deficiency of manganese, on the other hand, results in leg deformities with over-knuckling in calves, eggs not formed correctly, degeneration of testicles, offspring born dead, delayed heat periods. Shortage is created by excess of calcium and phosphorus.

9. zinc

An excess of zinc means decreased copper availability and interference with utilization of copper and iron, bringing about anemia. A zinc excess also shows up as bald patches and skin disorders (rough skin), a deficiency is created by excess of calcium.

Zinc is absolutely necessary for production of sperm. It also increases the need for vitamin A.

20. boron

An excess of boron means diarrhea, an increased flow of urine, and visual disturbances. A deficiency of boron reduces rate of growth and rumen bacteria.

21. *copper*

An excess of copper results in degeneration of the liver. It causes blood in urine and poor utilization of nitrogen.

A deficiency of copper is created by excess of molybdenum and cobalt. It produces anemia due to poor iron utilization. It depresses growth. Other symptoms . . . depigmentation of hair and abnormal hair growth; impaired reproductive performance and heat failure; scouring; fragile bones; retained placenta and difficulty in calving; and muscular incoordination in young lambs, and stringy wool.

22. *molybdenum*

An excess of molybdenum makes copper unavailable. It brings on depigmentation of hair, and also severe scouring.

A deficiency of molybdenum is created by excess of sulfur.

It slows down cellulose digestion. It accounts for calcium deposits in the kidneys. Chronic poisoning is also an observed result, depending on the level of copper. A deficiency of molybdenum slows down the conversion of nitrogen to protein.

23. *iodine*

An excess of iodine means secretion of mucus from the lungs and bronchial tubes. A rapid pulse and nervous tremors also accompany iodine excess.

A deficiency of iodine brings on unmistakable signs . . .

Young are born dead, or die soon after birth.

Abortion at any stage, or reabsorption of the fetus.

Retention of fetal membrane.

Irregular or suppressed heat period, infertility and sterility.

Decline in sex drive, deterioration in semen.

24. *cobalt*

An excess of cobalt reduces the availability of copper, aluminum, iron, manganese, molybdenum and iodine—if the excess is severe. Also, the ability of bacteria to convert nitrogen to protein is reduced.

Cobalt deficiency means . . .

Rumen bacteria fail to manufacture enough vitamin B-12.

Starved appearance with pale skin.

Decreased fertility, milk, or wool production.
Cobalt is necessary for utilization of propionic acid.
Cellulose digestion is sharply reduced.

25. nickel

An excess of nickel makes the ration unpalatable. Excess nickel can be reduced by chelated iron.

26. vitamin a

This is the anti-infection vitamin. An excess of vitamin A is stored in the liver and in fat tissues. It works against vitamin D.

On the other hand, a deficiency of vitamin A results in the following inventory of symptoms . . .

Nasal discharge, coughing, scouring and watering eyes due to crying and hardening of the mucous membranes which line the lungs, throat, eyes, and intestines.

Calves' horns are weak.

Severe diarrhea in young calves is observed.

Redness and swelling around dewclaws.

Stiffness in hock and knee joints and swelling in the brisket.

Increased incidence of mastitis and other udder problems due to drying hardening of the mucous membranes of the udder.

Decline in sexual activity. Sperm decrease in number and mobility.

Loss of appetite.

Zinc deficiency, nitrates, and low ash rations increase the need for vitamin A.

27. vitamin d

An excess of vitamin D will result in a deposit of calcium in the heart and kidneys. An excess of vitamin D works against vitamin A.

A deficiency of vitamin D means . . .

Joints and hocks swell and stiffen.

Back arches.

Increases need for calcium and phosphorus.

Stiffness of gait, dragging hind feet.

Rickets.

28. vitamin e

A deficiency of vitamin E usually occurs in young animals. Effects include muscular dystrophy (white muscle disease) and heart failure, paralysis varying in severity from light lameness to complete inability to stand.

29. *vitamin k*

A deficiency of vitamin K results in failure of blood to clot.

30. *a disease condition*

There isn't a disease condition in animal husbandry that isn't started or sustained by poor nutrition. Likewise, there need be no animal disease if the animals are fed plants that give good nutritional support.

The late W. P. Scott of Naremco, Inc., Springfield, Missouri, said it all when he noted that "virtually all disease causing organisms are everywhere in nature, and in all forms of life. Yet we never see infection caused by bacteria or fungi present in all people, all animals, or all birds." Why do some escape? More important, what are the conditions that make a bird or an animal unable to guard its own health?

John Whittaker, a veterinarian who writes a column for *Acres U.S.A.*, has followed viruses, parasites, protozoa, bacteria and fungi cause-effect relationships his entire professional life. During the last two decades he has written often and well about mold toxins, those unfriendly fungi that seem to get their go-ahead when friendly fungi in the soil are overwhelmed. Poor soil management helps underwrite this mycotoxin proliferation— *Aspergillus flavus*, for instance—but not half as much as increased use of toxic genetic chemistry.

31. *build a ration*

There are consultants afloat in agriculture who attempt to put all these nutritional requirements in a computer and spin out answers, often with remarkable results. About a decade ago, a group of veterinarians—wearied of their black bag full of medical tricks—moved into nutrition. They formed Animal Nutrition, Inc., at Belleville, Illinois. The basic approach was to take core samples and test the bulk feed available to the farmer, get a laboratory readout, then with a computer doing the computations "build a ration." The same general job is being done by Rocky Mountain Nutritional Consultants, Evans, Colorado. Both operate on the basis of premises suitable to those who pursue environmentally sound agriculture.

Basic foundation ingredients and the trace nutrients all figure when computer lights flash. Much of the fine-tuning relies on chelation, a basic phenomena in plant and animal life. When feedstuff production fails to deliver the full balance that health requires—which is frequently the case nowadays—then the next best thing is to turn to the chelated additives.

All this is not to suggest that sound principles in crop production won't produce feedstuff support for animal health. Sound principles will. Gene Poirot, whose words were quoted in the opening lines of this book, has kept a cow herd free from disease for decades. Environmental farmers everywhere are doing the same thing without computers or even in-depth knowledge of nature's chemistry. They know that they must reach back to the soil for nature's balance, rhythm and harmony, or the sad notes on which this lesson started become an assured legacy. And they act accordingly.

CHELATION

The word chelation surfaces routinely among eco-farmers. It describes how modern technology can work with nature. Animal Nutrition, Inc. has offered this explanation.

Chelation involves the coating or surrounding of an inorganic metal ion by a chelating or trapping agent. When metals are properly complexed or chelated before ingestion, the positive charges on the metals are absorbed or neutralized by the organic chelating substances carrying a negative charge. This results in a slightly negatively charged molecule which in unaltered form can transport itself through the intestinal wall into the blood stream. Chelation is commonly used to express this type of product. But the more scientific term would be that of organic metal compounds or amino acid chelates.

We are prone to assume that all the food factors that reach the upper intestine are absorbed uniformly and adequately. Of course such is not the case. On ingestion of inorganic or non-chelated nutrient salts, cations disassociate from anions, with cations taking a positive charge and anions taking on the negative charges. Anions pass through the pores of the intestinal wall quite readily because both are negatively charged. This provides enough electrical repulsion to keep anions from adhering to negative pore membranes. On the other hand, positive charges on the cations cause enough fixation with the negative pore walls to prevent absorption of a major part of the cations. When inorganic minerals are ingested, natural chelating factors must rush to the rescue to absorb and complex a certain portion of the metals.

The enzyme systems of the body depend on chelation whereby the metal is tied into or surrounded by the organic moiety of the enzyme molecule.

ANI's chelates are produced from amino acids, peptides and polypeptides, and are referred to as natural or organic chelates. This is opposed to synthetic or inorganic chelates such as EDTA. Since natural chelating agents are not foreign to the body, toxicity problems are avoided and overall efficiency is increased. Chelation is the most fundamental

principle of life. No cell in animal, human being or plant can function except by utilizing the chelation principle.

EDTA chelates are not compatible nor reliable for either plant or animal metabolic processes. EDTA is called an organic chelate, but the chelating agent does not dissipate, and may continually react with other exchangeable elements. This results in active disequilibrium and lower capacity to react to stress conditions. Plant and animal systems must always be balanced with properly chelated minerals exchangeable to the inner cellular systems, to keep the electrons tuned-up for optimum cell and genetic function in the system of life as nature intended.

Silage

A well-known farm consultant once said that the "silo is a monument to the stupidity of man. Why? You can't stabilize the nutrient values in feeding. The big monuments of our past mistakes or the status symbol tanks always lead to the nearest feed consultant. As a result of successful silo salesmanship, many technical feed consultants have come on the scene in the last 15 to 20 years, all building their business around silage problems."

Silage has always been a prime farm problem for the grower who has livestock. There is almost never a uniform way to provide feed throughout the course of the year. Temperature should control the rate at which ensilage is placed into storage, and conditions are almost always difficult for obtaining proper fermentation.

It is difficult to see how any farm can afford the kind of investment most silos require. The money is locked in. After four or five years, many farmers want to sell the farm to a doctor or investment house. A farmer ought to be able to bulldoze every five years and say, *this silo owes me nothing!*

Silage quality is directly influenced by many factors, each varied with respect to individual farm situations. And each such factor requires evaluated thought and immediate judgment to determine the methods that can be used to obtain properly fermented silage. There is a distinct art to making good silage, and good silage can be obtained from many forms of storage.

These variable objectives must be woven into the fabric of individual farm feed systems. They must include the following.

1. Access and availability to a uniform supply of feed for every day of the feeding season.

2. Simple and practical methods of handling, storing and feeding. Included here is the temperature goal of 105 F., 12 inches deep, before addition of more green chop material; lactic fermentation to stabilize nutrient values; moisture management of fresh ingredients as well as plant juices, liquid saps, chlorophyll and cellular protoplasm. These moisture forms are the essential ingredients which will govern the nutrient value and nutritional quality. They give permission to life for the various life forms invited to endure.

3. Capacity of storage must be flexible enough to salvage varied forms of bulk tonnage at different times of the year, and yet be practical in cost and function while relying on a minimum of mechanical devices. Mandatory is reliable access to choices of ensilage for frequent feeding each day, again with a minimum of mechanical limitations. All this translates into the final requirement—low investment and high capacity storage structured to equal low cost handling. Horizontal bunker silos and well built trench silos fill these requirements best, and at the same time permit principles of fermentation management to underwrite the art of making and keeping good feed.

1. silage

What is silage?

Sugar + $O_2 \rightarrow H_2O + CO_2$. Heat, and hot silage means lost energy.

Textbook definition has it that silage is "the product of fermentation of carbohydrates in green plant tissues by certain species of lactic acid bacteria." This is correct. Fermentation in fact means production of acids by lactic bacteria, and lactic bacteria are acid producers. As just about any eco-farmer knows by now, bacteria are not bacteria without vast distinction. The right bacteria are required in the soil, and the right bacteria are mandatory in silage making, beer brewing, wine fermentation, even whiskey making. The wrong ones are worse than wrong, they are impossible.

This fact has pretty much escaped modern agronomy. Chemical embalming fluids and preservatives won't make good silage, much less preserve it. Only the right bacteria handle this chore properly. How well this population is nurtured and fed will determine nutrient and caloric losses resulting from mold, plus losses accounted for by improper fermentation.

The minute a chopper moves down the corn rows, harvesting and reducing to small pieces the tons of ensiling materials, the mass becomes a microscopic battleground. The nutrients in the plant are what they are at this time, a fine-tuned result of tillage, fertilization, management, seed selection and timely harvest. But bacteria from the soil, even from the equipment, find a new territory to occupy and conquer. Juices suddenly exposed to air become a growth medium. Even good quality forage has plenty of nutrition for growth of butyric spoilage bacteria. These bacteria immediately declare war on lactic acid bacteria. Each gains a toehold and each goes to work producing various byproducts. If a farmer has judged correctly, harvested properly, and properly packed the ensiling materials, the battle eventually turns in favor of the lactic acid bacteria. Fermentation proceeds properly and the completed material becomes silage. There is an if in this procedure—if the lactic acid bacteria manage to crowd out the competing bacteria.

What if the lactic acid bacteria lose the battle? The first available evidence will be heat. This means the battle is going badly. Heat means that carbohydrates are being burned up by aerobic bacteria. This burning accounts for conversion of carbohydrates to carbon dioxide and water. Carbon dioxide is lost in the air. Water is run off, carrying many of the silage goodies into the nearest ditch. The second available evidence will be cold fermentation. This means a poor nutrient media to invite lactic acid processes even to begin.

2. good silage

The Bureau of Dairy Industry, USDA, has advised that the amount of silage available for feeding will average 20 to 40% loss in stacked silage, and 15 to 25% loss in trench ensiled forage even when silage is made by an expert. All this is based on the obsolete technology of wild fermentation. By way of contrast, even the apparent spoilage on properly treated silage materials remains as a high protein and desirable nutrient.

Good silage should be bright and not much darker in color than the original plant material. It should have a clean, sharp, but not heavily butyric aroma, and it should be acid enough to inhibit spoilage bacteria. There should be practically no runoff.

Fortunately, Bureau of Dairy Industry findings can now be negated because of developments in eco-agriculture. Several firms have developed silage makers that harness friendly bacteria and inhibit the dangerous *Candida albicans* mold. Proper fermentation is the key. It lowers both free oxygen and oxidation-reduction, and this lowering is a strike force against the growth of contaminating molds.

Molds mean spoilage. Molds may mean mycotoxin production, including the insidious *Aspergillus flavus* (aflatoxin). Yet friendly molds may assure good keeping quality. Once the wrong molds invade silage, the result is weight loss (or failure to gain) in livestock. At necropsy, veterinarians find diseased viscera brought on by this cancer causer, *A. flavus*, subclinical infections, and all the while the animal appears to be well on the outside.

The ecologically sound silage maker is not an embalming preservative. It is rather a select seeding material that does for the silage pulp what inoculants do for the manure pile, the beer ferment or bread dough. Friendly bacteria overwhelm unfriendly agents to control molds, reduce temperatures in the pit or silo, retain moisture while at the same time conditioning the crude fiber for easy digestion. The signs of good silage are easy to discern. Color and aroma are pleasing to the senses. Waste is virtually eliminated. Nitrate levels never assail the nostrils.

3. when should crops be cut

When should crops be cut for silage? It depends on the crop and the moisture content of that crop.

Testing for the proper condition does not require apparatus. Simply chop enough material for a double handful. Compact it between the hands. Moisture will be just right when a ball takes shape and unfolds slowly. If the material has a springy effect, moisture is already too low. If it compacts and holds its shape, and moisture oozes out copiously, the material is too wet. There is more tolerance for wet silage when eco-inoculants are used. Indeed, wet materials properly handled can end up as good feed, too.

Clover should be cut when half heads are in bloom and moisture is 70 to 80%. That's just below the runoff stage. Alfalfa is right when one-tenth heads are in bloom, moisture 70 to 80%. Cereal grains make good silage when the kernel has a milk to soft dough effect, moisture 75% plus. High moisture ear corn should also have that milk to soft dough kernel, moisture 40 to 42 %. Grass, of course, is at its protein stage when it first joints, and optimum harvest is at 4:30 in the morning. No farmer can comply with exacting nature in this instance, and in any case optimum lasts only a few minutes. When first stems head, grass should be cut for silage. Moisture content at this time will be 65 to 75%. These data refer to cutting when plant nutrients are at a certain level of protein or available energy content. But healthy plant growth may also permit more physiological maturity which can also be properly stored.

When moisture in grasses is low, say 55%, lactic bacteria are at a disadvantage, and butyric fermentation results. A fair grass silage can be made at 60% moisture when a good inoculum is used. Without an inoculum, such silage can do little more than become an unpalatable mess.

There is little evidence to suggest that tall silos are worth the money it takes to erect them. A pit silo is cheap to construct and usually owes a farmer nothing after a few years of use. Sidewalls improve yield and compactibility. As indicated in the drawing, a pit silo might slope out at the top

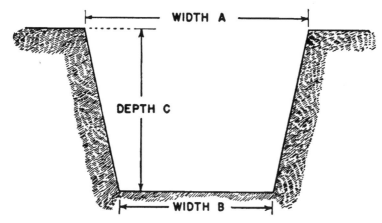

COWS	Trench Width		Depth	Feeding Period			Amount of silage per running foot
	A	B	C	90 days	120 days	150 days	
Number	Feet	Feet	Feet	Feet	Feet	Feet	Pounds
3	4	3	3	26	35	43	315
5	5	3	4	28	38	47	480
10	8	6	4	33	43	54	840
15	10	8	4	38	50	63	1080
25	12	10	5	41	55	69	1650
50	12	10	5	82	110	138	1650

TRENCH SILO CAPACITIES

Width	Top Width	Bottom Width	Approx. No. of Tons per Running Ft.
10 (Feet)	11 (Feet)	6 (Feet)	1.5 (Tons)
10	15	8	2.0
10	18	10	2.5
8	10	6	1.1
8	12	7	1.3
8	15	8	1.6
6	9	6	0.8
6	11	7	0.9
6	13	8	1.1
4	7	5	0.4
4	9	6	0.5
4	10	7	0.6

* Based on settled silage weighing 35 pounds per cubic foot.

a bit, about one inch per foot of height. This permits compacting equipment to work the edges and force chopped material tightly against the side as it settled. A regular wheel tractor is best for the compaction chore. As a matter of fact, the wheel does a better job than track type vehicles.

4. compaction

Compaction in upright silos is no problem. The material will handle its own, weight on top packing the lower levels. Some farmers make a foot job of it, but those who don't bother seem to fare as well.

Most good silage makers are added at the rate of one pound per ton of silage. This can be done by hand broadcasting, or via mechanical distribution, depending on availability of equipment and resources. Amino complexes may also be applied on the harvest machine. (Parenthetically, it should be noted that amino acid complexes applied at harvest both enhance the quality of hay, and permit instant baling.)

Is a silage maker absolutely necessary? We would have to say yes! It is true, bread can be made with the wild yeasts in the air rather than with controlled fermentation. Compost can be made the same way, relying on soil and air bacteria that float in to inoculate a pile of manure. Unfortunately you can't make a very good loaf of bread without yeast. And compost takes too long and results are too spotty when an inoculum is not used. The same is true of silage. It can be made without help, and many farmers do a fair job—albeit with the losses Bureau of Animal Industry has properly defined. It costs quite a bit to fertilize properly, prepare the rootbed, till and harvest a crop. It makes little sense to risk it all with seat-of-the-pants technology when science has offered so much for so little.

Should a pit silo be covered? Depending on experience, climate and other factors, it is usually best to cover and protect. Black plastic is available just about anywhere. It can be tucked in at the foot and covered with soil or other weights to prevent animal access. Animal runs between plastic and forage material are essentially destructive, and possible sources of disease problems, and should be avoided.

POST GRADUATE

LESSON 27

The New Agriculture

The new insight called eco-agriculture has now ranged well beyond pure organics and use of the homestead tools a man, his labor and nature might provide. It has given birth to all sorts of sophisticated products wherein nature has married up with science in a most compatible union.

1. the soil activators

The soil activators came first. One and all they claim to increase soil fertility by accelerating the growth of beneficial soil organisms, and also by speeding up the formation of humus. Humus, as we have seen, has the ability to absorb plant nutrients and hold them available for plant use. All of these products came on at a time when college professors were saying "there is no substitute for N, P and K." The wisest answered, "Of course there isn't, but N, P and K are no substitute for trace minerals, functional enzyme systems, chlorophyll capacity and sunlight utilization, tilth, and biotic life in the soil." The soil activators presumed to deal with many of the other variables, and to set up an equation that provided for better utilization of N, P and K in the soil bank account than had heretofore been possible. Primary ingredients in many activators were and are cultured microorganisms and the medium which supported the organisms. How well such technology gains a toehold on worn out, chemicalized soil is openly debated. The earliest workers with legume inoculants came to understand that a life structure could be imposed on soil only if the soil was capable of supporting it. William A. Albrecht put it this way. "You can't get a calf just because you've got a prize bull. The momma cow has something to say about it too." Nevertheless, the soil activators have worked. How they have worked has often depended on where and how.

It may be true that some entries in the new agriculture at first sounded like a secret formula whispered by a dying monk to a titled traveler, but this was so largely because too few in the institutions understood the grammar of the subject. Agronomist John Porter, for instance, found some genetically superior strains of aerobic bacteria, anaerobic bacteria, facultatively anaerobic and algae while working with coffee planters in Brazil. These strains were both symbiotic and nonsymbiotic in their relationship to the higher plants which shared the soils in which they lived. They were held in the soil activator he developed by a host medium composed of formulated and treated organic material designed to sustain the several strains of bacteria while they genetically adapted to the soil environment in which they were placed.

This may sound like high flown language for exactly what Fletcher Sims has been doing with compost, and what silage makers are doing with corn, alfalfa and oats.

The soil is a living ecosystem in which bacteria are an integral part. Without the proper bacterial activity to promote mineralization of nitrogenous organic matter, the soil cannot feed the plant for very long, if at all on a continuous basis.

It is equally true that when there is insufficient bacterial action in the soil's ecosystem, then each phase of the natural feeding cycle is occurring at a suboptimum level. There are bacteria which subsequently oxidize nitrites to nitrates. These various forms of bacteria ammonify, nitrify, and aid in conversion—and through increased enzymatic activity influence a variety of factors in the soil's makeup, including crumb structure, salt content, pH, aeration, and basic ionic exchange with regard to macro and micronutrients a plant needs. Indeed, the microorganisms are the all important entity in the soil's relationship to plants.

2. the foundation precepts

One of the foundation precepts at *Acres U.S.A.* has always been that a farmer is entitled to know what a product tool contains and what it does, and that this information must be communicated in language a literate farmer can understand. We may not always succeed in this, but we have to try. A good college try is most important as agriculture enters the fifth dimension. We'll call them catalytic growth regulators. We hinted at this in the lesson on the cell.

It is ironic that most of the development in the new agriculture has taken place with the land grant colleges knowing very little about it. Even today, according to agronomist Lee Fryer, very few colleges have people competent in the area of catalytic growth regulators. Some 30 years ago, college workers still scoffed at trace elements because they "didn't believe in them." To this John Porter replied, "Ah, but the plants do, and they didn't go to college."

3. enzyme systems

Plants rely on enzyme systems as regulated by their micronutrient keys. Here are a very few of these enzyme systems and their trace element keys.

Enzyme	Key
Nitrate reductase	Molybdenum
Gentanic dehydrogenase	Copper
Phospholipase	Manganese
Cytochrome	Iron
Starch phosphorylase	Boron
Auxin enzyme	Zinc, etc.

The enzymes are protein amino acid clusters. They speed up the rate of a specific chemical reaction in a plant. Each specific enzyme is tied to its individual chemical reaction. As a result, this necessary enzyme is produced in each plant cell at the time of cell division.

Each enzyme is inert and non-specific until activated by a specific trace element. This trace element is the key that unlocks enzymatic activity and fires it up so that it can do its work in the metabolic process. N, P and K are not taken up as plant nutrients except under enzymatic control. When this activity falters, the result is a cellular short circuit. This may mean less production, production without keeping quality, or a crop incapable of defending itself against insect, bacterial and fungal attack.

Trace minerals in a soil system can either vanish or become locked up as a result of imbalanced fertilization. This means return of trace minerals to the soil itself is fraught with unwanted possibilities, namely the new supply being promptly complexed and held unavailable.

4. catalytic growth regulators

The integration of the growth rates of different parts of the plant is controlled by hormones. Six micronutrients—molybdenum, copper, manganese, iron, boron, and zinc—are necessary in balanced proportions to assure ideal plant growth. The form of each element is quite strategic, more so than a mere quantity of water soluble individual elements can assure. The wire service that keeps each enzyme paced is the hormone system.

As a general class, the catalytic plant growth regulators are simply soil, seed, and plant foliar inoculums for maintaining optimum plant growth, yield, resistance to stress, storability and nutritional maturity. In addition to normal soil microorganisms, the new breed of growth regulators contain various aerobic and anaerobic nonsymbiotic nitrogen fixing bacteria along with good nutrient media capable of taking nitrogen from the air for plant use. These bacteria under ideal conditions can fix up to 80 pounds of slow release protein nitrogen per acre per year. It has been calculated that seed treatments, crop residue digesters and foliar sprays can assist the soil and plant system to utilize nitrogen more efficiently. The trick is to supply the bacteria with their proper nutrients for performance even though they are inoculated into low organic matter soils.

In this manner nature-science products create natural cooperation between soil microorganisms and the plants to supply a uniform, metered, readily available nitrogen with proper cutoff time. This is quite opposite from synthetic nitrogen. The upshoot of the new technology is quicker dry down, physiological maturity, and feeding of the plant through periods of stress and drought.

During the past several years, foliar mists of combinations of plant catalytic growth regulators with major, secondary and micronutrients have been created to accomplish several things. Those who sell them can take the responsibility for their specific performance—and directions for use. It is enough here to orient the farmer on what he can expect. In a general way—

1. They help set more blooms and increase the size of extra produce, flowers, fruit or grains.

2. They create natural cooperation between soil microorganisms with the plants to supply a uniform, metered, readily available nitrogen with the proper cutoff time, which means quicker dry down, physiological maturity and reduced growing degree days.

3. They increase the protein, ribonucleic acid (RNA) and deoxyribonucleic acid (DNA) contents of plants.

4. They extend the productive period and shelf life of plants and produce by retarding the aging process, or senescence, in plants. Crops sprayed with foliar nutrients stay in production longer and yield more and higher quality products.

5. They retard the loss of protein, chlorophyll and RNA in produce, fruits, vegetables, haylage, silage. Products grown with the foliars have a longer shelf life, ship far better and exhibit excellent cosmetic effect at the market as well.

There are numerous single item products under development or in the supply system available to farmers. It must be remembered that single factor ingredients—when used in conjunction with or when complimented by other essential growing nutrients—can give satisfactory function and performance. And in many cases when these are used alone, environment or chemical limitations present in unbalanced soils can actually degenerate or significantly reduce the actual benefits possible.

A product produced under the guidance and support of A. H. J. Rajamannan (a Ceylon-trained scientist) is a biological combination of bacteria nutrients and enzymes that when placed in the seedbed can function on its own and release or convert essential soil and plant nutrients, especially from the locked-up portions of the soil. It can also stimulate the carbon-oxygen cycle of decay and release of organic nutrients. Related type products are also available for seed, root or foliar application. These often can supply or stimulate plant response and recovery from water, temperature, air or other stress factors during the growing season.

5. growth hormones

Many deal with growth hormones. There are three groups of growth hormones, each of which can interact to affect plant changes and growth patterns.

There are the *gibberellins*. Unfortunately, these degrade rapidly and lack functional stability. They can easily be produced by plants if nutrition is ample and proper.

There are the *auxins*. These are very complex compounds related directly to environment and character of nutrition. They decompose or can be degenerated rapidly in each type of environment. As an example, a green manure crop plowed down in the spring rootbed may leave its auxins (root exudes) in the soil to keep weed seeds dormant. In some soils these auxins may endure 1 or 2 days, but in properly balanced soils the effect may hold back weed germination as long as 6 to 8 weeks. These complex hormone compounds are extremely powerful in a biological system. They are also quite fleeting in their functions, and can constantly be produced by all forms of living systems to meet and react to hormonal direction as needs and media require. Most herbicide compounds are auxin-like, albeit without nature's finely tuned governing capacity or application apparatus.

Last, there are *cytokinins*. These are produced in culture systems under laboratory control. They are stable, natural, and exchangeable to the biological energy field of soil and plant systems. They are basically cell dividers. That is to say, they stimulate cell division and enlargement. They cause differentiation of plant tissues and bud formation. They regenerate shoots of root segments, increase rate of germination, and induce flowering. They reverse the effect of plant inhibitors such as excess cold or heat, or improper light conditions. They offset cumarins which prevent germination and eliminate the need for infrared stimulus or magnetic effects on seed germination.

Cytokinins take on front burner attention when the products of the new agriculture are considered. These points, for instance, cannot be contradicted successfully.

Cytokinins used in all areas of life systems can increase the capacity of those systems to utilize fertilizers 20 to 30% better. Cytokinins stabilize and enhance chlorophyll efficiency, and tend to improve protein synthesis and function in leaves of plants. Thus they increase photosynthesis capacity. In relying on this fact the farmer must make certain that the soil's nutritional capacity is equal to the task of filling the increased demand for leaf production. Cytokinins mobilize plant nutrients and tend to draw elemental nutrients to the areas of application, either on seeds, transplants, or leaves. Finally, the cytokinin input from any good eco-farming product has the effect of delaying senescence or early die-back of plants because now the plant can overcome and correct nutritional deficiencies.

6. complimentarity in biology

Earlier in this primer, mention was made of a new frontier in biology (see page 75). Unfortunately, the concepts and realities hinted at are much too sophisticated for those who still believe that the biology of immunity lies in the first available can of toxic genetic chemicals. Yet as we near the end of this little primer, it is not too much to ask our special readers to walk with us through advanced expressions of principles of biological activity in plant and animal nutrition.

The age of atomic energy and nuclear science has, of course, probed into the vast energy pool of molecules of physical matter. It has given a similar window of understanding into the field of biological energy. For the first time since von Liebig and Pasteur clashed over their 19th century theories of chemistry and bacteriology, is it now possible to sort out the *principles of differences*—to begin the restructure of knowledge and recomprehension of basic facts that relate on the one hand to physical and laboratory chemistry, and to the biological chemistry of living systems. Atomic science has learned how to unleash the tremendous power and energy contained in the atom of physical matter. And now agriculture can learn how to release biological energy, a natural source of power present in plants, soil microbial systems and animal nutrition. Because of this, it is necessary for biology to become concerned with the complimentary descriptions of energy in stable action states of which the most stable forms are those produced by biological processes in the living units of molecules. James P. Issacs and John C. Lamb in *Complimentarity in Biology* describe these living units of molecular energy as bions and relate an updated viewpoint of the factors which may be called biological energy, a force perhaps even more powerful and swift than atomic energy.

Biological processes occur in all living matter as cell metabolism or in reproduction systems. Substances which act as triggers, catalysts or starters of these biological energy producing systems are: antigens, activators of physiological processes, hormonal moderators (chelates of minerals), vitamins, vital minerals (preferably amino acid chelates and enzyme complexes), and certain toxins.

All of these starters are able to extend their influence or action in very low concentrations—in other words the *Law of the Little Bit* is quite appropriate.

Examples may be suggested as follows: only one molecule of ragweed antigen E is necessary per cell for immunological action—a principle defined as recent as 1964. Only one molecule of Hageman factor is required for initiating blood clotting (1965). Only 2,000 molecules of botulism toxin are fatal to a mouse. Only an average of 0.2 atom of zinc per cell in human leucocytes is normal, yet it may fall to as low as 10% of this in patients with leukemia.

7. energy transfer

In the presence of specific amino acids, enzyme systems can exert speedy conversion of energy transfer from proteins and their surface, as well as convert, nitrogen with oxygen. This principle is basic in aerobic pre-digestion of organic residues, manure compost, etc. Within the cell-milieu in cytoplasm and nucleoplasm there are many systems and pathways, cross pathways, enzyme and catalyst systems, all of which are influenced and regulated by some of the following factors: colloidal pH—calcium, magnesium, potassium and sodium in balance to each other; redox potential—oxidation and reduction of oxygen; ionic strength and ability of ions to exchange energy within cellular systems; specific ions, especially P_2O_43-, Ca^{3+}, Mg^{2+} and SO_42-; lipids, protein lipids, cholesterol; hormones—the regulators of life cycles; atom levels of vital minerals and amino groups; and hormone systems as catalyzed by vital minerals. When cultures are overlaid with a medium lacking in one or more of the essential amino acids, there is no net synthesis of protein. Amino acids are the building blocks for all living systems. Light, darkness microwave energy sources, temperature, vital minerals and nutritional ingredients have a combined effect in multi-cellular and bion molecule pools which may determine the time for completion of energy-exchange or cellular function. Such processes may complete within as low as 20 minutes or can continue on for the full life of a multi-cellular system. Amino acids are absorbed into cellular functions at a high rate of .7 to 1.0% per hour, and may also be reversed, such as an amino can leave a protein at the same rate with much less energy, providing, however, that an amino acid is present on the cell surface or the culture medium. Another example was revealed by research of 1956 in which it was discovered that DNA synthesis occurs during about 5% of the life cycles of typical mammalian cells. Certain hormone molecules in effector cells, located in all plasma, plant globulin, or in colloidal solutions and fluids, as a result of energy response, can exert effects even with only a few such molecules per cell in a target organ or tissue. Scientists to date are not able to deal with such low molecule concentrates, and as a result, we have not been able to make day to day decisions to accomplish better soil, plant and animal nutrition. A few of the hormone molecules can be listed as follows: angiotensin, serotonin, aldosterone, catecholamines, thyroid, peptide hormones, dimboa (in corn plants). Yes, some hormones have recently been identified. However, such things were not revealed to von Liebig or Pasteur in the late 19th century.

Chromosome bands and nucleotide substances are associated with particular genetic characters. In all seeds and grains certain and specific genomes comprise entities of genetic influence which may be less than 50 characters, but remember even an infected cell or transformed cell can supply all the machinery to accomplish the function of the particle. The use of seed nutrients supplied at the time of planting, soaked into the seed, or those nutrients released in a germinating seedbed, all can have an impact on the energy exchange of chromosomes and nucleotides, especially by amino acid solutions in the presence of Ca^{2+} and ATP, the storehouse of light energy, Again the significance of calcium in all living systems, which includes the soil, is emphasized. The Ca^{3+} ions acts as a carrier of excitation and gives momentum to electrical stimulus, but also relates to the uptake of Ca^{2+} in the presence of ATP. These actions can occur in 16 milliseconds or more depending on the equilibrium of the medium present.

8. pools of large molecules

Biology should be dealing with pools of large molecules of many bions energized for interaction now that it has completed its concerns with description of the numerous forms (or phyla) of life. When viruses enter cells or living bions they do to the cells what alpha particles do when they are bombarded into the nuclei of atoms of physical matter. The atoms change energy states and eject some similar particle. We must continue the biological evaluation of the energy factors that compliment the stability of biological processes for optimum bion function. This may give a meaningful approach and insight into the diseases of man, as well as plants and animals.

Disease conditions considered here are the following in man: cancer, congenital malfunction, aging, hypertension, artherogenase, collagen-vascular hypersensitivity, and vicarious calcification. In plant: fungal and decay infections, poor pollination, physiological maturity, herbicide stress, vascular bundle constriction, accumulated toxin and nodal systems, photosynthesis limits, malnutrition and stress reaction. Diseases, insects and growth vitality are totally related to the above causes. In animals: Merricks disease (cancer in chickens), grass tetany, bloat, pink eye, retained placenta, brucellosis, breeding efficiency, acetonemia, milk fever, cholera, and foot rot.

A few biological functions that relate to energy—exchange efficiency and nutrient processes—can be compared to the following: riboflavin is produced by *Aerobacter aerogens* in a favorable nutrient media at an average of 1.4 to 2 molecules per bacteria per second. Biotin is produced at a rate of .08 to .34 molecules per cell per second. P-aminobenzoic acid is produced by *Escherichia coli* at 3.3 molecules per cell per second. Vitamin molecules are produced by enzymes in vitro systems at the rate of 100 molecules per enzyme per second. By estimates there are about 9 enzyme molecules per cell.

9. biological forms of nutrients

Malnutrition places a limit on the energy exchange potential of all living systems of soil, plants, animals and man. Poor quality foodstuffs cannot sustain the basic biological function of life, and synthetic substitutes cannot enter into the fundamental systems required.

Some molecules involved in the multiple pathways are those of energy-yielding substrates, i.e., glucose, fatty acids, amino acids, cofactors and energy rich intermediates, nucleotide phosphate, acetyl co-enzyme "A," folic acid of one carbon pool, ulvic and fulvic acid of simple carbon pools. These form multiple couplings and thus result in greater energy exchange or actual enzyme sensitivity.

Photosynthesis processes are able to distribute carbon to all synthetic storage, structural and energy supply systems (as in ATP). A single hormone can show multiple effects on a target cell. ACTH thyroxine and insulin influence gene activation and at the same time many other actions. Several hormones such as epenephrine and glucogen each produce identical effects in liver phosphorylase.

In summary, the energy releases from all living systems can be regulated, expanded, or limited by the biological forms of nutrients. Seed-treat, seed-zone nutrients, soil pH modifiers, calcium or dolomite lime, sulfur, mineral chelates, complexes of amino acid groups and trace minerals used on seed and foliage—all are products now available for the farmer's considerations. Any of such products which have been produced from cultured or fermented ingredients will be easily accepted by seeds and foliage for active and efficient energy exchange. Similar energy will be utilized by leaf photosynthesis for more efficient nutrition and production. The final result will be better quality nutrition for good vigor and health.

As this new technology gains a toehold, new cautions and sets of standards for evaluation become mandatory. Leaves will tend to carry more nutrient content than may actually be added. This is the reason foliar analysis takes on an entirely different set of standards than have been established for foliar study under regular fertilizer systems.

10. the suppliers

Data on supply pipelines in eco-agriculture are presented in *Acres U.S.A.* monthly. An index to products and services is updated each issue in the eco-directory of *Acres U.S.A.*

LESSON 28

Case Report

The big question for any farm is simply, how do you put it all together? In a simple primer such as this one, we can only hint and illustrate. Most farmers think in terms of problems when they contemplate a change. So let's look at a common problem, potato scabs.

1. *streptomyces scabies*

At eye level, potato scab is caused by a fungus called Streptomyces scabies. This is usually associated with improper decay of crop residues in the latter part of the season and is most active in warmer soils. Farmers who grow potatoes already know about the Russet Burbank, which was developed for its scab resistance. Yet the Russet Burbank will pick up scab. The conventional Extension viewpoint says use scab-free seed of resistant varieties and rotate. They readily admit that soil chemicals and chemical seed treatments fail in field trials.

So let's take what we have learned in this primer and peel back cause and effect in terms of nature's harsh demands. We will note that scab lesions usually start in or adjacent to pockets of foul smelling clumps of organic residues. As soil temperatures increase, especially during August and September, the scabies mold gets fired up. Since there is no competition from friendly decay organisms, Streptomyces scabies dominate. This is improper decay of organic matter speaking. Our lessons on management of organic matter, pH, calcium and cation balance come to the rescue. They tell us that improper decay occurs in soils where pH character is not balanced, and the chemical, physical and biological factors are in a state of disequilibrium.

2. disequilibrium

Now, just what brings about that disequilibrium? Our lessons on magnesium, calcium and potassium tell us that magnesium and calcium may be out of ratio to potassium. Let's recall the lessons on soil tilth and structure. These are directly related to pH character. Crusts on soil surface, sedimentary leaching of fine clay particles to bony layers below, compaction, slow water intake or poor capillary capacity—all are consequences of pH disequilibrium. Such conditions restrict the release and conversion of mineral nutrition which affects desired plant growth and soil microbial potential. Moreover, sodium and potash can disintegrate particle aggregation. And soils with low humus content cannot buffer the flushing capacity of large doses of nitrogen, which in turn may release excesses of certain heavy metals and complex other desirable minerals. A factor in disequilibrium might be that excess phosphates in a low calcium or imbalanced pH environment interferes with the function of zinc and its effect on nitrogen utilization. Finally, microbial systems are not possible in soils treated with fungicides, fumigants or other toxic genetic chemicals. Large doses of nitrogen can over-feed, over-stimulate or restrict necessary soil processes.

By now we've used well over half the lessons in this primer just to understand what happens when potatoes get scab. But there is more. For one thing, ordinary soil audits do not provide sufficient data for proper decision making. This shortfall is often complicated by the fact that pH management cannot be accomplished with the use of N, P and K fertilizers or the common forms of sulfur generally available.

3. nutritional support

Even the Extension specialists admit that crop rotations and pH modification and soil nutritional support alone will not cleanse the soil of undesirable molds, viruses and fungi. Desirable decay systems must be given nutritional support throughout the growing season. Frequently such nutrients can be flushed down into the plant root zone through sprinkler irrigation.

Many farmers turn to fumigants when potato scab appears. But these toxic agents cannot endure long enough to control scabies once Streptomyces scabies start to take a crop. As a matter of fact, scabies control must involve integration of almost all the factors discussed in this primer. The common denominator comes styled physical, chemical and biological equilibrium. No single control measure can be effective alone. The total living system has to be considered—and this means plant and soil nutrition. The same common denominators answer weed and insect problems in any crop, as we have seen.

4. to get specific

But to get specific while staying with a single crop and a single problem, potato scab.

1. Fields with a history of potato scab must be sampled and analyzed. Soil corrections and pH adjustments must be started at least a year prior to being replanted with potatoes.

2. The tillage and fertility system invoked must encourage and sustain total decay of lignated organic residues by desired types of actinomycetes molds instead of Streptomyces scabies.

3. Fields must be inoculated with bacterial activator materials or pre-digested manure. This step will set up appropriate decay systems for crop stubble.

4. A modest dose of nitrogen along with determined amounts of specific forms of pH modifiers—processed sulfurs, for instance, perhaps humic acids, and mineral nutrients. Wetting agents such as WEX and Basic-H give these treatments a mighty assist.

5. All crop residues must be mulched into the soil immediately after harvest. Quick fermentation and decay can be encouraged with irrigation water whenever possible.

6. It is absolutely mandatory that the spring seedbed be ripened and I cured out via early aeration and light tillage. In most cases an application of nitrogen and additional pH modifiers is indicated.

7. No eco-farmer needs to be told not to kill off his soil decay system with spring applications of fungicides or other toxic genetic chemicals. They do not endure into August and September in any case. In the meantime they offend the life system of the soil on which so much depends.

8. Finally, it is just plain common sense to plant seed from carefully selected seed stocks.

5. heads-up management

The battle against potato scab isn't over yet—not this year, certainly not before the entire rootbed has been cleaned out and repopulated with the right forms of soil life. In irrigated acres, it is possible to avoid moisture stress by heads-up management. Otherwise the potato farmer is more or less at the mercy of the gods and the men who play god with silver iodide.

Light and water shortfalls impose stress. However, timely applications of foliar nutrition can bypass the limitations imposed by imbalanced soils, inclement weather and other stress factors.

A potato scab soil is a soil in trouble, and this reality makes mandatory use of the best available tools. Timely small doses of the right nitrogen form fit into the management picture simply because the natural nitrogen cycle isn't working properly yet! Sprinkler application of the necessary materials—pH modifiers included—works well for irrigated acres. This treatment is needed in late July and early August to compliment the maintenance of desired decay while warmer soil temperatures prevail. This is when tubers begin to mature. This is when microbial nutritional balance is to be maintained.

Immediately after harvest of the potato crop, composted manure should be mulched into the soil. If such materials are not available, a bacterial inoculant and the necessary materials for pH adjustment—plus the necessary nitrogen—must be used. The cleansing program is now near completion. The soil that once infected an entire crop with scab will be ready for return to potatoes after one inter-crop season.

6. almost every lesson

Almost every lesson in this primer comes into play when a problem is faced with the precepts of eco-agriculture in tow. It is not possible to be a specialist in roots or leaves or bugs or weeds or soils. All relate to each other and everything is connected to everything else.

If everything revealed here is clear, give yourself an "A." If not, remember that a primer is merely the first step in getting an education. It isn't definitive. Indeed, for agriculture only the Creator has all the answers.

LESSON 29

Summary

At the head of the biotic pyramid stands man. His health and permission to life depend on the welfare of life forms beneath him, and nature's balance of that life and death. Unlike plants, composed chiefly of sunshine, air and water—with only a trace of earth minerals—man is a superb package of proteins. Everything that shows—muscle, skin, hair, eyes, nails—all are protein tissues. Even teeth contain a little protein. Those component parts kept well out of sight—blood and lymph, heart and lungs, tendons and ligaments, brain and nerves—rest largely on a foundation of protein. Squeeze the water out of a human being, and the half that's left would be largely protein.

The human life process itself requires protein as a prime mover of sorts. That's what the Greek root word for protein means—first place. A Dutch physician turned chemist first announced this fact to the world in 1838 as a conclusion to studies of living plants and animals. All, he said, contain a substance without which life cannot proceed. He couldn't put a handle on it at the time, but he knew it was vital.

1. the "first place" substance

He knew that all foodstuffs contain carbon, hydrogen and oxygen in varying ratios. The first place substance also contained these elements— and nitrogen as well. This made the nitrogen component of proteins of special power and importance.

The chromosomes that blueprint a human life at conception are protein in character. Offended by toxic genetic chemicals, these same chromosomes dictate the birth of a deformed or mentally retarded child. The full scope of heredity is controlled by a very special kind of protein. All the hormones that regulate human and plant life, and the enzyme spark plugs of biological reactions, are proteins.

The text books usually call proteins the building blocks of cells, and they are. There is a dilemma here, because proteins in turn have to be made by living cells. They do not exist in the air. They do not float in from the sun, like energy. They are built and in turn they build.

The splendid opening chapter of *Yearbook of Agriculture, 1959*, hits the nail on the head quite perfectly, and explains why nature has to be biological, not just laboratory synthetic. "Most plants make their own protein by combining the nitrogen from nitrogen-containing materials in the soil with carbon dioxide from the air and with water. The energy they need comes from the sun. Legumes, which include beans, peas and peanuts, can use the nitrogen directly from the air for combining with other substances to make protein.

"Animals and people cannot use such simple raw materials for building proteins. We must get our proteins from plants and other animals. Once eaten, these proteins are digested into smaller units and rearranged to form the many special and distinct proteins we need."

Proteins do supply energy, much as do starches, sugars and fats. Proteins give a mighty assist in the exchange of nutrients between cells and intercellular fluids, also between blood and lymph. In turn proteins are assisted. Proteins that are in the process of change can release three to six times more energy than was required in the re-synthesis process.

A symbiont in the human cell is the mitochondria. Other small beings are also living in human cells, sorting, holding, working, fighting. Cells in fact must rate attention as miniature eco-systems, as complex as an estuary and as detailed as spaceship earth itself.

2. strangers in our midst

Plant life is not plant life as such. It too is rented out and occupied. Chloroplasts manage the photosynthetic business in plant cells, working with the sweep of the sun to manufacture the oxygen we breathe. They're individualists, these chloroplasts. They have their own genomes. They speak their own language.

These strangers in our midst are hostile only on occasion. Even viruses have a function hardly understood by the people who seek to kill them. They seem to shadow the whole of the biotic pyramid, dancing from organism to organism, from mammal to insect to plant and vice versa. They carry along strings of genes, graft a DNA here, a heredity trait there, trucking genomes all the way, possibly presiding over mutations that keep life on this planet afloat. The virus that brings disease is probably a mistake in the game plan called life.

Laboratory science has never understood the interrelatedness of the biotic pyramid. It has refused to admit its mistakes and to retrace its steps. As the late G. H. Earp-Thomas, a physician, so succinctly put it, "Science tries to harmonize the latest discoveries with the incongruities of its false position."

An even bigger mistake is the one made routinely by the man on the street or farm. Every attempt is made to destroy the little creatures that rent space in the corn crop or in the human body on the premise that destruction of friendly organisms is a small price to pay for getting at the harmful organisms. The well individual, corn crop or human, is the one that has less harmful renters than beneficial ones.

Balance of power is the name of the game. Among the beneficial creatures in the human being are the red corpuscles. Born in the marrow of bones and in the spleen, these little fellows live about 30 to 90 days. They come and go the way human beings live and die. They collect oxygen from the air by tripping to the lungs, and then deliver it to all parts of the body. Just as a fire needs air, nourishment on tap by cells must have oxygen to burn.

Workaday partners with the little red caddies of oxygen are the white corpuscles, a veritable phalanx of Coldstream Guardsmen ready to do battle at a moment's notice. It is their function to hunt and destroy. The enemy is any harmful germ.

We live in a veritable cesspool, from a white corpuscle's point of view. The air we breathe is filled with pathogenic invaders. The water we drink is likely to be loaded. Pathogens lodge themselves on our skin, in our clothing, amid molecules of odor, everywhere. Should the invaders find a weak spot in the Maginot line of health defense, they set up shop and proliferate. Just about every ailment can be characterized in this way.

White corpuscles don't take these invasions sitting down. When a breath of air brings harmful bacteria or viruses into the lungs, and these enter the blood stream, white defenders tear them to pieces. There is nothing in nature as vicious as a white cell taking apart an invader pathogen. These cells hunt in packs like Jack London's wolves. They surround, attack, counterattack and annihilate the enemy—generally! They can't lose if arms and ammunition are kept in good repair. If, however, the body is allowed to run down, if nutrients are either imbalanced or deficient, the white defenders fight a losing battle. Disease follows.

Thus, as if by one leap, the head of the biotic pyramid is bound to the soil and the nutrition he takes from the soil, and eco-agriculture.

3. organic - inorganic

It has been a century since most physicians believed several types of minerals in trace form were necessary for human health. Calcium for the bones, phosphorus for the nerves were copybook maxims at one time. Not so well understood was the fact that minerals were not necessarily minerals as far as the human body was concerned. Calcium was excellent for a whitewashed fence, but it wasn't food when eaten. First, it had to be changed into organic calcium either via the biology of a plant or the microorganism of a gut. "Animals are not mineral eaters," the late Bill Albrecht used to say. "It is only a symptom of desperation when they take to the mineral box."

It was routinely demonstrated some few years ago that dead iron injected into the blood stream produced a definite reaction. Blood became excited, frustrated—if one can be permitted that term—but a decade later that same dead iron hung on like an anchor, unassimilated. As a natural element it obeyed the law of homeostasis and eliminated itself as waste material. It was also routinely demonstrated that organic iron easily passed into the blood. Yet you can't eat a shingle nail and expect it to be used as body nourishment, not this year, not in 20 years.

The organic-inorganic debate raged for a time because institutional business could synthesize solubility for an element, whereas genuine organic nutrients depended on agriculture. Plants became the final arbiter of this debate, but they didn't have credentials and they hadn't gone to college.

Still, the live iron out of a plant cannot be formed into a nail again. It will not melt. No art of science can return it to its inorganic state. It is not by accident that every part of the United States fields a definite lack: calcium and phosphorus. Yet most of the soils heavily fertilized by factory acidulated nutrients—those rich in calcium and phosphorus—are producing foods deficient in those very minerals.

When foods lack enough calcium, various respiratory diseases enjoy an upswing—pneumonia, bronchitis, sinus trouble.

"The truth is," wrote Earp-Thomas, "that the germ would probably be rendered harmless if there was enough calcium in our body . . . As an example—the germ of tuberculosis uses an acid to destroy a small spot in the human body in order to get a proper place to start, as it lives on destroyed tissue. When there is plenty of calcium in that tissue on which the germ settles—when the human body is well stocked with this essential mineral—not only in the bones and teeth but in the blood and every cell, this calcium—being lime—will neutralize the acid used by the germ.

Therefore the colony the germ would start is stopped before it really gets started. This neutralized combination of germ acid and calcium is called *formate of lime.* There is no doubt that extra calcium in the body helps fight—and conquer—tuberculosis."

4. alkaline creatures

Human beings, in fact, are alkaline creatures. If the bloodstream should become even slightly acid, death results. A dead person is instantly acid and negative. A living person is alkaline and positive. This is very likely a law of nature, and like gravity, it is immutable. It was not by accident that the physician Ira Allison and William Albrecht could cure chronic diseases with minerals in concentrated application. Blood became unfit as food for so-called disease germs, and those germs died off rapidly.

We are what we eat. And we are what our farms produce. The steady march of metabolic degenerative disease across the population of the nation and the world does not speak well of the prevailing form of agriculture. The course of human events has ordered change, and yet change is kept an arm's length away.

C. J. Fenzau, writing in an early issue of *Acres U.S.A.,* best expressed the crossroad at which we wait: "When living life emerges into an ultimate objective of our nation, only then can we begin to contruct the understanding of the ecology of the soil and be able to benefit from its minuteness and wondrous patience. Nature has no limit on time. She is patient and forgiving. She is able to repair herself from the ignoble treatment of man in spite of his tremendous physical capacity for destruction. As we continue to replace nature, we assuredly prevent the development of our mental capacity to learn and fully compliment nature—a requirement expected from us—in permission for life. Apparently the growing interest of many farmers in searching for answers reflects the beginning of an intellectual attitude in conflict with both agronomic science and industry. This virgin knowledge is finally emerging as a test and challenge to the introverted and integrated industrial exploiters, and absconders of our most precious resource."

A soil system for nutrient and energy production is a living system in which bacteria and the many soil organisms must receive nutrition and energy from proteins, carbohydrates and cellulose and lignin—all organic materials—in a soil that has a managed supply of both air and water within a balanced chemical environment. This chemical balance involves more than just N, P and K. It requires an equilibrium of pH, calcium, magnesium, sodium, potassium, humus and a nutritional balance of sulfur to nitrogen, nitrogen to calcium, calcium to magnesium, magnesium to potassium and sodium. Without this balanced equilibrium, no chemical system and no organic system can have an enduring potential for soil building, or plant nutrition. This total equilibrium is even more essential for the vital nourishment of man.

There are few trained soil scientists who understand and present any comprehensive judgment, other than lip service, to the standard viewpoints that are important to human health—namely soil bacteria, earthworms, soil drainage, tilth, structure, and the true function of organic material in the soil system. After all, organic matter converts, regulates and releases nutrients to soil life which, in turn, makes available biological forms of nutrients to the biological system of the soil from which man was originally evolved. His creation and permission to life is indeed profoundly linked to this biotic beginning.

So let's leave the peak of the biotic pyramid for a moment and take a more informed look at what we've learned in this little primer.

5. aaa

Few can deny that the fossil fuel industry had developed a need to sell soluble fertilizers by the end of World War II, and the chemical firms (largely owned by the oil companies) saw great sales potential in the war against insect and fungal crop destroyers. That farmers were surviving, even prospering without this harsh technology had become troublesome reality in the late 1920s. And so, during the 1930s, the government moved in to give the friends of bureaudom a helping hand. An AAA experience provides a good "for instance."

When the Agricultural Adjustment Act came into being in 1933, its announced aim was to help farmers reclaim their land. Farmers were told to lime their soils to the neutral point—"Use all the lime you want," read the slogans, "you can't overdo it." One after another the intellectual advisers in the colleges, the Extension, the USDA modified their teachings to bring them into line with government edict. A few diehard professors objected and found themselves drummed from the classrooms.

Circa 1933, not all of America's acres required calcium by any stretch of the imagination, taking the state of the arts into consideration. The general lime requirement—where it was a requirement—was not to cure an acid soil condition, but to supply a missing nutrient. A few farmers objected, "Governments should stick to the two things they know best, stealing and killing, and leave the farmer alone." Little did they realize that it was stealing that this neo-science was all about.

But there was an even bigger sales ticket in the wings—toxic rescue chemistry. Even in the early 1930s it was known that pH achieved by an equilibrium of calcium, magnesium, potassium and sodium governed hormonal systems of plants, and plants with balanced hormone and enzyme systems provided their own defense against insect, bacterial and fungal attack.

With the pH thrown out of whack so that soluble salt fertilizers could be sold, the managers of scientific "findings" took out an insurance policy for the sale of lots of toxic rescue chemistry and herbicides.

The organic folks revolted. They had a pragmatic feeling that something was wrong, but they couldn't put their finger on it.

This was years before Rachel Carson wrote: "We are rightly appalled by the genetic effects of radiation; how then can we be indifferent to the same effect in chemicals that we disseminate widely in our environment?" In May 1963, Jerome Wiesner—a science counselor to President John F. Kennedy—told a Senate Commission assembled to evaluate the report, *Use of Pesticides*, that "using agricultural pesticides is more dangerous than atomic fallout."

The dangers continue to hang over agriculture and over the end product of agriculture—human beings capable of thought and reason. In the meantime, this little primer has—it is hoped—set forth "an intellectual attitude in conflict with both agronomic science and industry." So be it. As we prevail, both agronomic science and industry will be recast. Our permission to life depends on it.

GLOSSARY OF TERMS
FIELD NOTES
INDEX

GLOSSARY

Absolute temperature. The degree of the body in Centigrade plus 273.16 degrees. Absolute zero is -273.16 C.

Acid soils. Soils with a pH below 7.

Adventitious roots. Roots that grow from the tap root either immediately above or below the soil line.

Aeration. Making air available to a material.

Aerobic. Living and functioning in the presence of air or free oxygen.

Aflatoxin. The mold toxin caused by Aspergillus flavus. This is a coined word—"a" coming from the Aspergillus, "fla" from flavus, and toxin is added on—thus, aflatoxin.

Agar. A gelatinous substance extracted from seaweed. It is used in making culture media to study growth characteristics of microorganisms.

Alkaline soils. Soils with a high pH, or one over 7. Alkalinity is usually the consequence of heavy sodium and potassium concentration.

Anaerobic. Living and functioning without air or oxygen.

Angstrom. A unit for measuring light. An angstrom is one-ten-billionth of a meter.

Anion. An acid forming element. It is negatively charged and therefore migrates to the anode in an electrolyzing solution. It is the opposite of cation.

Annual. A plant that has a complete life cycle in one year.

Anther. The sac-like structure in which the pollen is formed in the flower. Anthers commonly have two lobes or cavities, which open by long-itudinal slits or by terminal pores and release the pollen.

Asexual. Denotes reproduction by purely vegetative means, or without the function of two sexes.

Auxins. Plant hormones which govern the elongation and differentiation of newly formed cells. Auxins also regulate water movement in plant tissues, causing turgor changes. Turgor refers to the fullness of vessels and capillaries. Auxin-like hormones given off by roots during early germination are important in regulating competing weeds and seedzone bacterial systems.

Bertrand's Law. After Gabriel Bertrand, a French scientist. More specifically, Bertrand's Law states that a plant cannot live in the absence of an essential element; it thrives on the right amount; and an excess is toxic.

Biennial. Plant that produces vegetative growth during the first year of growing season. After a period of storage or overwintering out of doors, flowers, fruits, and seeds are produced during the second year; then the plant dies.

Bioclimatics. Specifically, the study of geographical location and elevation in terms of plant response. The study has now evolved into evaluation of location and elevation as they affect the elements of climate.

Biodynamics. A school of agriculture founded by the Swiss, Rudolf Steiner. It gained a foothold in the U.S. under the auspices of Ehrenfried Pfeiffer, and is practiced by many Americans as a valid form of eco-agriculture.

Biological control. Using either insects or bacteria, viruses or other natural organisms to control bacterial, insect and fungal attack on farm crops. Hormones, pheromones and other chemicals that alter insect behavior are rapidly entering the bug control field and are more basic to both insect and disease systems of plants.

Biotic geography. The study of geography in terms of climax crops produced by nature in a certain geographical area.

Biotic pyramid. A conceptualization used by W. A. Albrecht to explain the inter-relatedness of nature.

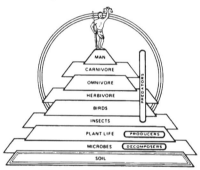

The Biotic Pyramid

Blackbody. A surface that efficiently absorbs radiant (IR) energy of any wavelength, and effectively radiates IR energy also. It does not reflect energy.

Botanical insecticides. These are poisons extracted from plants. Pyrethrins, rotenone and nicotine are the most common types. During recent years these poisons have been synthesized in the laboratory close enough to be called "relatives." Botanical insecticides are approved by organic gardening people because they break down quickly and cause little environmental damage. Rotenone can kill fish quite easily.

Bud. A plant structure that contains an undeveloped shoot or flower.

Budding. The process of transferring a live bud from one plant to another, usually by insertion under the bark. The term also means the process of forming buds. Also, asexual reproduction characteristic of acellular organisms. The new individual develops as a bud from the older or larger amount.

Calorie. Heat required to raise 1 kilogram of water 1 C. at a pressure of 1 atmosphere.

Calyx. Outer floral leaves.

Cambium layer. A one cell thick layer of tissue which foundations the growth of new tissue in stems and roots of dicot trees and shrubs. The cambium layer is responsible for growth rings in trees.

Carbamates. Technically, an ester of carbamic acid. In trade, it is characterized as a broad-spectrum poison that is highly toxic to bees and fish, among other things. It is said to be less toxic to man—less being judged on the basis of make-it-to-the-door toxicology. The long range effects in terms of cellular mischief have never been evaluated properly.

Carbon dioxide; (CO2). A gas compound that forms when 1 carbon combines with 2 oxygens. In the case of human beings, it leaves the body chiefly when air is exhaled from the lungs.

Carbohydrate. An organic compound in which hydrogen and carbon typically occur in a ratio of 2 to 1.

Catalyst. A chemical or biologically-formed substance which speeds up the rate of chemical and energy-exchange reactions without entering the end product or being changed itself. See enzyme.

Cation. Positively charged base elements, either alkali metals or alkaline earths, which migrate to the cathode in an electrolyzing solution. A cation is the opposite of anion.

Cation exchange capacity (CEC). The capacity of soil components which exhibit surface-active adsorbing properties to hold positively charged nutrients as expressed in milliequivalents (ME). 1 ME of total cation exchange capacity represents the colloidal energy necessary to adsorb and hold to the colloidal system in the top 7 inches of 1 acre the following: 400 pounds of calcium; or 240 pounds magnesium; or 780 pounds of potassium; or 20 pounds of exchangeable hydrogen, etc.

Cause. As generally used, it means precedence. Every phenomenon, act, whatever, preceding another could be considered a cause. In terms of experimentation, cause is little more than an order of succession in a time frame. Is a cannon fired by the percussion cap or by the soldier who pulled the firing mechanism? Or is the cause the charge of powder? Without the soldier's hand on the percussion cap, that powder could remain inert forever. What if the soldier's hand is replaced by a small ray of light that started its way to earth from the star Arcturus 40 years earlier? Is the star responsible for the damage done by a shell landing some 30 miles distant? Or are the workers who made the powder, or the men who built the factory, or their parents or their grandparents—or the hand that rocked any of those cradles—are they responsible? Responsibility fades away, and yet it never disappears entirely, before reaching back to the origin of the world. Thus we arrive at the First Cause. (See Creator.)

Cell. The basic structural unit of living organisms. In plants, it is comprised of protoplasm enclosed in a cell wall. the protoplasm consists of a nucleus and a semifluid matrix, the cytoplasm, which contains plastids and many other smaller bodies. Mature plant cells usually contain a large cavity or vacuole filled with a water solution of sugars, salts, acids and other substances.

Cellulose. Insoluble polysaccharide that makes up the cell wall of plant cell.

Chelate. A chemical or a biologically synthesized compound in which a metallic ion is bound up within the chelating molecule.

Chlorinated hydrocarbons. A family of insecticides and herbicides, including DDT, chlordane, aldrin, dieldrin, endrin and methoxychlor, and thousands more. Many are long-lasting and efficient in killing weeds and insect pests. They are also capable of entering the food chain. Almost all have been linked with degenerative metabolic disease. According to many of the world's top scientists, they have no safe level and they have no tolerance level. In agriculture they are used by farmers who have mastered the precepts of eco-agriculture.

Chloroplasts. Specialized chlorophyll-containing units of life that trap light and become donors of electrons in the photosynthesis process.

Chromosome. A rod-like body in the nucleus of the plant cell. It carries the hereditary message.

Climax. As related to plant life, it means the final stage and condition of equilibrium of vegetation after a series of progressional stages which have developed—without serious interruption—under the influence of a given complex of environmental factors.

Clone. In horticulture, it means a group of individuals of common ancestry which have been propagated vegetatively, usually by cuttings or natural multiplication of bulbs or tubers.

Colloidal position. A designation of how available nutrients are positioned, whether on the colloid of clay or humus or in leachable water solution.

Companion planting. Introducing plants into a crop situation that differ from each other, yet grow side by side. Auxin effects from one hand offer an assist to the other in governing weed and insect populations.

Compound. Two or more different ions or atoms in union forming a different substance when held in chemical bond.

Corn silk. The stigma and style of the female corn flower through which the pollen tube grows to reach the embryo sac.

Creator. As used in this *Acres U.S.A. Primer*, it means First Cause, and the author of all causes. To explain: all scientific laws rest on chance, on the concept of absolute disorder at the base. That molecules, atoms and electrons are in perfectly disordered motion is a prime requirement for the statistical reasoning used to frame scientific laws, laws that enable us to foresee. This seems odd since the laws of nature exhibit harmony in terms of the human scale of observation. Odder still is the fact that the calculus of probabilities used as a foundation for science makes it impossible to explain the birth of life as a consequence of pure chance. An illustration might be in order. The probability of throwing the number 5 twice in succession in a dice game is 1/6 x 1/6 = 1/36, or .0277. The probability of throwing the same number five times in a row is only 1/7,776, or .00013. Numbers soon become astronomical. To make an end run around this problem, science uses power figures. There are 30,000,000,000,000,000,000, molecules in a cubic centimeter of gas. It strains the eye to read such a figure. Science writes this number as 3 x 10^{19}, meaning 3 x 10 to the power of 19, and 19—the exponent—merely expresses the number of zeros after the last significant figure. It is estimated that the earth is about 2,000 million years old—or 2 x 10^9 years. In terms of centuries, this would be 2 x 10^7. There are 100,000 days in a century, or 105 days. This means that less than 2 x 10^{12} days have gone by since the beginning. Nature doesn't cheat. At the scale of observation being considered here, there is no dissymmetry. Now we are faced with the problem of calculating the appearance by chance along of life essentials—large molecules of proteins, for example. Professor Charles-Eugene Guye once calculated an imaginary protein molecule. He found that the chance for a single molecule to be formed by the action of chance and normal thermic agitation—supposing 500 trillion shakings per second, which corresponds to the magnitude of light frequencies—would be practically nil, or one chance in 10^{242} billions of years. We can't forget, however, that the earth has been around for no more than 2 billion years, and life has existed only since the earth cooled, about half that time. A single molecule is of little use. Life requires hundreds of millions of identical molecules, and a cell—born on the basis of probability—would call for figures that make the foregoing calculations negligible. If we want to be honest about it, we have to admit that it is totally impossible to explain the phenomena pertaining to life using our scientific laws. As the scientist Eddington put it, to study life we are forced to call on an anti-chance, a cheater who routinely and systematically violates the laws of large numbers, those statistical laws that deny individuality to the particles of the universe. As the French philosopher Lecomte du Nouy so aptly put it, "for a man of science there is no difference between the meaning of the words anti-chance and God." Obviously, Carnot-Clausius Law (sometimes called the second law of thermodynamics) does not apply to living organisms. The second law of thermodynamics states that an isolated material system cannot pass twice identically through the same state. Every successive state entails a definitive decrease in its available energy. Hence its irreversibility. Again to quote du Nouy: "To account for what has taken place since the appearance of life, we are obliged to call in an anti-chance which orients this immense series of phenomena in a progressive, highly improbable direction (incompatible with chance) . . ." We call this principle of creation and life the Creator—or God.

Cross-pollinate. To apply pollen of one flower to the stigma of another. The term commonly refers to the pollination of flowers of one plant with pollen for another.

Cytokinin. A plant hormone capable of modifying plant development either by stimulating or altering the cellular RNA. Most growth hormones—such as gibberellins and auxin-like hormones—are more fleeting and unstable when compared to the cytokin group of plant hormones.

Cytoplasm. Contents of a cell other than the nucleus or chromosome bearing portion. In reproduction, the male parent usually contributes only chromosomes. Both nuclear and cytoplasmic constituents of the female parent become a part of the offspring.

Damping-off. A disease of seeds or young seedlings caused by fungi. It is sometimes confused with germination failure, in which the seeds is invaded during early stages of germination by fungi and fails to sprout. Pre-emergence damping-off means young seedlings were attacked by fungi before it pushed its way through the surface of the soil.

Detassel. To remove the tassel or pollen-producing organ of a corn plant before pollen is released.

Diapause. In insects, the state during which growth and development are temporarily arrested.

Dicotyledon (dicot). A plant with two seed leaves.

Dioecious. Plants that have the male reproductive system on one plant, the female on another.

DNA (deoxyribonucleic acid). A nucleic acid found in all living cells. Geneticists believe that genes such as are in chromosomes, consist of complex organic molecules composed mainly of this acid. Genes differ from each other according to their molecular structure.

Double-cross hybrid. A type of hybrid corn that is the result of mating two pairs of inbred lines to produce two single crosses which are then mated to produce a double-cross.

Ecdysone. A molting hormone secreted by the prothoracic glands in arthropods. Ecdysone and ecdysine are two juvenile hormones that are significant to the molting, early maturation or death of most arthropods, caterpillar, etc.

Ecology. The study of living systems in relation to environment and their effects upon each other.

Electrogenic. An energy form that contributes to delivering nutrients through cell membranes.

Electron. A negatively charged particle that orbits the atomic nucleus.

Elements. Naturally occurring fundamental kinds of matter. A total of 92 is listed in the Periodic Table.

Endosperm. Tissue of seeds that nourishes the embryo.

Enzyme. In nature, a catalyst produced in living matter, each related to specific combinations of minerals. It is a specialized protein capable of aiding in bringing about chemical changes. Enzymes promote a reaction without entering into the reaction, undergoing change themselves, or being destroyed.

Epiphytes. A plant that sends out stems and roots that cling to other plants, albeit not for the purpose of nourishment.

Equivalent. The weight in grams of an ion or element that combines with or replaces 1 gram of hydrogen. This means the atomic weight or formula weight divided by the valence of the element.

Eukaryotic cells. Having a nucleus, with nuclear material surrounded by a membrane and containing organelles, or particles of living substance. Organelles serve as the site for cellular respiration.

Far-red light. Radiant energy near the long wave length side of the visible spectrum. This is the place in the spectrum where the average eye begins to fail to detect radiation. It is important in germinating seeds, insect systems and bacterial processes.

Fermentation. An anaerobic process wherein carbohydrates or derivatives are decomposed by living microorganisms.

Gametophyte. A plant that generates via two sexes.

Gel. A solid form of colloidal suspension, an important byproduct of the proper decay or fermentation of organic materials.

Gene pool. The many genes that exist in all the individuals in a particular interbreeding population.

Gibberellins. Plant hormones capable of affecting growth, flowering and pollen development, but usually unstable or unreliable if nutritional weaknesses are present.

Glucose. A simple—or monosaccharide—sugar. It is the prime fuel substance for most organisms.

Graft. Transferring a piece of stem, with buds attached, to another plant. Cambium layers are placed adjacent so that union will occur.

Hermaphrodite. Two sexes on the same plant.

Homeostasis. The ability of a system to maintain internal stability in the face of external change or altered intake. Regulation of body temperature, adjustment of circulation, elimination of toxins—all are examples.

Hormone. In nature, a biologically energized chemical substance produced in one part of a plant that induces a growth response in another

part at very low concentrations. Hormones also have been made synthetically. The most common are 2,4-D and 2,4,5-T, herbicides which literally cause certain plants to grow to exhaustion and death.

Humate. A salt form of humic acids. Writing in Economic Botany, Everette M. Burdick put it this way. "Humus is the organic matter of soils that has decayed sufficiently to have lost its identity with regards to its origin. The most important and biochemically active group of the many degradation products of soil organic materials is the alkalisoluble fraction commonly called the humic acids. The salts of these humic acids are known as the humates." Burdick's other paragraphs on humates help explain the definition cited above. "The ability of the humates to poise or regulate water-holding capacity or content is probably their most significant property so far as agriculture is concerned, since from a quantitative point water is the most important plant material derived from the soil. In conjunction with this water regulating effect, the humates possess extremely high ion exchange capacities, and it is this property that makes possible better retention and utilization of fertilizers by preventing excessive leaching away from the root zones and ultimately releasing them to the growing plants as needed. The humates reduce soil erosion by increasing the cohesive forces of the very fine soil particles. The desirable friable character of fertile soils is maintained through the formation of colloidal mineral complexes, which assist in aeration and the prevention of large clods and stratification. Very low concentrations of purified humates have been shown to stimulate seed germination and viability, root growth, especially lengthwise. Significant increased yields have been reported for many crops, such as cotton, potatoes, wheat, tomatoes, mustard, and nursery stock. They have also been shown to stimulate growth and proliferation of desirable soil microorganisms as well as algae and yeasts. A number of workers have reported that the humic acids can solubilize and make available to plants certain materials that are otherwise unavailable, such as rock phosphates. The humates seem to play an important role in plant utilization and metabolism of the phosphates. The humic acids apparently can liberate carbon dioxide from soil calcium carbonates and thus make it available to the plant through the roots for photosynthesis. The humates are known to stimulate plant enzymes. The humates . . . are nature's soil conditioners par excellence." Humate materials are also known as Leonardite. Humate concentrates in liquid form are now also marketed, reducing the shipping problem.

Humus. The end product from the decomposition of plant and animal residue under proper environmental and biological conditions. This does not occur accidentally when plant or animal residues are left to rot or ferment in soils of hostile bacterial conditions.

Hybridization. Using seed parents of differing genetic makeups to engineer crop performance. The term hybrid technically means a cross between different species. It is the first generation seed of a cross produced by controlling the pollination and by combining two, three or four inbred lines; one inbred of a single cross with an open-pollinated variety; or two varieties or species, except open pollinated varieties of corn. Commercial use of the term hybrid is restricted to the first generation following hybridization.

Hydrophyte. Plants that grow either under water or with part of the plant structure above water or floating on water. These are water loving plants.

Inbred. Successively self-fertilized. It also means a plant or progeny resulting from successive self-fertilization.

Indole. The substance, C_8H_6N. It is usually found by decomposition of proteins containing tryptophan. One of the phenolic wastes found in animal excrements.

Inorganic insecticides. Usually lead arsenate, Bordeaux mixture, sulfur and the like. Lead is a heavy metal and very dangerous. Bordeaux mixture is copper sulfate and lime, and probably provides more in terms of nutritional support than it does fungal properties. Bordeaux mixture provides nutritional support to grape plants, and is strictly not a poison any more than most metallic nutrients. It is generally characterized as a fungicide, however. Needless to say, calcium lime is a nutrient. Sulfur is both dangerous and a requirement in any form of agriculture.

Internode. The portion of a plant stem between the places (nodes) from which leaves grow.

Ion. An atom or atoms with an electrical charge.

Isotope. Elemental substances having identical chemical properties, but differing atomic weights. Most of the common elements have been shown to consist of a mixture of two or more isotopes. Isotopes of any one element have the same number of extra-nuclear electrons, and the same nuclear charge, but have differing nuclear masses. The importance in soil science lies in use of radioisotopes. These are basic elements that emit radioactive particles that can be detected by electronic apparatus or film. Radioisotopes of carbon, iodine, cobalt, phosphorus, and other nutrients are widely used in plant and animal research to study translocation of chemicals and how they react in the tissue systems.

Kelvin's K, or K. As used in physics, K is the symbol for absolute temperature. It is thus named for Lord Kelvin.

Larva. The wormlike, albeit immature form of certain insects.

Lateral. Located on or developing from the side, as a bud located on the side of a shoot.

Lateral roots. Smaller roots that form from the swellings along the sides of larger roots.

LD50. This symbol, which is used to designate toxicity, means that the cited milligrams of poison per kilogram of test animal body weight will kill 50% of the test animals. *Acres U.S.A.* calls this the make-it-to-the-door test. No consideration is given to damage that may surface later—in human beings, many years later. Routes for administration of test animals are generally oral or dermal. LC_{50} means lethal concentration (as in water).

Lipids. Organic substances insoluble in water, but soluble in alcohol, chloroform and other fat solvents. Generally speaking, the term means fatty acids.

Macronutrients. Essential plant nutrients required in large amounts.

Male sterile cytoplasm. This designates a type of male sterility conditioned by the cytoplasm rather than by nuclear genes. It is transmitted only through the female parent.

Mass numbers. Total number of protons and neutrons in a given atom's nucleus.

Mendel's Law. It states that inherited traits are not intermediate between the parents, but come from one or the other of the parents. Dominance of one parent over the other is the determining factor.

Mesophytes. Plants that require a medium amount of water. Most agricultural crops are mesophytes.

Metabolism. Changes within a cell that provide the energy required by plant or animal. The term takes in both constructive and destructive changes.

Micron. A unit of microscopic measurement. It is equal to 1/1,000th part of a millimeter. The symbol for micron is now MU as decreed by physicists.

Micronutrient. An element required only in trace amounts by living systems. See trace element.

Midrib. The rib section of a leaf that holds the leaf blade in position.

Milliequivalent (ME). One thousandth part of an equivalent.

Mitochondria. A particle of living substance in cells that serves as a site for cellular respiration.

Mixture. Two or more differing substances in close proximity to each other, but chemically not bonded.

Monocotyledon (monocot). A seed plant with a single seed leaf.

Monoecious. Plants that have both male and female reproductive organs.

Mycelium. Thread-like growth emanating from fungal spores, such as actinomycetes molds which act as food-taking structures for the fruiting body.

Mycorrhiza. Fungi in close contact with and entering into plant roots growing in biotically active soil.

Nematodes. Microscopic eel-like worms that attack and destroy plant roots in poorly managed and imbalanced soil. There are also many species complementary to plant and soil systems.

Neutral soils. Soils with a pH at 7.

Neutron. A particle within the nucleus of an atom which has no charge.

Node. A place along a stem where leaves and buds form.

Nucleic acid. Organic acid composed of joined nucleotide complexes. The best known are deoxyribose nucleic acid (DNA) and ribose nucleic acid (RNA).

Open-pollinated. Generally refers to corn that is pollinated naturally by wind and insects, and not engineered genetically as are hybrid seed varieties.

Organophosphorus insecticides. Compounds with very high acute toxicity. These compounds are better known as organic phosphates, a happenstance that has brought problems to the old organic gardening movement. Included among the organophosphates are parathion, malathion, mevinphos, and too many more. Using stone age testing methods, scientists have concluded that some organophosphorus materials are extremely toxic (acute) to most living things, others less so. Few tests have ever been conducted to cover several generations of test animals. Questions regarding the genetic effect on human beings over several generations have never been answered. Indeed, they have never been asked in most testing circles. Parathion poisoning brings on a roster of acute ailments that can stagger the imagination, if not the person. Included are general weakness, headache, sweating, nausea and vomiting, salivation, miosis, dyspnea, difficulty in walking, diarrhea, muscular fasciculation, disturbance in speech and consciousness, abdominal pain, fever, bronchophyaryngeal secretion, increased blood pressure, loss of pupillary reflex, cramps and cyanosis, among other things.

Osmosis. A process by which water diffuses through a semi-permeable membrane from a side with less concentration. Diffusion continues until particle concentration is the same on both sides.

Outcross. Mating a hybrid with a third parent. Also, an offtype plant resulting from pollen of a different kind contaminating a seed field.

Ovary. The part of the pistil that contain a ovule. It ripens to form the fruit.

Oxalate. A poisonous, crystalline dibasic acid. It is produced by certain plants and is not digestible.

Petiole. The narrow leaf stalk.

pH modifiers. Prescription processed nutrient elements used to bring a soil's pH to the proper level for crop production with either calcium, magnesium, potassium or sodium in combination with sulfurs, humic acid and other specific solubilizers as soil conditions may suggest.

Pheromone. An outside hormone. The chemical a female insect releases to sexually stimulate and attract males. Pheromones also figure in other forms of insect communication.

Phloem. Designating cylindrical cells which specialize in taking dissolved organic nutrients to various parts of a plant.

Photosynthesis. Nature's process for manufacturing carbohydrates out of carbon dioxide and water with the use of light energy and the green light trapping pigment called chlorophyll.

Pistil. An ovule bearing seed plant organ.

Pollen. Microscopic bodies borne in the anthers of flowers. Pollen contains the male generative cells.

Pollen tube. A microscopic tube by which the sperm cells move to the embryo sac of the ovule.

Polysaccharide. A carbohydrate which yields more than ten monosaccharides on hydrolysis.

Porosity. As used in eco-farming, soil that is mellow and open and porous and capable of taking water and air.

Protein. Complex nitrogenous substances made up of many combinations of amino acid molecules.

Proton. Fundamental particles of the nuclei of all atoms. The proton carries a positive electric charge.

Protoplasm. The essential, complex, living substance of cells upon which all the vital functions of nutrition, secretion, growth and reproduction depend.

Radicle. The first root to emerge when a seed germinates.

Radiomimetic chemicals. Chemicals that provide the same effects as radioactivity. The farm chemicals are radiomimetic since they have the same effect as alpha, beta and gamma rays evident in atomic fallout.

Rhizobia. Bacteria that live in symbiotic harmony with legumes. They exist in nodules on plant roots and fix nitrogen for plant use.

Rhizomes. Stems which grow under the soil and account for propagation of certain plants.

Rod. A sensitive visual cell that registers light, albeit not color.

Root. The mineral and water absorbing underground part of a plant. True roots do not bear leaves, scales, flowers or buds.

Saccharide. One of a number of carbohydrates, including the sugars.

Scientific system. Most of what is generally called the "scientific system" is not science at all, but merely procedural. The procedural aspect calls for setting up experiments that eliminate other possibilities, or it deals with making instruments that enable the investigator to find what he is looking for. The backbone of the scientific system has to do with asking the right questions. A scientist can only ask the right questions after his life has absorbed the experiences that lead him to a vision of the Creator's handiwork, hence the right question. In the final analysis new discovery is accomplished by the mind and soul of the whole person and cannot be a mechanical scientific procedure. It stands to reason that you can't get the answers if you don't know the questions. When science falters, it is because no one is asking the right questions. (See cause).

Scion. A portion of the shoot of a plant which is grafted upon a plant having a root system.

Sepal. A division of the calyx.

Single-cross parent. The F_1 offspring of two inbred parents, which in turn is used as a parent—usually with another single cross parent to produce a double-cross hybrid. The term is generally used in corn breeding.

Skatole. The compound CgHgN found in intestines and feces, and in several plants. A phenolic waste.

Spectrometry. The measurement of absorption or emission of light by a substance at specific wavelengths.

Sperm. The male generative cell that fertilizes the egg cell.

Spermatophytes. Seed producing plants.

Sporophyte. The asexual or vegetative part of the plant as opposed to the gametophyte or sexual part.

Stamen. The part of the flower bearing the male reproductive cells, the pollen sac (anther).

Stigma. The part of the pistil that receives the pollen.

Stomata. Plant leaf pores that permit breathing. These specialized openings are surrounded by two guard cells. They are scattered over the epidermal surface of leaves of all vascular plants for the process of gas exchange called breathing.

Style. The stalk of the pistil between stigma and ovary.

Symbiosis. Life together of two dissimilar organisms with resultant mutual benefit, each depending on the other.

Tap root. The anchor root in dicot plants.

Terminal bud. The growing point at the end of each stem, branch, shoot and sucker.

Tissue. Specialized cells similar in structure and function.

Toxic genetic chemistry. A term used by *Acres U.S.A.* to designate manmade molecules of poisons used by some farmers to rescue crops from insect, fungal and bacterial crop destroyers. The word genetic is inserted here because the most commonly used molecules offend the genesis of life at the cellular level in plant, animal and man. Properly managed soil systems produce crops that do not require rescue with toxic genetic chemicals. The term also includes herbicides, which presume to rescue crops from weed takeover in the wake of management mistakes. Under U.S. regulations, materials with an acute oral LD_{50} value of 0 to 50 mg./kg. must carry signal words "DANGER" and "POISON" as well as skull and crossbones. There must be an antidote statement, and "Call Physician Immediately" and "Keep Out of Reach of Children" language must appear on the label. A poison with a LD_{50} value of 50 to 500 mg./kg. is considered moderately toxic by the government. It has to have the signal word "WARNING," but no antidote statement is required. The federal people figure a victim will live long enough to call a Poison Control Center as listed in the yellow pages of a telephone directory. A moderately toxic substance must nevertheless carry "Keep Out of Reach of Children" language. If it takes 500 to 5,000 milligrams per kilogram of body weight to bowl over 50% of the test animals, a poison is considered of low order toxicity. The next word in the signal word pecking order is "CAUTION." No antidote statement is required. The usual "children" language is mandatory. A 5,000 plus LD_{50} rating is considered comparatively free from instant danger. It may deliver a cancer or a scrambled child—but that won't be this week or this year. No warnings, no cautions are required. The government frowns on unqualified claims of safety, and it requires the "Keep Out of Reach of Children" mandatory line. Toxic genetic chemical poisoning can mimic brain hemorrhage, heat stroke, heat exhaustion, hypoglycemia, gastroenteritis, pneumonia and other severe respiratory infections, and asthma.

Trace elements. Elements such as manganese, iron, copper, zinc, boron, etc., which are found as mere traces in soils and are used sparingly by plants. They are of vital importance to the growth and functioning of plants, as well as to the soil biological systems.

Translocation. Movement of water through the vascular tissues of plants.

Transpiration. The loss of water vapor as a consequence of gas exchange in the plant. This in turn makes constant water replacement mandatory

Tuber. A thickened underground branch or stem structure loaded with stored reserves of food and so modified as to serve as a vegetative reproductive structure. A potato is a tuber.

Valence. A measure of the capacity of any element to combine with another. Specifically the number of atoms of hydrogen or chlorine which one atom of an element will combine with or displace. Monovalent means one valence. Divalent, two valence. Trivalent, three valences. Polyvalent, many valences.

van't Hoff's Law. Each species and variety has a temperature below which growth is no longer possible, usually called the minimum growth temperature, or a temperature beyond which growth ceases. From somewhere above the minimum growth temperature, the rate of growth follows van't Hoff's Law. van't Hoff's Law states that for every 10 C. (18 F.) rise in temperature, the rate of growth approximately doubles. Above the optimum, the growth rate falls off rapidly until the maximum temperature is reached. Beyond that growth stops.

Vernalization. A process for hurrying the flowering and fruiting of plants. This is usually accomplished by treating seeds or bulbs or seedlings to shorten the vegetative period. In grain, to partially germinate a seed and then hold it as near freezing temperature before spring sowing. It is sometimes used to spring sow a winter wheat crop.

Villi. Very small, finger-like vascular processes on the mucous membranes of small intestines. Villi aid in taking nutrients from the slurry that comes by.

von Liebig's Law of the Minimum. It states that the limiting factor in plant growth is determined by the absence of any essential nutrient, or its availability as an inadequate supply.

Weinberg's Principle. After Eugene D. Weinberg. It is an extension of Bertrand's Law. It states that amounts of a nutrient adequate for growth are not necessarily adequate for optimal function.

Wien's Constant. After Wilhelm Wien, a figure used to compute the wavelength of radiation associated with the temperature of any heated object. The constant figure is 2,897. This number divided by the absolute temperature of an object gives the wavelength at peak temperature. See absolute temperature.

Xylem. Cells that bring water and minerals from roots to leaves. These cells die and their bulk makes up the wood in trees.

Xerophytes. Plants highly resistant to drought conditions. All xerophytes survive under conditions of extreme water shortage.

Zygote. Cell produced by union of two gametes. A fertilized egg.

CONVERSION FACTORS

Multiply	by	to obtain
acres	.404687	hectares
acres	4.04687×10^3	square kilometers
degrees angular	.0174533	radians
degrees, F. (less 32 F.)	.5556	degrees C.
degrees, C.	1.8	degrees, F. (less 32 F.)
foot pounds	.13826	kilogram meters
feet	30.4801	centimeters
feet	.304801	meters
feet	304.801	millimeters
feet	1.64468×10^4	miles, nautical
gallons, British	1.20091	gallons, U.S.
gallons, U.S.	.832702	gallons, British
gallons, U.S.	.13368	cubic feet
gallons, U.S.	231.	cubic inches
gallons, U.S.	3.78543	liters
grams, metric	2.20462×10^3	pounds, avoirdupois
hectares	2.47104	acres
hectares	1.076387×10^5	square feet
hectares	3.86101×10^3	square miles
horsepower, metric	.98632	horsepower, U.S.
horsepower, U.S.	1.01387	horsepower, metric
inches	2.54001	centimeters
inches	2.54001×10^2	meters
inches	25.4001	millimeters
miles, statute	1.60935	kilometers
miles, statute	.8684	miles, nautical
miles, nautical	6080.204	feet
miles, nautical	1.85325	kilometers
miles, nautical	1.1516	miles, statute
pounds, avoirdupois	453.592	grams, metric
pounds, avoirdupois	.453592	kilograms
pounds, avoirdupois	4.464×10^4	tons, long
pounds, avoirdupois	4.53592×10^4	tons, metric
pounds per foot	1.48816	kilograms per meter
pounds per square foot	4.88241	kilograms per square m.
pounds per square inch	7.031×10^2	kilograms per square cm.
pounds per square inch	7.301×10^4	kilograms per square mm.
pounds per cubic foot	16.0184	kilograms per cubic meter.
radians	57.29578	degrees, angular

Multiply	by	to obtain
square feet	9.29034×10^6	hectares
square feet	.0929034	square meters
square inches	6.45163	square centimeters
square inches	645.163	square millimeters
square miles	259.0	hectares
square miles	2. 590	square kilometers
square yards	.83613	square meters
tons, long	1016.05	kilograms
tons, long	2240.	pounds
tons, long	1.01605	tons, metric
tons, long	1.120	tons, short
tons, metric	2204.62	pounds
tons, metric	.98421	tons, long
tons, metric	1.10231	tons, short
tons, short	907.185	kilograms
tons, short	.892857	tons, long
tons, short	.907185	tons, metric

VEGETABLE SEEDS

Asparagus (Asparagus officinalis), a perennial, 700 seeds per ounce. Germination in 7 to 21 days at 68 to 86 F. Tolerates cool soil. Plant 1 to 1 1/2 inches deep in 8 x 10 inch deep furrows; cover periodically during first year. Tractor cultivated rows, 4 to 5 feet apart; hand cultivated rows, 1 1/2 to 2 feet apart. Plants in the row, 18 inches apart.

Beans, garden (Phaseolus vulgaris), an annual, 100 to 125 seeds per ounce. Germination in 5 to 8 days at 68 to 86 F. Requires warm soil. Plant 1 to 11/2 inches deep. Both tractor and hand cultivated rows, 2 1/2 to 3 feet apart. Plants in the row—3 to 4 inches apart.

Beans, dry edible (Phaseolus vulgaris), an annual, 100 to 125 seeds per ounce. Germination in 5 to 8 days at 68 to 86 F. Requires warm soil. Plant 1 to 1/2 inches deep . Both tractor and hand cultivated rows, 2 1/2 to 3 feet apart. Plants in the row—3 to 4 inches apart.

Beans, lima (Phaseolus lunatus), an annual, 25 to 75 seeds per ounce. Germination in 5 to 9 days at 68 to 86 F. Requires warm soil. Plant 1 to 1 1/2 inches deep. Tractor cultivated rows, 3 to 4 feet apart for pole varieties; bush types, 2 1/2 to 3 feet apart. Hand cultivated rows, 2 feet apart. Plants in the row 3 to 4 inches apart.

Beans, runner (Phaseolus coccineus), an annual, 25 to 30 seeds per ounce. Germination in 5 to 9 days at 68 to 86 F. Requires warm soil. Plant 1 to 1 1/2 inches deep. Both tractor and hand cultivated rows, 3 1/2 to 4 feet apart. Plants in the row—3 to 4 inches apart.

Beet (Beta vulgaris), a biennial, 1,600 seeds per ounce. Germination in 3 to 14 days at 68 to 86 F. Tolerates cool soil. Plant 1 inch deep. Tractor cultivated rows, 2 to 2 1/2 feet apart; hand cultivated rows, 14 inches apart. Plants in the row—2 to 3 inches apart.

Broadbean (Vicia faba), an annual, 20 to 50 seeds per ounce. Germination in 4 to 14 days at 68 to 86 F. Tolerates cool soil. Plant 1 to 1 1/2 inches deep. Both tractor and hand cultivated rows, 2 1/2 to 3 feet apart. Plants in the row, 3 to 4 inches apart.

Broccoli (Brassica oleracea var. botrytis), an annual or biennial, 9,000 seeds per ounce. Germination in 3 to 10 days at 68 to 86 F. Tolerates cool soil. Plant 1/2 inch deep. Both tractor and hand cultivated rows, 2 to 2 1/2 feet apart. Plants in the row—14 to 24 inches apart.

Brussels sprouts (Brassica oleracea var. gemmifera), a biennial, 9,000 seeds per ounce. Germination in 3 to 10 days at 68 to 86 F. Tolerates cool soil. Plant 1/2 inch deep. Tractor cultivated rows, 2 1/2 to 3 feet apart; hand cultivated rows, 2 to 2 1/2 feet apart. Plants in the row, 14 to 24 inches apart.

Cabbage (Brassica oleracea var. capitata), a biennial, 9,000 seeds per ounce. Germination in 3 to 10 days at 68 to 86 F. Tolerates cool soil. Plant 1/2 inch deep. Tractor cultivated rows, 2 1/2 to 3 feet apart; hand cultivated rows, 2 to 2 1/2 feet apart. Plants in the row—14 to 24 inches apart.

Cabbage, Chinese (Brassica pekinensis), and annual or biennial, 18,000 seeds per ounce. Germination in 3 to 7 days at 68 to 86 F. Tolerates cool soil. Plant 1/2 inch deep. Tractor cultivated rows, 2 to 2 1/2 feet apart. Hand cultivated rows, 18 to 24 inches apart. Plants in the row—8 to 12 inches apart.

Carrot (Daucus carota), a biennial, 23,000 seeds per ounce. Germination in 6 to 21 days at 68 to 86 F. Tolerates cool soil. Plant 1/2-inch deep. Tractor cultivated rows, 2 to 2 1/2 feet apart. Hand cultivated rows, 14 to 16 inches apart. Plants in the row—2 to 3 inches apart.

Cauliflower (Brassica oleracea var. botrytis), an annual or biennial, 9,000 seeds per ounce. Germination in 3 to 10 days at 68 to 86 F. Tolerates cool soil. Plant 1/2 inch deep. Tractor cultivated rows, 2 to 2 1/2 feet apart; hand cultivated row, 2 to 2 1/2 feet apart. Plants in the row, 14 to 24 inches apart.

Celeriac (Apium graveolens var. rapaceum), a biennial, 72,000 seeds per ounce. Germination in 10 to 21 days at 50 to 68 F. Requires cool soil. Plant 1/8 inch deep. Tractor cultivated rows, 2 1/2 to 3 feet apart; hand cultivated rows, 18 to 24 inches apart. Plants in the row—4 to 6 inches apart.

Celery (Apium graveolens var. dulce), a biennial, 72,000 seeds per ounce; Germination in 10 to 21 days at 50 to 68 F. Requires cool soil. Plant one inch deep. Tractor cultivated rows, 2 1/2 to 3 feet apart; hand cultivated rows.

Chard, Swiss (Beta vulgaris var. cicla), a biennial, 1,600 seeds per ounce. Germination in 3 to 14 days at 68 to 86 F. Tolerates cool soil. Plant 1 inch deep. Tractor rows, 2 to 2% feet apart; hand rows, 18 to 24 inches apart. Plants in the row, 6 inches apart.

Chicory (Cichorium intybus), a perennial, 27,000 seeds per ounce. Germination in 5 to 14 days at 68 to 86 F. Tolerates cool soil. Plant 1/2 inch deep. Tractor cultivated rows, 2 to 2 1/2 feet apart; hand cultivated rows, 18 to 24 inches apart. Plants in the row—6 to 8 inches apart.

Collards (Brassica oleracea var. acephala), a biennial, 9,000 seeds per ounce. Germination in 3 to 10 days at 68 to 86 F. Tolerates cool soil. Plant 1/2 inch deep. Tractor cultivated rows, 3 to 3 1/2 feet apart; hand cultivated rows, 18 to 24 inches apart. Plants in the row—18 to 24 inches apart.

Corn, sweet (Zea mays), an annual, 120 to 180 seeds per ounce. Germination in 4 to 7 days at 68 to 86 F. Requires warm soil Plant 2 inches deep. Tractor cultivated rows, 3 to 3 1/2 feet apart; hand cultivated row, 2 to 3 feet apart. Plants in the row—drilled, 6 to 16 inches; hills, 2 1/2 to 3 feet apart.

Cornsalad (Valerianella locusta var. olitoria), an annual or biennial. Germination in 7 to 28 days at 68 F. Tolerates cool soil. Plant 1/2 inch deep. Tractor cultivated rows, 2% to 3 feet apart; hand cultivated rows, 14 to 16 inches apart. Plants in the row, 1 foot apart.

Cress, garden (Lepidium sativum), an annual, 12,000 seeds per ounce. Germination in 4 to 10 days at 68 F. Light sensitive. Plant 1/8 to 1/2 inch deep, 2 to 3 inches apart for upland cress. Tractor cultivated rows, 2 to 2 1/2 feet apart; hand cultivated rows, 14 to 16 inches apart. Plants in the row—2 to 3 inches apart.

Cress, water (Rorippa nasturtium-aquaticum), a perennial, 150,000 seeds per ounce. Germination in 4 to 14 days at 68 to 86 F. Tolerates cool soil. Plant 1/8 to 1/2 inch deep. Tractor cultivated rows, 2 to 2 1/2 feet apart; hand cultivated rows, 18 to 24 inches apart. Plants in the row—4 to 6 inches apart.

Cucumber (Cucumis sativus), an annual, 1,100 seeds per ounce. Germination in 3 to 7 days at 68 to 86 F. Requires warm soil. Plant 1/2 inch deep. Both tractor and hand cultivated rows, 6 to 7 feet apart. Plants in the row—drills, 3 feet; hills, 6 feet apart.

Dandelion (Taraxacum officinale), a biennial or perennial, 35,000 seeds per ounce. Germination in 7 to 21 days at 68 to 86 F. Tolerates cool soil. Plant % inch deep, 8 to 12 inches apart. Tractor cultivated rows, 2% to 3 feet apart; hand cultivated rows, 14 to 16 inches apart. Plants in the row—8 to 12 inches apart.

Eggplant (Solanum melongena var. esculentum), an annual, 6,500 seeds per ounce. Germination in 7 to 14 days at 68 to 86 F. Requires warm soil.

Plant 1/2 inch deep. Tractor cultivated, 3 feet apart; hand cultivated, 2 to 2 1/2 feet apart. Plants in the row—3 feet apart.

Endive (Cichorium endivia), an annual or biennial, 27,000 seeds per ounce. Germination in 5 to 14 days at 68 to 86 F. Tolerates cool soil. Plant 1/2 inch deep. Tractor cultivated rows, 2 1/2 to 3 feet apart; hand cultivated rows, 18 to 24 inches apart. Plants in the row—12 inches apart.

Kale (Brassica oleracea var. acephala), a biennial, 9,000 seeds per ounce. Germination in 3 to 10 days at 68 to 86 F. Tolerates cool soil. Plant 1/2 inch deep. Tractor cultivated rows, 2 1/2 to 3 feet apart; hand cultivated rows, 18 to 24 inches apart. Plants in the row—12 to 15 inches apart.

Kohlrabi (Brassica oleracea var. gongylodes), a biennial, 9,000 seeds per ounce. Germination in 3 to 10 days at 68 to 86 F. Tolerates cool soil. Plant 1/2 inch deep. Tractor cultivated rows, 2 1/2 to 3 feet apart; hand cultivated rows, 14 to 16 inches apart. Plants in the row—5 to 6 inches apart.

Leek (Allium porrum), a biennial, 11,000 seeds per ounce. Germination in 6 to 14 days at 68 F. Requires cool soil. Plant 1/2 to 1 inch deep. Both hand and tractor cultivated rows, 2% to 3 feet apart. Plants in the row— 2 to 3 inches apart.

Lettuce (Lactuca sativa), an annual, 25,000 seeds per ounce. Germination in 7 days at 68 F. Requires cool soil. Some varieties are light sensitive. Plant 1/2 inch deep. Both tractor and hand cultivated rows, 2 1/2 to 3 feet apart. Plants in the row—12 to 15 inches apart.

Muskmelon (Cucumis melo), an annual, 1,300 seeds per ounce. Germination in 4 to 10 days at 68 to 86 F. Requires warm soil. Plant 1 inch deep. Both tractor and hand cultivated rows, 6 to 7 feet apart. Plants in hills, 6 feet apart.

Mustard (Brassica juncea), an annual, 18,000 seeds per ounce. Germination in 3 to 7 days at 68 to 86 F. Tolerates cool soil. Plant 1/2 inch deep. Tractor cultivated rows, 2 1/2 to 3 feet apart; hand cultivated rows, 14 to 16 inches apart. Plants in the row—12 inches apart.

Mustard, spinach (Brassica perviridis), an annual, 15,000 seeds per ounce. Germination in 3 to 7 days at 68 to 86 F. Tolerates cool soil Plant 1/2 inch deep. Tractor cultivated rows, 2 1/2 to 3 feet apart; hand cultivated, 14 to 16 inches apart. Plants in the row—12 inches apart.

Okra (Hibiscus esculentus), an annual, 500 seeds per ounce. Germination in 4 to 14 days at 68 to 86 F. Requires warm soil. Plant 1 to 1 1/2 inches deep. Both tractor and hand cultivated rows, 3 to 3 1/2 feet apart. Plants in the row—2 feet apart.

Onion (Allium cepa), a biennial, 9,500 seeds per ounce. Germination in 6 to 10 days at 68 F. Requires cool soil. Plant 1 to 2 inches deep. Tractor cultivated rows, 2 to 2 1/2 feet apart; hand cultivated seeds and set rows, 14 to 16 inches apart. Plants in the row—2 to 3 inches apart.

Onion, Welsh (Allium fistulosum), a biennial. Germination in 6 to 12 days at 68 F. Requires cool soil. Plant 1 to 2 inches deep Tractor cultivated rows,

2 to 2 1/2 feet apart; hand cultivated seeds and set rows, 14 to 16 inches apart. Plants in the row—2 to 3 inches apart.

Parsley (Petroselinum hortense; P. crispum), a biennial, 18,000 seeds per ounce. Germination in 11 to 28 days at 68 to 86 F. Tolerates cool soil. Plant 1/8 to 1/4 inch deep. Tractor cultivated rows, 2 to 2 1/2 feet apart; hand cultivated rows, 14 to 16 inches apart. Plants in the row—2 to 3 inches apart.

Parsnip (Pastinaca sativa), a biennial, 12,000 seeds per ounce. Germination in 6 to 28 days at 68 to 86 F. Tolerates cool soil. Plant 1/2 inch deep. Tractor cultivated rows, 2 to 2 1/2 feet apart. Plants in the row—2 to 3 inches apart.

Pea (Pisum sativum), an annual, 90 to 175 seeds per ounce. Germination in 5 to 8 days at 68 F. Requires cool soil. Plant 2 to 3 inches deep. Tractor cultivated rows, 2 to 4 feet apart; hand cultivated rows, 1 1/2 to 3 feet apart. Plants in the row—1 inch apart.

Pepper (Capsicum spp), an annual, 4,500 seeds per ounce. Germination in 6 to 14 days at 68 to 86 F. Requires warm soil. Plant 1 inch deep. Tractor cultivated rows, 3 to 4 feet apart; hand cultivated rows, 2 to 3 feet apart. Plants in the row—18 to 24 inches apart.

Potato (Solanum tuberosum), a perennial. Germination at 68 F. Tolerates cool soil. Plant 4 inches deep. Tractor cultivated rows, 2 1/2 to 3 feet apart; hand cultivated rows, 2 to 2 1/2 feet apart. Plants in the row—10 to 18 inches apart.

Pumpkin (Cucurbita pepo), an annual, 100 to 300 seeds per ounce. Germination in 4 to 7 days at 68 to 86 F. Requires warm soil. Plant 1 to 2 inches deep. Both tractor and hand cultivated, 5 to 8 feet apart. Plants in the row—3 to 4 feet apart.

Radish (Raphanus sativus), an annual, 2,000 to 4,000 seeds per ounce. Germination in 4 to 6 days at 68 F. Requires cool soil. Plant 1/2 inch deep. Tractor cultivated rows, 2 to 2 1/2 inches apart; hand cultivated rows, 14 to 16 inches apart. Plants in the row—1 inch apart.

Rhubarb (Rheum rhaponticum), a perennial, 1,700 seeds per ounce. Germination in 7 to 21 days at 68 to 86 F. Tolerates cool soil. Both tractor and hand cultivated rows, 3 to 4 feet apart. Plants in the row —3 to 4 feet apart.

Rutabaga (Brassica napus var. napobrassica), a biennial, 12,000 seeds per ounce. Germination in 3 to 14 days at 68 to 86 F. Tolerates cool soil. Plant 1/4 to 1/2 inch deep. Both tractor and hand cultivated rows, 2 to 2 1/2 feet apart. Plants in the row—2 to 3 inches apart.

Salsify (Tragopogon porrifolius), a biennial, 1,900 seeds per ounce. Germination in 5 to 10 days at 68 F. Requires cool soil. Plant 1/2 inch deep. Tractor cultivated rows, 2 to 2 1/2 feet apart; hand cultivated rows, 18 to 26 inches apart. Plants in the row—2 to 3 inches apart.

Sorrel (Rumex acetosa), a perennial, 30,000 seeds per ounce. Germination in 3 to 14 days at 68 to 86 F. Tolerates cool soil. Plant 1/2 inch deep. Tractor cultivated rows, 2 to 2 1/2 feet apart; hand cultivated rows, 18 to 24 inches apart. Plants in the row—5 to 8 inches apart.

Soybean (Glycine max), an annual, 175 to 350 seeds per ounce. Germination in 5 to 8 days at 68 to 86 F. Requires warm soil. Plant 1 to 1 1/2 inches deep. Tractor cultivated rows, 2 1/2 to 3 feet apart; hand cultivated rows, 24 to 30 inches apart. Plants in the row—3 inches apart.

Spinach (Spinacea oleracea), an annual, 2,800 seeds per ounce. Germination in 7 to 21 days at 59 F. Requires cool soil. Plant 1/2 inch deep. Tractor cultivated rows, 2 to 2 1/2 feet apart; hand cultivated rows, 14 to 16 inches apart. Plants in the row—3 to 4 inches apart.

Spinach, New Zealand (Tetragonia expansa), an annual, 350 seeds per ounce. Germination in 5 to 28 days at 50 to 86 F. Germinates irregularly. Plant 1 to 1 1/2 inches deep. Tractor cultivated rows, 3 to 3 1/2 feet apart: hand cultivated rows, 3 feet apart. Plants in the row—18 inches apart.

Sweetpotato (Ipomoea batatas), a perennial. Germination at 77 F. Break or remove seedcoat. Plant 2 to 3 inches deep. Both tractor and hand cultivated rows, 3 to 3 1/2 feet apart. Plants in the row, 12 to 14 inches apart.

Squash (Cucurbita moschata; C. maxima), an annual, 200 to 400 seeds per ounce. Germination in 4 to 7 days at 68 to 86 F. Requires warm soil. Plant both varieties 1 to 1 1/2 inches deep. Both tractor and hand cultivated rows of bush squash, 4 to 5 feet apart; vine squash rows, 8 to 12 feet apart. Bush plants in the row by drills, 15 to 18 inches apart; in hills, 4 feet apart. Vine types, drills, 2 to 3 feet apart; hills, 4 feet apart.

Tomato (Lycopersicon esculentum), an annual, 11,500 seeds per ounce. Germination in 5 to 14 days at 68 to 86 F. Requires warm soil. Plant seeds 1/2 inch deep. Tractor cultivated rows, 3 to 4 feet apart; hand cultivated rows, 2 to 3 feet apart. Plants in the row—1 1/2 to 3 feet apart.

Tomato, husk (Physalis pubescens), an annual, 35,000 seeds per ounce. Germination in 7 to 28 days at 68 to 86 F. Requires warm soil. Plant seeds 1/2 inch deep. Tractor cultivated rows, 3 to 4 feet apart; hand rows, 2 to 3 feet apart. Plants in the row—1 1/2 to 3 feet apart.

Turnip (Brassica rapa), a biennial, 15,000 seeds per ounce. Germination in 3 to 7 days at 68 to 86 F. Tolerates cool soil. Plant 1/4 to 1/2 inch deep. Tractor cultivated rows, 2 to 2 1/2 feet apart; hand cultivated rows, 14 to 16 inches apart. Plants in the row—2 to 3 inches apart.

Watermelon (Citrullus vulgaris), an annual, 200 to 300 seeds per ounce. Germination in 4 to 14 days at 68 to 86 F. Requires warm soil. Plant 1 to 2 inches deep. Both tractor and hand cultivated rows, 8 to 10 feet apart. Plants in the row with drills, 2 to 3 feet apart; in hills, 8 feet apart.

SEEDS AND SEEDING

Crop to be seeded	Amount Per Acre	Weight Per Bushel (lbs)
Alfalfa	10-12 lbs.	60
Barley	7-10 pecks	48
Clover, red	8-10 lbs.	60
Clover, Alsike	4-6 lbs.	60
Clover, White in mixtures	2-5 lbs.	60
Clover, mammoth	8-10 lbs.	60
Clover, sweet	10-12 lbs.	60
Corn, broom	4-6 lbs.	45
Corn, shelled	6-10 lbs.	56
Corn, on the ear		70
Lespedeza, (Japanclover)	20-25 lbs.	25
Oats	8-10 pecks	32
Peanuts	50 lbs.	22
Potatoes	10-20 cwt.	60
Rye	6-8 pecks	56
Sorghum	8-15 lbs.	50
Soy Beans, drilled solid	6-8 pecks	60
Wheat, winter	6-8 pecks	60
Wheat, spring	6 pecks	60

GERMINATION

Many new seedings of grass and legumes fail because of faulty seeding practices. Each figure recorded here represents the percentage planted which will grow and establish new plants, according to University of Wisconsin research. In other words, at 2 inches depth, only 1% of the bluegrass seed will grow. At a half inch, 43% will grow. The rest of the table is self explanatory.

Depth	Blue grass	Timothy	Brome-grass	Alsike clover	Red clover	White clover	Sweet clover	Alfalfa
½"	43	89	78	53	56	47	51	64
1"	27	81	69	49	62	28	45	53
1½"	4	39	51	9	22	2	26	45
2"	1	12	24	4	14	0	14	19

DISTANCE EQUALING 1/100TH OF AN ACRE

Between two 40-inch rows	131 feet
Between two 38-inch rows	138 feet
Between two 36-inch rows	145 feet
Between two 30-inch rows	174 feet
Between two 20-inch rows	261 feet

ROWS IN AN ACRE

One fairly accurate way to determine the number of acres in a cornfield—or portion of a cornfield—is by computing the length of the rows and the distance between rows. The following table shows the number and length of rows in one acre.

Length of Rows	If distance between rows is:					
	15"	20"	30"	38"	40"	42"
40 rods	52.8	39.6	26.4	20.8	19.8	18.8
60 rods	35.2	26.4	17.6	13.9	13.2	12.6
80 rods	26.4	19.5	13.2	10.4	9.9	9.4
100 rods	21.0	15.8	10.5	8.3	7.9	7.5
120 rods	17.4	13.2	8.7	6.9	6.5	6.2
140 rods	15.0	11.3	7.5	5.9	5.6	5.3
160 rods	13.2	9.8	6.6	5.2	4.9	4.7

PLANT POPULATION PER ACRE

Plants Per Acre By Row Width				Average Plants Per 50" Row
20"	30"	38"	40"	
56,500	37,600	29,700	28,200	108
54,400	36,200	28,600	27,200	104
52,300	34,800	27,500	26,100	100
50,200	33,400	26,400	25,100	96
48,100	32,100	25,300	24,000	92
46,000	30,700	24,200	23,000	88
44,800	29,900	23,600	22,400	85

20"	30"	38"	40"	50" Row
42,900	28,600	22,600	21,400	82
41,800	27,900	22,000	20,900	80
39,700	26,500	20,900	19,900	76
39,200	26,100	20,600	19,600	75
37,600	25,100	19,800	18,800	72
36,600	24,400	19,300	18,300	70
35,600	23,600	18,700	17,800	68
34,500	23,000	18,200	17,300	66
33,500	22,300	17,600	16,700	64
33,000	22,000	17,400	16,500	62
31,400	20,900	16,500	15,700	60
30,300	20,200	16,000	15,200	58
29,800	19,900	15,700	14,900	57
29,300	19,500	15,400	14,600	56
28,500	19,000	15,000	14,300	55
27,200	18,100	14,300	13,600	52
26,100	17,400	13,800	13,100	50
25,100	16,700	13,200	12,500	48
24,000	16,000	12,700	12,000	46
23,200	15,500	12,200	11,600	44
22,400	14,900	11,800	11,200	42
20,900	13,900	11,000	10,500	40

DETERMINING PLANT POPULATION AT HARVEST

When comparing various planting rates, it is helpful to know the actual plant population at harvest time (this may differ considerably from the "planted" population and will often be somewhat changed from a seeding stand count).

Mark off several 50 foot row lengths in various parts of the field and count the number of plants (tillers should not be counted as separate plants). Determine the approximate number of plants per acre by using the average number of plants per 50 feet and your row width, as shown in the table on page 442. Or, pace off 1,000th of an acre depending on row width—13.1 feet for a 39 inch row, or 17.4 feet for a 30 inch row. Then count plants or ears and multiply by 1,000. This will yield population or ear count.

CONVERTING HARVEST YIELD TO NO. 2 CORN

To calculate yield, multiply weight of ears obtained by 100 and divided by pounds of ear corn per bushel (from conversion table) determined by moisture percentage of grain sample.

Moisture in kernels	lbs. ear corn per bushel	lbs. shelled corn per bushel
11%	63.3 lbs.	53.17 lbs.
12	64.2	53.77
13	65.2	54.39
14	66.2	55.02
15	67.3	55.67
16	68.4	56.33
17	69.6	57.01
18	70.8	57.71
19	72.1	58.42
20	73.4	59.15
21	74.8	59.90
22	76.2	60.67
23	77.7	61.45
24	79.2	62.26
25	80.7	63.09
26	82.2	63.95
27	83.7	64.82
28	85.2	65.72
29	86.7	66.65
30	88.2	67.60
31	89.9	68.58
32	91.7	69.59
33	93.6	70.63
34	95.6	71.70
35	97.7	72.80
36	99.9	73.94

ESTIMATING CORN YIELDS

Hand harvested ear samples are useful for making quick yield estimates or for comparing small plots. By picking 1/100 of an acre, you can closely approximate corn yield. Here is a table giving the feet in a row required to equal 1/100 acre.

Row width	Length of row	Row width	Length of row
20"	262 ft.	34"	154 ft.
28"	186 ft.	36"	145 ft.
30"	174 ft.	38"	137 ft.
32"	163 ft.	40"	131 ft.

CORN LOST PER ACRE AS WHOLE KERNELS

Average number whole Kernels left per 10 square feet	Bushels per acre
2	0.1
4	0.2
6	0.3
8	0.4
10	0.5
20	1.0
30	1.5
40	2.0
50	2.5
100	5.0
200	10.0

A wooden frame can be made to facilitate the count. To measure 10 square feet, inside dimensions must be 2' x 5', 3' x 3'4", 4' x 2'6", etc. A circle or hoop measuring approximately 3'7" across includes a 10 square foot area.

Main Reasons for Lost Kernels	Prescribed Correction
• Worm snapping rolls	• Replace them
• Snapping roll and/or stripper plates not set right	• Adjust
• Kernels left on cob	• Adjust
• Sieve setting not right	• Adjust
• Engine or ground speed too fast	• Operate at proper speeds

Add bushels lost as ears to bushels lost as kernels to determine total field loss.

FACTS ABOUT SORGHUM

If you dry 30% moisture grain at harvest down to 15%, you'll lose 17.6 pounds hundredweight. Read the rest of this table to determine grain sorghum shrinkage.

Original Moisture %	Final Moisture Content—%			
	15	14	13	12
30	17.6	18.6	19.6	20.5
29	16.5	17.4	18.4	19.4
27	14.1	15.1	16.1	17.0
25	11.8	12.8	13.8	14.8
23	9.4	10.4	11.5	12.5
21	7.0	8.1	9.2	10.2
19	4.7	5.8	6.9	8.0
17	2.4	3.5	4.6	5.7
15		1.2	2.3	3.4

FACTS ABOUT CORN

Kernels /acre at various spacings in 30" and 40" rows

avg. inches per Kernel in Row		Kernals per acre	Final Stand with—	
30	40"		10% loss	20% loss
17.3	13	12,062	10,856	9,650
16.7	12 ½	12,545	11,291	10,036
16.0	12	13,068	11,754	10,448
15.3	11 ½	13,636	12,272	10,909
14.7	11	14,256	12,830	11,405
14.0	10 ½	14,934	13,441	11,947
13.3	10	15,681	14,113	12,545
12.7	9 ½	16,506	14,855	13,205
12.0	9	17,424	15,682	13,939
11.3	8 ½	18,448	16,603	15,078
10.7	8	19,602	17,642	15,682
10.0	7 ½	21,908	19,717	17,526
9.3	7	22,402	20,161	17,922
8.7	6 ½	24,134	21,721	19,307
8.0	6	26,136	23,522	20,909

If hill-dropping or power-checking, you can plant 21,908 kernels per acre by dropping one kernel every 7 ½ inches. From this, you can normally expect a final stand of from 17,500 to 19,700 plants per acre. Planting 38 inch rows increases stand by about 5% above figures given for 40 inch rows. The same average kernel spacing in 30 inch rows as you used in 40 inch rows gives you 33% higher population. You'll probably want to adjust the average kernel spacing when switching to narrower rows.

CORN HARVEST LOSSES

Bushels left in the field at harvest are 100% profit loss. Here are suggestions for determining the extent of harvest loss and for correcting the most common causes. You can't avoid some loss. Agricultural engineers say a 3 to 5% loss cannot be avoided.

Whole ears left on the stalk plus loose ones (husked or unhusked) lying in a harvested field. Count whole ears in several 1/100 acre areas and determine average.

Av. number ears left in 1/100 Acre	Average dry ear weight, pound			
	0.7	0.6	0.5	0.4
1	1.0 bu.	0.9 bu.	0.7 bu.	0.6 bu.
2	2.0 bu.	1.7 bu.	1.4 bu.	1.1 bu.
3	3.0 bu.	2.6 bu.	2.1 bu.	1.7 bu.
4	4.0 bu.	3.4 bu.	2.9 bu.	2.3 bu.
5	5.0 bu.	4.3 bu.	3.6 bu.	2.9 bu.
6	6.0 bu.	5.1 bu.	4.3 bu.	3.4 bu.
7	7.0 bu.	6.0 bu.	5.0 bu.	4.0 bu.
8	8.0 bu.	6.9 bu.	5.7 bu.	4.6 bu.
9	9.0 bu.	7.7 bu.	6.4 bu.	5.2 bu.
10	10.0 bu.	8.5 bu.	7.1 bu.	5.7 bu.

Main Reasons for Lost Ears	Prescribed Correction
• Ears fell off before harvest	• Select hybrids that hold their ears; control nutrition; harvest early
• Improper stripper plate and/or snapping bar adjustment	• Adjust
• Excess ground speed	• Slow down

• Gathering points too high	• Run them as low as practical
• Mismatched row	• Planter and harvester row spacing should be the same
• Corn lodged before harvest	• Select hybrids that stand well; harvest early; correct soil nutrition

Whole kernels—count the loose ones as well as those left attached to cobs or pieces of cob in several 10 square foot areas and determine average. Divide the average kernels per 10 square feet by 20 to get bushels per acre lost.

FIELD LAYOUT SHEET

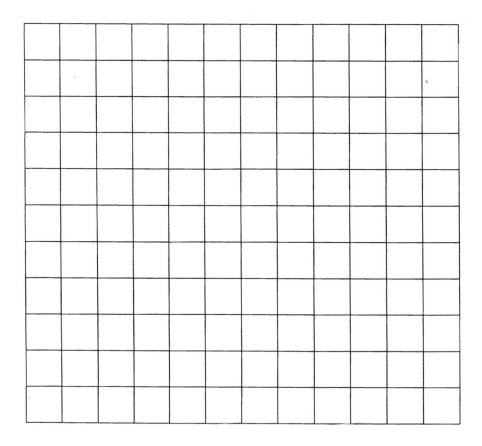

CROP RECORDS

Field # . . . . Acres . . . .

    Planting date . . . . Seed variety . . . .

    Soil materials . . . .

    Tillage . . . .

    Weed management . . . .

    Date harvested . . . .

    Per acre yield . . . .

Field # . . . . Acres . . . .

    Planting date . . . . Seed variety . . . .

    Soil materials . . . .

    Tillage . . . .

    Weed management . . . .

    Date harvested . . . .

    Per acre yield . . . .

Field # . . . . Acres . . . .

    Planting date . . . . Seed variety . . . .

    Soil materials . . . .

    Tillage . . . .

    Weed management . . . .

    Date harvested . . . .

    Per acre yield . . . .

INDEX

Carson, Rachel, 304, 415
catalyst, 421
cation, 421
cation exchange capacity (CEC), 133-138, 421
cattle, data on, 367
cause, 421
Cavendish, Henry, 152
celandine, 311
celery, sodium chloride in, 204
cell, 421, 422
cell, eukaryotic, 424
Cell, The, 71-80
cellulose, 422
Chang, Sing Chen, 214
charlock, 311
cheat, 332-334
chelate, 422
chelation, 97, 384, 385
chickweed, 311, 343-346
China, Taichi, 68
chlorinated hydrocarbons, 208, 422
chlorine, 148; in feed ration, 379
chlorophyll, 29, 31
chloroplasts,13-15, 49, 422
chlorosis,211
cholera, 369, 370
chromosome, 422
climax, 422
clone, 422
Clostridium aerobe, 95
cobalt, 38, 210, 211; in feed ration, 381
coccidiosis, 369
cocklebur, 122, 361, 362
codling moth, 293
cold damage, 61
cold injury, 61
colloidal position, 140,141, 422
Commoner, Barry, 164
companion planting, 422

Complementarity in Biology, 75; explained, 400-403
compost, 231-242; technique of making, 236-239
compost starters, 235
compound, 422
Connecticut Agricultural Experiment Station, 42
Conquest of the Land Through 7,000 Years, 102, 104
conversion factors, 437, 438
cool season crops, 58
copper, 148, 211; in feed ration, 380
corn, anatomy of, 27; estimating yields, 447; facts about, 449, 450; harvest losses (as ears), 450, 451; hybridization, 38-43; lost per acre as whole kernels, 448; moisture requirement, 66, 67; No. 2, converting harvest yield to, 446, 447
corn borer, 288, 289
corn silk, 422
corn smut, 287
Cornelian cherry, 312
cotton fleahopper, 291
Cousteau, Jacques, 203
cover crop principles, 252
crabgrass, 360, 361
Creator, 422, 423
cross-pollinate, 423
cutworms, 287
cycads, 24
cytokinin, 399, 400, 424
cytoplasm, 424
cytoplasmic male sterility, 42

damping-off, 424
dandelion, 311, 346-348
Davey, Humphrey, 152
DDT, 276

fertilizer, arithmetic of, 158, 159;
   the United States of (map), 151
fertilizers, mixed, 156
field madder, 312
field speedwell, 311
fire, and weeds, 316
fleahopper, cotton, 291
flowered asters, 312
flowers, anatomy of, 34-37
fluoridation, 115
fluorides, 115
fluorine, 114, 115
fly control, 304
fly parasites, 299
foliar fertilization, 31
foliar nutrition, 263-269
folklore of fossil fuel chemistry,
   149
Food Power from the Sea, 266
Ford, Brian J., 82, 83
forest belts, 118
Foundation, Wisconsin Alumni
   Research, 375
fox glove, 312
foxtail, 312; barley, 353, 354; giant,
   319-321
Friar, Thorpe, 195
Friends of the Land, 131
frosts, killing, 56, 57
Fryer, Lee, 266, 268
fulvic acids, 233
fungi, 88, 104; as botany division,
   22
fungus, predacious, 86; stinkhorn,
   23
fusarium, 88

gametophyte, 425
garden sorrel, 312
gas, natural, for fertilizer, 266
gasless bean, 42
gel, 425

gene pool, 425
genetics, 34-43
Geological Survey Offices, 178
gibberellins, 399, 425
gliocladium, 88
glucose, 8, 425
goldenrod, 312
graft, 425
grafting, 36
grape leaf folder, 303
grape leaf hopper, 303
grass, 19, 20
grass tetany, 198
grasshopper, 294
green lacewings, 299
Grew, Nehemiah, 4
groundwater in the U.S., 70
*Growing Degree Units for Corn in
   the North Central Region*, 60
growing point, 31
growth regulators, what they do,
   398
growth ring, 32
gypsum, source of sulfur, 222

Halbeib, Ernest M., 38
Hales, Stephen, 5, 7
Hanway's Iowa Research, 264
hare's ear mustard, 312
Haughley Experiment, the, 142
Hawaiian Entomological Society,
   304
Health and Light, 48
*Heat Considered as a Mode of
   Motion*, 280
hedge bindweed, 312
hedge nettle, 312
hemorrhagic septicemia, 372
Hendricks, S. B., 47
hermaphrodite, 425
homeostasis, 199, 425
Hood, S. C., 86, 87

Wiesner, Jerome, 415
wild buckwheat, 312
wild garlic, 312
wild mustard, 311, 341-343
wild pansy, 312
wild radish, 311
wild oats, 328-330
willamette mite, 303
willow, anatomy of, 29
Winogradsky, Serge N.,153
winterburn, 61
wireworm, 290
Wisconsin Alumni Research
    Foundation, 375
Wittwer, Sylvan H., 31, 264, 265
woody aster, 312
wormseed, 312

X disease, 373
xerophytes, 68, 433
xylem, 433

Yearbook of Agriculture, 1938, 113,
    221
Yearbook of Agriculture, 1959, 410
yellow camomile, 312
yellow toad flax, 312
yield, converting to No. 2 corn,
    446, 447; corn, estimating, 447

Zaderej, Andrew, 72-76
zero, absolute, 282
zinc, 148,211; in feed ration, 380
zygote, 433

Agriculture in Transition

BY DONALD L. SCHRIEFER

Now you can tap the source of many of agriculture's most popular progressive farming tools. Ideas now commonplace in the industry, such as "crop and soil weatherproofing," the "row support system," and the "tillage commandments," exemplify the practicality of the soil/root maintenance program that serves as the foundation for Schriefer's highly-successful "systems approach" farming. A veteran teacher, lecturer and writer, Schriefer's ideas are clear, straightforward, and practical. *Softcover, 238 pages. ISBN 978-0-911311-61-7*

From the Soil Up

BY DONALD L. SCHRIEFER

The farmer's role is to conduct the symphony of plants and soil. In this book, learn how to coax the most out of your plants by providing the best soil and removing all yield-limiting factors. Schriefer is best known for his "systems" approach to tillage and soil fertility, which is detailed here. Managing soil aeration, water, and residue decay are covered, as well as ridge planting systems, guidelines for cultivating row crops, and managing soil fertility. Develop your own soil fertility system for long-term productivity. *Softcover, 274 pages. ISBN 978-0-911311-63-1*

Science in Agriculture

BY ARDEN B. ANDERSEN, PH.D., D.O.

By ignoring the truth, ag-chemical enthusiasts are able to claim that pesticides and herbicides are necessary to feed the world. But science points out that low-to-mediocre crop production, weed, disease, and insect pressures are all symptoms of nutritional imbalances and inadequacies in the soil. The progressive farmer who knows this can grow bountiful, disease- and pest-free commodities without the use of toxic chemicals. A concise recap of the main schools of thought that make up eco-agriculture — all clearly explained. Both farmer and professional consultant will benefit from this important work. *Softcover, 376 pages. ISBN 978-0-911311-35-8*

Bread from Stones

BY JULIUS HENSEL

This book was the first work to attack Von Liebig's salt fertilizer thesis, and it stands as valid today as when first written over 100 years ago. Conventional agriculture is still operating under misconceptions disproved so eloquently by Hensel so long ago. In addition to the classic text, comments by John Hamaker and Phil Callahan add meaning to the body of the book. Many who stand on the shoulders of this giant have yet to acknowledge Hensel. A true classic of agriculture. *Softcover, 102 pages. ISBN 978-0-911311-30-3*

To order call 1-800-355-5313 or order online at www.acresusa.com

How to Grow World Record Tomatoes

BY CHARLES H. WILBER

 For most of his 80+ years, Charles Wilber has been learning how to work with nature. In this almost unbelievable book he tells his personal story and his philosophy and approach to gardening. Finally, this Guinness world record holder reveals for the first time how he grows record-breaking tomatoes and produce of every variety. Detailed step-by-step instructions teach you how to grow incredible tomatoes — and get award-winning results with all your garden, orchard, and field crops! Low-labor, organic, bio-intensive gardening at its best. *Softcover, 132 pages. ISBN 978-0-911311-57-0*

Weeds: Control Without Poisons

BY CHARLES WALTERS

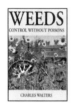 For a thorough understanding of the conditions that produce certain weeds, you simply can't find a better source than this one — certainly not one as entertaining, as full of anecdotes and homespun common sense. It contains a lifetime of collected wisdom that teaches us how to understand and thereby control the growth of countless weed species, as well as why there is an absolute necessity for a more holistic, eco-centered perspective in agriculture today. Contains specifics on a hundred weeds, why they grow, what soil conditions spur them on or stop them, what they say about your soil, and how to control them without the obscene presence of poisons, all cross-referenced by scientific and various common names, and a new pictorial glossary. *Softcover, 352 pages. ISBN 978-0-911311-58-7*

The Biological Farmer
A Complete Guide to the Sustainable
& Profitable Biological System of Farming

BY GARY F. ZIMMER

 Biological farmers work with nature, feeding soil life, balancing soil minerals, and tilling soils with a purpose. The methods they apply involve a unique system of beliefs, observations and guidelines that result in increased production and profit. This practical how-to guide elucidates their methods and will help you make farming fun and profitable. *The Biological Farmer* is the farming consultant's bible. It schools the interested grower in methods of maintaining a balanced, healthy soil that promises greater productivity at lower costs, and it covers some of the pitfalls of conventional farming practices. Zimmer knows how to make responsible farming work. His extensive knowledge of biological farming and consulting experience come through in this complete, practical guide to making farming fun and profitable. *Softcover, 352 pages. ISBN 978-0-911311-62-4*

To order call 1-800-355-5313 or order online at www.acresusa.com

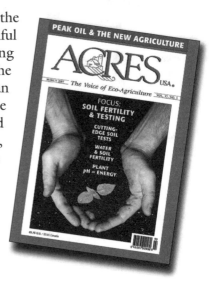